SCHAUM'S OUTLINE OF

THEORY AND PROBLEMS

OF

TRIGONOMETRY

Second Edition

WITH CALCULATOR-BASED SOLUTIONS

•

FRANK AYRES, JR., Ph.D.
Former Professor and Head, Department of Mathematics
Dickinson College

ROBERT E. MOYER, Ph.D.
Associate Professor of Mathematics
Fort Valley State College

•

SCHAUM'S OUTLINE SERIES
McGRAW-HILL PUBLISHING COMPANY

New York St. Louis San Francisco Auckland Bogotá Caracas
Hamburg Lisbon London Madrid Mexico Milan Montreal
New Delhi Oklahoma City Paris San Juan São Paulo
Singapore Sydney Tokyo Toronto

FRANK AYRES, Jr., Ph. D., was formerly professor and Head of the Department of Mathematics at Dickinson College, Carlisle, Pennsylvania. He is the author of eight Schaum's Outlines, including CALCULUS, DIFFERENTIAL EQUATIONS, 1st YEAR COLLEGE MATH, and MATRICES.

ROBERT E. MOYER has been teaching mathematics at Fort Valley State College in Fort Valley, Georgia since 1985. Prior to joining the FVSC faculty, he served for seven years as the Mathematics and Computer Consultant for the Middle Georgia Cooperative Educational Service Agency serving five county-wide public school systems, and he taught high school mathematics for twelve years in Carmi, IL and Rantoul, IL. He received his Doctor of Philosophy in Mathematics Education from the University of Illinois in 1974. From Southern Illinois University, he received his Master of Science in 1967 and his Bachelor of Science in 1964, both in Mathematics Education.

AUSTIN COMMUNITY COLLEGE
LEARNING RESOURCE SERVICES

Schaum's Outline of Theory and Problems of
TRIGONOMETRY

Copyright © 1990, 1954 by McGraw-Hill, Inc. All rights reserved. Printed in the United States of America. Except as permitted under the Copyright Act of 1976, no part of this publication may be reproduced or distributed in any form or by any means, or stored in a data base or retrieval system, without the prior written permission of the publisher.

2 3 4 5 6 7 8 9 10 11 12 13 14 15 16 17 18 19 20 SHP SHP 8 9 2 1 0 9

ISBN 0-07-002659-9

Sponsoring Editor, John Aliano
Production Supervisor, Janelle Travers
Editing Supervisor, Meg Tobin

Library of Congress Cataloging-in-Publication Data

Ayres, Frank.
    Schaum's outline of theory and problems of trigonometry
  Frank Ayres, Jr., Robert E. Moyer. — 2nd ed.
      p.   cm. — (Schaum's outline series)
  Includes index.
  ISBN 0-07-002659-9
    1. Trigonometry—Outlines, syllabi, etc. I. Moyer, Robert E.
  II. Title. III. Title: Theory and problems of trigonometry.
QA531.A97 1990
516.2'4—dc19   89-31294

Preface

In revising this book, the strengths of the first edition were retained while reflecting the changes in the study of trigonometry since the first edition was written. This edition focuses entirely on plane trigonometry, deemphasizes the use of logarithms, includes the use of a calculator, provides all tables necessary to do the problems without a calculator, and provides a summary of geometry properties and theorems that are helpful in solving trigonometry problems.

The book is complete in itself and can be used equally well by those who are studying trigonometry for the first time and those who wish to review the fundamental principles and procedures of trigonometry.

Each chapter contains a summary of the necessary definitions and theorems followed by a set of solved problems. These solved problems include the proofs of theorems and derivations of formulas. The chapters end with a set of supplementary problems and their answers.

Procedures using trigonometric tables and those using a calculator are included as needed to solve problems. The choice of whether to use the tables provided or to use a calculator is left to the student. Problems that are specifically intended for just one solution procedure are clearly labeled; otherwise, either procedure is appropriate. The work with logarithms is entirely optional. Examples and problems fully demonstrate how to use logarithms in trigonometry but may be omitted without loss of continuity of material by those who elect not to use these procedures.

Triangle solution problems, trigonometric identities, and trigonometric equations require a knowledge of elementary algebra. The problems have been carefully selected and their solutions have been spelled out in detail and arranged to illustrate clearly the algebraic processes involved as well as the use of the basic trigonometric relations.

Robert E. Moyer
Fort Valley, Georgia
November 1988

Frank Ayres
Carlisle, Pennsylvania

Contents

CONTENTS

Chapter 1

Angles and Applications

1.1 INTRODUCTION

Trigonometry, as the word implies, is concerned with the measurement of the parts sides, and angles, of a triangle. Plane trigonometry, which is the topic of this book, is restricted to triangles lying in a plane. Trigonometry is based on certain ratios, called trigonometric functions, to be defined in the next chapter. The early applications of the trigonometric functions were to surveying, navigation, and engineering. These functions also play an important role in the study of all sorts of vibratory phenomena—sound, light, electricity, etc. As a consequence, a considerable portion of the subject matter is concerned with a study of the properties of and relations among the trigonometric functions.

1.2 PLANE ANGLE

The plane angle XOP, Fig. 1-1, is formed by the two rays OX and OP. The point O is called the *vertex* and the half lines are called the sides of the angle.

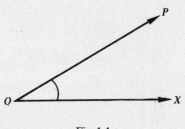

Fig. 1-1

More often, a plane angle is to be thought of as generated by revolving (in a plane) a ray from the initial position OX to a terminal position OP. Then O is again the vertex, $\overrightarrow{OX}$ is called the *initial side*, and $\overrightarrow{OP}$ is called the *terminal* side of the angle.

An angle, so generated, is called *positive* if the direction of rotation (indicated by a curved arrow) is counterclockwise and *negative* if the direction of rotation is clockwise. The angle is positive in Fig. 1-2(a) and (c), negative in Fig. 1-2(b).

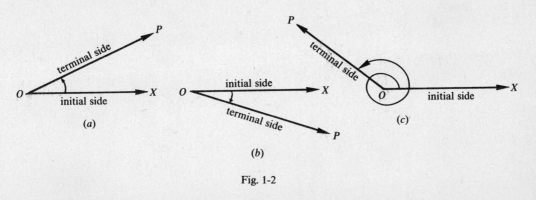

(a)

(b)

(c)

Fig. 1-2

1

1.3 MEASURES OF ANGLES

A *degree* (°) is defined as the measure of the central angle subtended by an arc of a circle equal to 1/360 of the circumference of the circle.

A *minute* (′) is 1/60 of a degree; a *second* (″) is 1/60 of a minute, or 1/3600 of a degree.

EXAMPLE 1.1   (a) $\frac{1}{4}(36°24') = 9°6'$ (b) $\frac{1}{2}(127°24') = \frac{1}{2}(126°84') = 63°42'$

                     (c) $\frac{1}{2}(81°15') = \frac{1}{2}(80°75') = 40°37.5'$ or $40°37'30''$

                     (d) $\frac{1}{4}(74°29'20'') = \frac{1}{4}(72°149'20'') = \frac{1}{4}(72°148'80'') = 18°37'20''$

When changing angles in decimals to minutes and seconds, the general rule is that angles in tenths will be changed to the nearest minute and all other angles will be rounded to the nearest hundredth and then changed to the nearest second. When changing angles in minutes and seconds to decimals, the results in minutes are rounded to tenths and angles in seconds have the results rounded to hundredths.

EXAMPLE 1.2   (a) $62.4° = 62° + 0.4(60') = 62°24'$

                     (b) $23.9° = 29° + 0.9(60') = 23°54'$

                     (c) $29.23° = 29° + 0.23(60') = 29°13.8' = 29°13' + 0.8(60'')$
                                       $= 29°13'48''$

                     (d) $37.47° = 37° + 0.47(60') = 37°28.2' = 37°28' + 0.2(60'')$
                                         $= 37°28'12''$

                     (e) $78°17' = 78° + 17°/60 = 78.3°$ (rounded to tenths)

                     (f) $58°22'16'' = 58° + 22°/60 + 16°/3600 = 58.37°$ (rounded to hundredths)

A *radian* (rad) is defined as the measure of the central angle subtended by an arc of a circle equal to the radius of the circle. (See Fig. 1-3.)

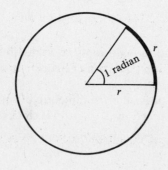

Fig. 1-3

The circumference of a circle = 2π(radius) and subtends an angle of 360°. Then 2π radians = 360°; therefore

$$1 \text{ radian} = \frac{180°}{\pi} = 57.296° = 57°17'45''$$

and

$$1 \text{ degree} = \frac{\pi}{180} \text{ radian} = 0.017453 \text{ rad,}$$

where $\pi = 3.14159$.

EXAMPLE 1.3 (a) $\dfrac{7}{12}\pi \text{ rad} = \dfrac{7\pi}{12} \cdot \dfrac{180°}{\pi} = 105°$ (c) $-\dfrac{\pi}{6} \text{ rad} = -\dfrac{\pi}{6} \cdot \dfrac{180°}{\pi} = -30°$

(b) $50° = 50 \cdot \dfrac{\pi}{180} \text{ rad} = \dfrac{5\pi}{18} \text{ rad}$ (d) $-210° = -210 \cdot \dfrac{\pi}{180} \text{ rad} = -\dfrac{7\pi}{6} \text{ rad}$

(See Probs. 1.1 and 1.2.)

1.4   ARC LENGTH

On a circle of radius r, a central angle of θ radians, Fig. 1-4, intercepts an arc of length

$$s = r\theta$$

that is, arc length = radius × central angle in radians

(NOTE: s and r may be measured in any convenient unit of length but they must be expressed in the same unit.)

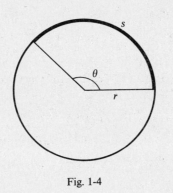

Fig. 1-4

EXAMPLE 1.4 (a) On a circle of radius 30 in, the length of the arc intercepted by a central angle of $\frac{1}{3}$ rad is

$$s = r\theta = 30(\tfrac{1}{3}) = 10 \text{ in}$$

(b) On the same circle a central angle of 50° intercepts an arc of length

$$s = r\theta = 30\left(\frac{5\pi}{18}\right) = \frac{25\pi}{3} \text{ in}$$

(c) On the same circle an arc of length $1\frac{1}{2}$ ft subtends a central angle

$$\theta = \frac{s}{r} = \frac{18}{30} = \frac{3}{5} \text{ rad} \qquad \text{when } s \text{ and } r \text{ are expressed in inches}$$

or

$$\theta = \frac{s}{r} = \frac{3/2}{5/2} = \frac{3}{5} \text{ rad} \qquad \text{when } s \text{ and } r \text{ are expressed in feet}$$

(See Probs. 1.3-1.8.)

1.5   LENGTHS OF ARCS ON A UNIT CIRCLE

The correspondence between points on a real number line and the points on a unit circle, $x^2 + y^2 = 1$, with its center at the origin is shown in Fig. 1-5.

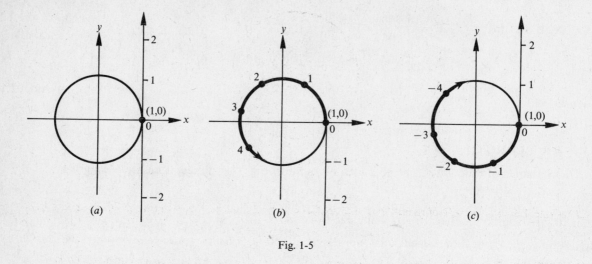

Fig. 1-5

The zero, 0, on the number line is matched with the point (1, 0) as shown in Fig. 1-5(a). The positive real numbers are wrapped around the circle in a counterclockwise direction, Fig. 1-5(b), and the negative real numbers are wrapped around the circle in a clockwise direction, Fig. 1-5(c). Every point on the unit circle is matched with many real numbers, both positive and negative.

The radius of a unit circle has length 1. Hence, the circumference of the circle, given by $2\pi r$, is 2π. The distance halfway around is π and 1/4 the way around is $\pi/2$. Each positive number is paired with a length of an arc s, and since $s = r\theta = 1 \cdot \theta = \theta$, each real number is paired with an angle θ in radian measure. Likewise, each negative real number is paired with the negative of the length of an arc and, therefore, with a negative angle in radian measure. Figure 1-6(a) shows points corresponding to positive angles, and Fig. 1-6(b) shows points corresponding to negative angles.

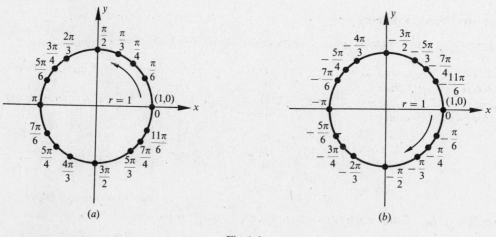

Fig. 1-6

1.6   AREA OF A SECTOR

The area K of a sector of a circle, the shaded part of Fig. 1-7, with radius r and central angle θ radians is

$$K = \tfrac{1}{2}r^2\theta$$

that is, the area of a sector = $\frac{1}{2}$ × the radius × the radius × the central angle in radians.

(NOTE: K will be measured in the square unit of area that corresponds to the length unit used to measure r.)

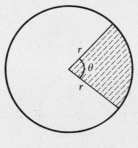

Fig. 1-7

EXAMPLE 1.5   For a circle of radius 30 in, the area of a sector intercepted by a central angle of $\frac{1}{3}$ rad is

$$K = \tfrac{1}{2}r^2\theta = \tfrac{1}{2}(30)^2(\tfrac{1}{3}) = 150 \text{ in}^2$$

EXAMPLE 1.6   For a circle of radius 18 cm, the area of a sector intercepted by a central angle of 50° is

$$K = \tfrac{1}{2}r^2\theta = \tfrac{1}{2}(18)^2\,\frac{5\pi}{18} = 45\pi \text{ cm}^2 \text{ or } 141 \text{ cm}^2 \text{ (rounded)}$$

(NOTE: 50° = $5\pi/18$ rad.)

(See Probs. 1.9 and 1.10.)

1.7 ANGULAR VELOCITY

The relationship between the linear velocity v and the angular velocity ω (the Greek letter *omega*) for a object with radius r is

$$v = r\omega$$

where ω is measured in radians per unit of time and v is distance per unit of time.

(NOTE: v and ω use the same unit of time and r and v use the same linear unit.)

EXAMPLE 1.7   A bicycle with 20-in wheels is traveling down a road at 15 mi/h. Find the angular velocity of the wheel in revolutions per minute.

Because the radius is 10 in and the angular velocity is to be in revolutions per minute (r/min), change the linear velocity 15 mi/h to units of in/min.

$$v = 15\,\frac{\text{mi}}{\text{h}} = \frac{15\text{ mi}}{1\text{ h}}\cdot\frac{5280\text{ ft}}{1\text{ mi}}\cdot\frac{12\text{ in}}{1\text{ ft}}\cdot\frac{1\text{ h}}{60\text{ min}} = 15{,}840\,\frac{\text{in}}{\text{min}}$$

$$\omega = \frac{v}{r} = \frac{15{,}840}{10}\,\frac{\text{rad}}{\text{min}} = 1{,}584\,\frac{\text{rad}}{\text{min}}$$

To change ω to r/min, we multiply by $1/2\pi$ revolution per radian (r/rad).

$$\omega = 1{,}584\,\frac{\text{rad}}{\text{min}} = \frac{1{,}584\text{ rad}}{1}\,\frac{1}{\text{min}}\cdot\frac{1}{2\pi}\,\frac{\text{r}}{\text{rad}} = \frac{792}{\pi}\,\frac{\text{r}}{\text{min}} \text{ or } 252 \text{ r/min}$$

EXAMPLE 1.8   A wheel that is drawn by a belt is making 1 revolution per second (r/s). If the wheel is 18 cm in diameter, what is the linear velocity of the belt in cm/s?

$$1\,\frac{\text{r}}{\text{s}} = \frac{1}{1}\cdot\frac{2\pi\text{ rad}}{1\text{ r}} = 2\pi \text{ rad/s}$$

$$v = r\omega = 9(2\pi) = 18\pi \text{ cm/s or } 57 \text{ cm/s}$$

(See Probs. 1.11 to 1.15.)

Solved Problems

1.1 Express each of the following angles in degree measure:
(a) 30°,   (b) 135°,   (c) 25°30′,   (d) 42°24′35″,   (e) 165.7°,   (f) −3.85°

(a) $30° = 30(\pi/180)$ rad $= \pi/6$ rad or 0.5236 rad

(b) $135° = 135(\pi/180)$ rad $= 3\pi/4$ rad or 2.3562 rad

(c) $25°30′ = 25.5° = 25.5(\pi/180)$ rad $= 0.4451$ rad

(d) $42°24′35″ = 42.41° = 42.41\ (\pi/180)$ rad $= 0.7402$ rad

(e) $165.7° = 165.7(\pi/180)$ rad $= 2.8920$ rad

(f) $-3.85° = -3.85(\pi/180)$ rad $= -0.0672$ rad

1.2 Express each of the following angles in degree measure:
(a) $\pi/3$ rad,   (b) $5\pi/9$ rad,   (c) 2/5 rad,   (d) 4/3 rad,   (e) $-\pi/8$ rad,

(a) $\pi/3$ rad $= (\pi/3)(180°/\pi) = 60°$

(b) $5\pi/9$ rad $= (5\pi/9)(180°/\pi) = 100°$

(c) $2/5$ rad $= (2/5)(180°/\pi) = 72°/\pi = 22.92°$ or 22°55.2′ or 22°55′12″

(d) $4/3$ rad $= (4/3)(180°/\pi) = 240°/\pi = 76.39°$ or 76°23.4′ or 76°23′24″

(e) $-\pi/8$ rad $= -(\pi/8)(180°/\pi) = -22.5°$ or 22°30′

1.3 The minute hand of a clock is 12 cm long. How far does the tip of the hand move during 20 min?

During 20 min the hand moves through an angle $\theta = 120° = 2\pi/3$ rad and the tip of the hand moves over a distance $s = r\theta = 12(2\pi/3) = 8\pi$ cm $= 25.1$ cm.

1.4 A central angle of a circle of radius 30 cm intercepts an arc of 6 cm. Express the central angle θ in radians and in degrees.

$$\theta = \frac{s}{r} = \frac{6}{30} = \frac{1}{5} \text{ rad} = 11°27′33″$$

1.5 A railroad curve is to be laid out on a circle. What radius should be used if the track is to change direction by 25° in a distance of 120 m?

We are required to find the radius of a circle on which a central angle $\theta = 25° = 5\pi/36$ rad intercepts an arc of 120 m. Then

$$r = \frac{s}{\theta} = \frac{120}{5\pi/36} = \frac{864}{\pi} \text{ m} = 275 \text{ m}$$

1.6 A train is moving at the rate 8 mi/h along a piece of circular track of radius 2500 ft. Through what angle does it turn in 1 min?

Since 8 mi/h $= 8(5280)/60$ ft/min $= 704$ ft/min, the train passes over an arc of length $s = 704$ ft in 1 min. Then $\theta = s/r = 704/2500 = 0.2816$ rad or 16°8′.

1.7   Assuming the earth to be a sphere of radius 3960 mi, find the distance of a point 36°N latitude from the equator.

   Since $36° = \pi/5$ rad, $s = r\theta = 3960(\pi/5) = 2488$ mi.

1.8   Two cities 270 mi apart lie on the same meridian. Find their difference in latitude.

$$\theta = \frac{s}{r} = \frac{270}{3960} = \frac{3}{44} \text{ rad} \qquad \text{or} \qquad 3°54.4'$$

1.9   A sector of a circle has a central angle of 50° and an area of 605 cm². Find the radius of the circle.

   $K = \frac{1}{2}r^2\theta$, therefore $r = \sqrt{2K/\theta}$.

$$r = \sqrt{\frac{2K}{\theta}} = \sqrt{\frac{2(605)}{(5\pi/18)}} = \sqrt{\frac{4356}{\pi}} = \sqrt{1386.56}$$

$$= 37.2 \text{ cm}$$

1.10   A sector of a circle has a central angle of 80° and a radius of 5 m. What is the area of the sector?

$$K = \frac{1}{2}r^2\theta = \frac{1}{2}(5)^2\left(\frac{4\pi}{9}\right) = \frac{50\pi}{9} \text{ m}^2 = 17.5 \text{ m}^2$$

1.11   A wheel is turning at the rate of 48 r/min. Express this angular speed in (a) r/s, (b) rad/min, and (c) rad/s.

   (a) $48 \dfrac{\text{r}}{\text{min}} = \dfrac{48}{1} \dfrac{\text{r}}{\text{min}} \cdot \dfrac{1}{60} \dfrac{\text{min}}{\text{s}} = \dfrac{4}{5} \dfrac{\text{r}}{\text{s}}$

   (b) $48 \dfrac{\text{r}}{\text{min}} = \dfrac{48}{1} \dfrac{\text{r}}{\text{min}} \cdot \dfrac{2\pi}{1} \dfrac{\text{rad}}{\text{r}} = 96\pi \dfrac{\text{rad}}{\text{min}}$ or $301.6 \dfrac{\text{rad}}{\text{min}}$

   (c) $48 \dfrac{\text{r}}{\text{min}} = \dfrac{48}{1} \dfrac{\text{r}}{\text{min}} \cdot \dfrac{1}{60} \dfrac{\text{min}}{\text{s}} \cdot \dfrac{2\pi}{1} \dfrac{\text{rad}}{\text{r}} = \dfrac{8\pi}{5} \dfrac{\text{rad}}{\text{s}}$ or $5.03 \dfrac{\text{rad}}{\text{s}}$

1.12   A wheel 4 ft in diameter is rotating at 80 r/min. Find the distance (in ft) traveled by a point on the rim in 1 s, that is, the linear speed of the point (in ft/s).

$$80 \frac{\text{r}}{\text{min}} = 80\left(\frac{2\pi}{60}\right) \frac{\text{rad}}{\text{s}} = \frac{8\pi}{3} \frac{\text{rad}}{\text{s}}$$

   Then in 1 s the wheel turns through an angle $\theta = 8\pi/3$ rad and a point on the wheel will travel a distance $s = r\theta = 2(8\pi/3)$ ft $= 16.8$ ft. The linear velocity is 16.8 ft/s.

1.13   Find the diameter of a pulley which is driven at 360 r/min by a belt moving at 40 ft/s.

$$360 \frac{\text{r}}{\text{min}} = 360\left(\frac{2\pi}{60}\right) \frac{\text{rad}}{\text{s}} = 12\pi \frac{\text{rad}}{\text{s}}$$

   Then in 1 s the pulley turns through an angle $\theta = 12\pi$ rad and a point on the rim travels a distance $s = 40$ ft.

$$d = 2r = 2\left(\frac{s}{\theta}\right) = 2\left(\frac{40}{12\pi}\right) \text{ ft} = \frac{20}{3\pi} \text{ ft} = 2.12 \text{ ft}$$

1.14 A point on the rim of a turbine wheel of diameter 10 ft moves with a linear speed 45 ft/s. Find the rate at which the wheel turns (angular speed) in rad/s and in r/s.

In 1 s a point on the rim travels a distance $s = 45$ ft. Then in 1 s the wheel turns through an angle $\theta = s/r = 45/5 = 9$ rad and its angular speed is 9 rad/s.

Since 1 r = 2π rad or 1 rad = $1/2\pi$ r, 9 rad/s = $9(1/2\pi)$ r/s = 1.43 r/s.

1.15 Determine the speed of the earth (in mi/s) in its course around the sun. Assume the earth's orbit to be a circle of radius 93,000,000 mi and 1 year = 365 days.

In 365 days the earth travels a distance of $2\pi r = 2(3.14)(93,000,000)$ mi.

In 1 s it will travel a distance $s = \dfrac{2(3.14)(93,000,000)}{365(24)(60)(60)}$ mi = 18.5 mi. Its speed is 18.5 mi/s.

Supplementary Problems

1.16 Express each of the following in radian measure:
(a) 25°, (b) 160°, (c) 75°30′, (d) 112°40′, (e) 12°12′20″, (f) 18.34°

Ans.   (a) $5\pi/36$ or 0.4363 rad   (c) $151\pi/360$ or 1.3177 rad   (e) 0.2130 rad
         (b) $8\pi/9$ or 2.7925 rad   (d) $169\pi/270$ or 1.9664 rad   (f) 0.3201 rad

1.17 Express each of the following in degree measure:
(a) $\pi/4$ rad, (b) $7\pi/10$ rad, (c) $5\pi/6$ rad, (d) 1/4 rad, (e) 7/5 rad

Ans.   (a) 45°, (b) 126°, (c) 150°, (d) 14°19′26″ or 14.32°, (e) 80°12′51″ or 80.21°

1.18 On a circle of radius 24 in, find the length of arc subtended by a central angle of (a) 2/3 rad, (b) $3\pi/5$ rad, (c) 75°, (d) 130°.

Ans.   (a) 16 in, (b) 14.4π or 45.2 in, (c) 10π or 31.4 in, (d) $52\pi/3$ or 54.5 in

1.19 A circle has a radius of 30 in. How many radians are there in an angle at the center subtended by an arc of (a) 30 in, (b) 20 in, (c) 50 in?

Ans.   (a) 1 rad, (b) $\frac{2}{3}$ rad, (c) $\frac{5}{3}$ rad

1.20 Find the radius of the circle for which an arc 15 in long subtends an angle of (a) 1 rad, (b) $\frac{2}{3}$ rad, (c) 3 rad, (d) 20°, (e) 50°.

Ans.   (a) 15 in, (b) 22.5 in, (c) 5 in, (d) 43.0 in, (e) 17.2 in

1.21 The end of a 40-in pendulum describes an arc of 5 in. Through what angle does the pendulum swing?

Ans. $\frac{1}{8}$ rad or 7°9′43″ or 7.16°

1.22 A train is traveling at the rate 12 mi/h on a curve of radius 3000 ft. Through what angle has it turned in 1 min?

Ans.   0.352 rad or 20°10′ or 20.17°

1.23 A reversed curve on a railroad track consists of two circular arcs. The central angle of one is 20° with radius 2500 ft and the central angle of the other is 25° with radius 3000 ft. Find the total length of the two arcs

Ans. $6250\pi/9$ or 2182 ft

1.24 Find the area of the sector determined by a central angle of $\pi/3$ rad in a circle of diameter 32 mm.

Ans. $128\pi/3$ or 134.04 mm^2

1.25 Find the central angle necessary to form a sector of area 14.6 cm^2 in a circle of radius 4.85 cm.

Ans.   1.24 rad or 71.05° or 71°3′

1.26 Find the area of the sector determined by a central angle of 100° in a circle with radius 12 cm.

Ans. 40π or 125.7 cm^2

1.27 If the area of a sector of a circle is 248 m^2 and the central angle is 135°, find the diameter of the circle.

Ans.   diameter = 29.0 m

1.28 A flywheel of radius 10 cm is turning at the rate 900 r/min. How fast does a point on the rim travel in m/s?

Ans. 3π or 9.4 m/s

1.29 An automobile tire is 30 in in diameter. How fast (r/min) does the wheel turn on the axle when the automobile maintains a speed of 45 mi/h?

Ans.   504 r/min

1.30 In grinding certain tools the linear velocity of the grinding surface should not exceed 6000 ft/s. Find the maximum number of revolutions per second of (*a*) a 12-in (diameter) emery wheel and (*b*) an 8-in wheel.

Ans.   (*a*) $6000/\pi$ r/s or 1910 r/s,   (*b*) $9000/\pi$ r/s or 2865 r/s

1.31 If an automobile wheel 78 cm in diameter rotates at 600 r/min, what is the speed of the car in km/h?

Ans.   88.2 km/h

Chapter 2

Trigonometric Functions of a General Angle

2.1 COORDINATES ON A LINE

A *directed line* is a line on which one direction is taken as positive and the other as negative. The positive direction is indicated by an arrowhead.

A *number scale* is established on a directed line by choosing a point O (see Fig. 2-1) called the *origin* and a unit of measure $OA = 1$. On this scale B is 4 units to the right of O (that is, in the positive direction from O) and C is 2 units to the left of O (that is, in the negative direction from O). The directed distance $OB = +4$ and the directed distance $OC = -2$. It is important to note that since the line is directed, $OB \neq BO$ and $OC \neq CO$. The directed distance $BO = -4$, being measured contrary to the indicated positive direction, and the directed distance $CO = +2$. Then $CB = CO + OB = 2 + 4 = 6$ and $BC = BO + OC = -4 + (-2) = -6$.

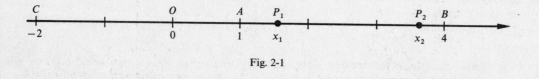

Fig. 2-1

2.2 COORDINATES IN A PLANE

A *rectangular coordinate system* in a plane consists of two number scales (called axes), one horizontal and the other vertical, whose point of intersection (*origin*) is the origin on each scale. It is customary to choose the positive direction on each axis as indicated in the figure, that is, positive to the right on the horizontal axis or x axis and positive upward on the vertical or y axis. For convenience, we shall assume the same unit of measure on each axis.

By means of such a system the position of any point P in the plane is given by its (directed) distances, called *coordinates*, from the axes. The x coordinate or *abscissa* of a point P (see Fig. 2-2) is the directed distance $BP = OA$ and the y coordinate or *ordinate* is the directed distance $AP = OB$. A point P with abscissa x and ordinate y will be denoted by $P(x, y)$.

The axes divide the plane into four parts, called *quadrants*, which are numbered in a counter clockwise direction I, II, III, and IV. The numbered quadrants, together with the signs of the coordinates of a point in each, are shown in Fig. 2-3.

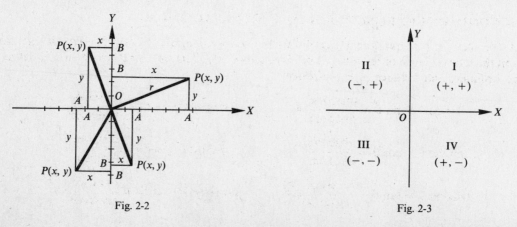

Fig. 2-2

Fig. 2-3

The undirected distance r of any point $P(x, y)$ from the origin, called the *distance of P* or the *radius vector of P*, is given by

$$r = \sqrt{x^2 + y^2}$$

Thus, with each point in the plane, we associate three numbers: x, y, and r.

(See Probs. 2.1 to 2.3.)

2.3   ANGLES IN STANDARD POSITION

With respect to a rectangular coordinate system, an angle is said to be *in standard position* when its vertex is at the origin and its initial side coincides with the positive x axis.

An angle is said to be a *first-quadrant angle* or to be *in the first quadrant* if, when in standard position, its terminal side falls in that quadrant. Similar definitions hold for the other quadrants. For example, the angles 30°, 59°, and −330° are first-quadrant angles [see Fig. 2-4(a)]; 119° is a second-quadrant angle; −119° is a third-quadrant angle; and −10° and 710° are fourth-quadrant angles [see Fig. 2-4(b)].

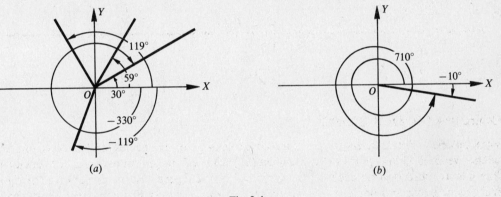

Fig. 2-4

Two angles which, when placed in standard position, have coincident terminal sides are called *coterminal angles*. For example, 30° and −330°, and −10° and 710° are pairs of coterminal angles. There are an unlimited number of angles coterminal with a given angle. Coterminal angles for any given angle can be found by adding integer multiples of 360° to the degree measure of the given angle.

(See Probs. 2.4 and 2.5.)

The angles 0°, 90°, 180°, and 270° and all the angles coterminal with them are called *quadrantal angles*.

2.4   TRIGONOMETRIC FUNCTIONS OF A GENERAL ANGLE

Let θ be an angle (not quadrantal) in standard position and let $P(x, y)$ be any point, distinct from the origin, on the terminal side of the angle. The six trigonometric functions of θ are defined, in terms of the abscissa, ordinate, and distance of P, as follows:

$$\text{sine } \theta = \sin \theta = \frac{\text{ordinate}}{\text{distance}} = \frac{y}{r} \qquad \text{cotangent } \theta = \cot \theta = \frac{\text{abscissa}}{\text{ordinate}} = \frac{x}{y}$$

$$\text{cosine } \theta = \cos \theta = \frac{\text{abscissa}}{\text{distance}} = \frac{x}{r} \qquad \text{secant } \theta = \sec \theta = \frac{\text{distance}}{\text{abscissa}} = \frac{r}{x}$$

$$\text{tangent } \theta = \tan \theta = \frac{\text{ordinate}}{\text{abscissa}} = \frac{y}{x} \qquad \text{cosecant } \theta = \csc \theta = \frac{\text{distance}}{\text{ordinate}} = \frac{r}{y}$$

As an immediate consequence of these definitions, we have the so-called *reciprocal relations*:

$$\sin \theta = 1/\csc \theta \qquad \tan \theta = 1/\cot \theta \qquad \sec \theta = 1/\cos \theta$$

$$\cos \theta = 1/\sec \theta \qquad \cot \theta = 1/\tan \theta \qquad \csc \theta = 1/\sin \theta$$

Because of these reciprocal relationships, one function in each pair of reciprocal trigonometric functions has been used more frequently that the other. The more frequently used trigonometric functions are sine, cosine, and tangent.

It is evident from the diagrams in Fig. 2-5 that the values of the trigonometric functions of θ change as θ changes. In Prob. 2.6 it is shown that the values of the functions of a given angle θ are independent of the choice of the point P on its terminal side.

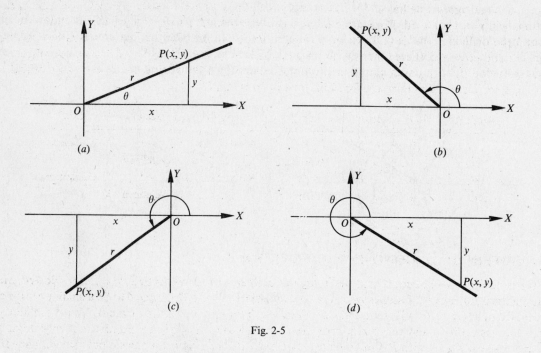

Fig. 2-5

2.5   QUADRANT SIGNS OF THE FUNCTIONS

Since r is always positive, the signs of the functions in the various quadrants depend on the signs of x and y. To determine these signs, one may visualize the angle in standard position or use some device as shown in Fig. 2-6 in which only the functions having positive signs are listed.

(See Prob. 2.7.)

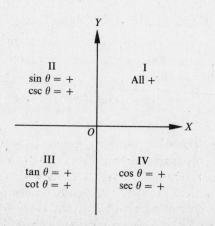

Fig. 2-6

When an angle is given, its trigonometric functions are uniquely determined. When, however, the value of one function of an angle is given, the angle is not uniquely determined. For example, if $\sin \theta = \frac{1}{2}$, then $\theta = 30°, 150°, 390°, 510°, \ldots$. In general, two possible positions of the terminal side are found for example, the terminal sides of $30°$ and $150°$ in the above illustration. The exceptions to this rule occur when the angle is quadrantal.

(See Probs. 2.8 to 2.16.)

2.6 TRIGONOMETRIC FUNCTIONS OF QUADRANTAL ANGLES

For a quadrantal angle, the terminal side coincides with one of the axes. A point P, distinct from the origin, on the terminal side has either $x = 0$ and $y \neq 0$, or $x \neq 0$ and $y = 0$. In either case, two of the six functions will not be defined. For example, the terminal side of the angle $0°$ coincides with the positive x axis and the ordinate of P is 0. Since the ordinate occurs in the denominator of the ratio defining the cotangent and cosecant, these functions are not defined. In this book, *undefined* will be used instead of a numerical value in such cases, but some authors indicate this by writing $\cot 0° = \infty$ and others write $\cot 0° = \pm \infty$. The following results are obtained in Prob. 2.17.

Angle θ	$\sin \theta$	$\cos \theta$	$\tan \theta$	$\cot \theta$	$\sec \theta$	$\csc \theta$
$0°$	0	1	0	Undefined	1	Undefined
$90°$	1	0	Undefined	0	Undefined	1
$180°$	0	-1	0	Undefined	-1	Undefined
$270°$	-1	0	Undefined	0	Undefined	-1

2.7 UNDEFINED TRIGONOMETRIC FUNCTIONS

It has been noted that $\cot 0°$ and $\csc 0°$ are not defined since division by zero is never allowed, but the values of these functions for angles near $0°$ are of interest. In Fig. 2-7(a), take θ to be a small positive angle in standard position and on its terminal side take $P(x, y)$ to be at a distance r from O. Now x is slightly less than r and y is positive and very small; then $\cot \theta = x/y$ and $\csc \theta = r/y$ are positive and very large. Next let θ decrease toward $0°$ with P remaining at a distance r from O. Now x increases but is always less than r while y decreases but it remains greater than 0; thus $\cot \theta$ and $\csc \theta$ become larger and larger. (To see this, take $r = 1$ and compute $\csc \theta$ when $y = 0.1, 0.01, 0.001, \ldots$.) This state of affairs is indicated by "If θ approaches $0°^{+}$, then $\cot \theta$ approaches $+\infty$," which is what is meant when writing $\cot 0° = +\infty$.

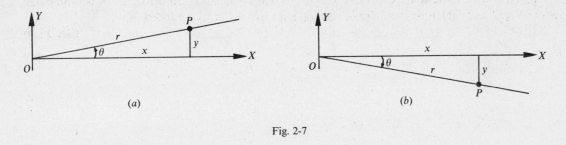

(a)   (b)

Fig. 2-7

Next suppose, as in Fig. 2-7(b), that θ is a negative angle but close to $0°$ and take $P(x, y)$ on its terminal side at a distance r from O. Then x is positive and slightly smaller than r while y is negative and has a small absolute value. Both $\cot \theta$ and $\csc \theta$ are negative with large absolute values. Next let θ increase toward $0°$ with P remaining at a distance r from O. Now x increases but is always less than r, while y remains negative with an absolute value decreasing toward 0; thus $\cot \theta$ and $\csc \theta$ remain negative but have absolute values that get larger and larger. This situation is indicated by "If θ approaches $0°^{-}$, then $\cot \theta$ approaches $-\infty$," which is what is meant when writing $\cot 0° = -\infty$.

In each of these cases, $\cot 0° = +\infty$ and $\cot 0° = -\infty$, the use of the $=$ sign does not have the standard meaning of "equals" and should be used with caution since $\cot 0°$ is undefined and ∞ is not a number. The notation is used as a short way to describe a special situation for trigonometric functions.

The behavior of other trigonometric functions that become undefined can be explored in a similar manner. The following chart summarizes the behavior of each trigonometric function that becomes undefined for angles from 0° up to 360°.

Angle θ	Function Values
$\theta \to \;\;\;0°^{+}$	$\cot \theta \to +\infty$ and $\csc \theta \to +\infty$
$\theta \to \;\;\;0°^{-}$	$\cot \theta \to -\infty$ and $\csc \theta \to -\infty$
$\theta \to \;\;90°^{-}$	$\tan \theta \to +\infty$ and $\sec \theta \to +\infty$
$\theta \to \;\;90°^{+}$	$\tan \theta \to -\infty$ and $\sec \theta \to -\infty$
$\theta \to 180°^{-}$	$\cot \theta \to -\infty$ and $\csc \theta \to +\infty$
$\theta \to 180°^{+}$	$\cot \theta \to +\infty$ and $\csc \theta \to -\infty$
$\theta \to 270°^{-}$	$\tan \theta \to +\infty$ and $\sec \theta \to -\infty$
$\theta \to 270°^{+}$	$\tan \theta \to -\infty$ and $\sec \theta \to +\infty$

(NOTE: The $^{+}$ means the value is greater than the number stated; $180°^{+}$ means values greater than 180°. The $^{-}$ means the value is less than the number stated; $90°^{-}$ means values less than 90°.)

2.8 COORDINATES OF POINTS ON A UNIT CIRCLE

Let s be the length of an arc on a unit circle $x^2 + y^2 = 1$ and each s is paired with an angle θ in radians (see Sec. 1.4). Using the point $(1, 0)$ as the initial point of the arc and $P(x, y)$ as the terminal point of the arc, as in Fig. 2-8, we can determine the coordinates of P in terms of the real number s.

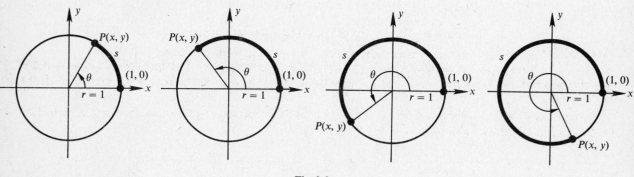

Fig. 2-8

For any angle θ, $\cos \theta = x/r$ and $\sin \theta = y/r$. On a unit circle, $r = 1$ and the arc length $s = r\theta = \theta$ and $\cos \theta = \cos s = x/1 = x$ and $\sin \theta = \sin s = y/1 = y$. The point P associated with the arc length s is determined by $P(x, y) = P(\cos s, \sin s)$. The wrapping function W maps real numbers s onto points P of the unit circle denoted by

$$W(s) = (\cos s, \sin s)$$

Some arc lengths are paired with points on the unit circle whose coordinates are easily determined. If $s = 0$, the point is $(1, 0)$; for $s = \pi/2$, one-fourth the way around the unit circle, the point is $(0, 1)$; $s = \pi$ is paired with $(-1, 0)$; and $s = 3\pi/2$ is paired with $(0, -1)$. (See Sec. 1.5.) These values are summarized in the following chart.

s	$P(x, y)$	$\cos s$	$\sin s$
0	(1, 0)	1	0
$\pi/2$	(0, 1)	0	1
π	(−1, 0)	−1	0
$3\pi/2$	(0, −1)	0	−1

2.9 CIRCULAR FUNCTIONS

Each arc length s determines a single ordered pair $(\cos s, \sin s)$ on a unit circle. Both s and $\cos s$ are real numbers and define a function $(s, \cos s)$ which is called the *circular function cosine*. Likewise, s and $\sin s$ are real numbers and define a function $(s, \sin s)$ which is called the *circular function sine*. These functions are called *circular functions* since both $\cos s$ and $\sin s$ are coordinates on a unit circle. The circular functions $\sin s$ and $\cos s$ are similar to the trigonometric functions $\sin \theta$ and $\cos \theta$ in all regards since, as shown in Chap. 1, any angle in degree measure can be converted to radian measure and this radian-measure angle is paired with an arc s on the unit circle. The important distinction for circular functions is that since $(s, \cos s)$ and $(s, \sin s)$ are ordered pairs of real numbers, all properties and procedures for functions of real numbers apply to circular functions.

The remaining circular functions are defined in terms of $\cos s$ and $\sin s$.

$$\tan s = \frac{\sin s}{\cos s} \qquad \text{for } s \neq \frac{\pi}{2} + k\pi \text{ where } k \text{ is an integer}$$

$$\cot s = \frac{\cos s}{\sin s} \qquad s \neq k\pi \text{ where } k \text{ is an integer}$$

$$\sec s = \frac{1}{\cos s} \qquad \text{for } s \neq \frac{\pi}{2} + k\pi \text{ where } k \text{ is an integer}$$

$$\csc s = \frac{1}{\sin s} \qquad \text{for } s \neq k\pi \text{ where } k \text{ is an integer}$$

It should be noted that the circular functions are defined everywhere that the trigonometric functions are defined and that the values left out of the domains correspond to values where the trigonometric functions are undefined.

In any application there is no need to distinguish between trigonometric functions of angles in radian measure and circular functions of real numbers.

Solved Problems

2.1 Using a rectangular coordinate system, locate the following points and find the value of r for each:
$A(1, 2)$, $B(-3, 4)$, $C(-3, -3\sqrt{3})$, $D(4, -5)$ (see Fig. 2-9).

For A: $r = \sqrt{x^2 + y^2} = \sqrt{1 + 4} = \sqrt{5}$

For B: $r = \sqrt{9 + 16} = 5$

For C: $r = \sqrt{9 + 27} = 6$

For D: $r = \sqrt{16 + 25} = \sqrt{41}$

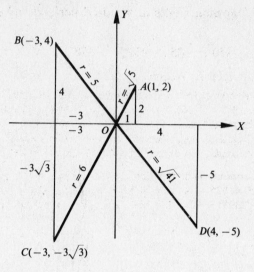

Fig. 2-9

2.2 Determine the missing coordinate of P in each of the following:

(a) $x = 2, r = 3, P$ in the first quadrant

(b) $x = -3, r = 5, P$ in the second quadrant

(c) $y = -1, r = 3, P$ in the third quadrant

(d) $x = 2, r = \sqrt{5}, P$ in the fourth quadrant

(e) $x = 3, r = 3$

(f) $y = -2, r = 2$

(g) $x = 0, r = 2, y$ positive

(h) $y = 0, r = 1, x$ negative

(a)   Using the relation $x^2 + y^2 = r^2$, we have $4 + y^2 = 9$; then $y^2 = 5$ and $y = \pm\sqrt{5}$.
Since P is in the first quadrant, the missing coordinate is $y = \sqrt{5}$.

(b)   Here $9 + y^2 = 25$, $y^2 = 16$, and $y = \pm 4$.
Since P is in the second quadrant, the missing coordinate is $y = 4$.

(c)   We have $x^2 + 1 = 9$, $x^2 = 8$, and $x = \pm 2\sqrt{2}$.
Since P is in the third quadrant, the missing coordinate is $x = -2\sqrt{2}$.

(d) $y^2 = 5 - 4$ and $y = \pm 1$. Since P is in the fourth quadrant, the missing coordinate is $y = -1$.

(e)   Here $y^2 = r^2 - x^2 = 9 - 9 = 0$ and the missing coordinate is $y = 0$.

(f) $x^2 = r^2 - y^2 = 0$ and $x = 0$.   (g) $y^2 = r^2 - x^2 = 4$ and $y = 2$ is the missing coordinate.

(h) $x^2 = r^2 - y^2 = 1$ and $x = -1$ is the missing coordinate.

2.3 In what quadrants may $P(x, y)$ be located if

(a) x is positive and $y \neq 0$?   (c) y/r is positive?   (e) y/x is positive?

(b) y is negative and $x \neq 0$?   (d) r/x is negative?

(a)   In the first quadrant when y is positive and in the fourth quadrant when y is negative

(b)   In the fourth quadrant when x is positive and in the third quadrant when x is negative

(c)   In the first and second quadrants

(d)   In the second and third quadrants

(e)   In the first quadrant when both x and y are positive and in the third quadrant when both x and y are negative

2.4 (*a*) Construct the following angles in standard position and determine those which are coterminal:

$$125°, \quad 210°, \quad -150°, \quad 385°, \quad 930°, \quad -370°, \quad -955°, \quad -870°$$

(*b*) Give five other angles coterminal with 125°.

(*a*) The angles in standard position are shown in Fig. 2-10. The angles 125° and −955° are coterminal since −955° = 125° + 3·360° (or since 125° = −955° + 3·360°). The angles 210°, −150°, 930°, and −870° are coterminal since −150° = 210° − 1·360°, 930° = 210° + 2·360°, and −870° = 210° − 3·360°. From Fig. 2-10, it can be seen that there is only one first-quadrant angle, 385°, and only one fourth-quadrant angle, −370°, so these angles can not be coterminal with any of the other angles.

(*b*) Any angle coterminal with 125° can be written in the form 125° + k·360° where k is an integer. Therefore, 485° = 125° + 1·360°, 845° = 125° + 2·360°, −235° = 125° − 1·360°, −595° = 125° − 2·360°, and −2395° = 125° − 7·360° are angles coterminal with 125°.

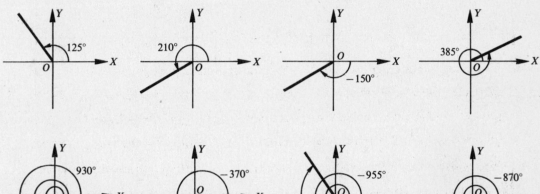

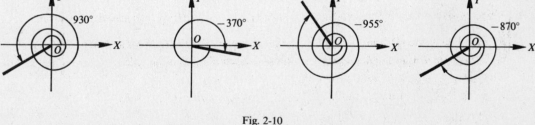

Fig. 2-10

2.5 State a positive angle and a negative angle coterminal with each radian-measure angle:
(*a*) π/6, (*b*) 5π/4, (*c*) 0, (*d*) −17π/6, (*e*) −10π/3, (*f*) 7π/2

$$k \cdot 360° = k(2\pi \text{ radians}) = 2k\pi \qquad \text{where } k \text{ is an integer}$$

(*a*) π/6 + 2π = 13π/6; π/6 − 2π = −11π/6

(*b*) 5π/4 + 2π = 13π/4; 5π/4 − 2π = −3π/4

(*c*) 0 + 2π = 2π; 0 − 2π = −2π

(*d*) −17π/6 + 4π = 7π/6; −17π/6 + 2π = −5π/6

(*e*) −10π/3 + 4π = 2π/3; −10π/3 + 2π = −4π/3

(*f*) 7π/2 − 2π = 3π/2; 7π/2 − 4π = −π/2

2.6 Show that the values of the trigonometric functions of an angle θ do not depend on the choice of the point P selected on the terminal side of the angle.

On the terminal side of each of the angles of Fig. 2-11, let P and P′ have coordinates as indicated and denote the distances OP and OP′ by r and r′ respectively. Drop the perpendiculars AP and A′P′ to the x axis.

In each of the diagrams in Fig. 2-11, the triangles OAP and $OA'P'$, having sides a, b, r and a', b', r' respectively, are similar; thus, using Fig. 2-11(a),

(1) $b/r = b'/r'$ $a/r = a'/r'$ $b/a = b'/a'$ $a/b = a'/b'$ $r/a = r'/a'$ $r/b = r'/b'$

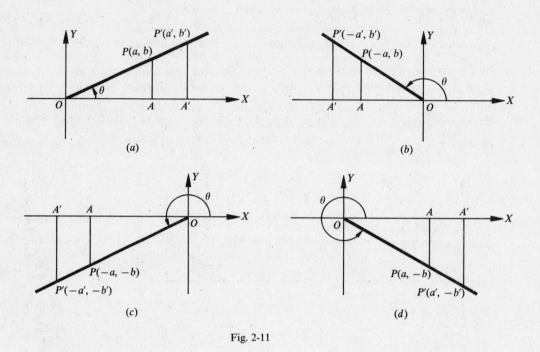

Fig. 2-11

Since the ratios are the trigonometric ratios for the first-quadrant angle, the values of the functions of any first-quadrant angle are independent of the choice of P.

From (1) and Fig. 2-11(b) it follows that

$b/r = b'/r'$ $-a/r = -a'/r'$ $b/-a = b'/-a'$ $-a/b = -a'/b'$ $r/-a = r'/-a'$ $r/b = r'/b'$

Since these are the trigonometric ratios for the second-quadrant angle, the values of the functions of any second-quadrant angle are independent of the choice of P.

It is left for the reader using Fig. 2-11(c) and (d) respectively, to consider the cases,

$$-b/r = -b'/r', \quad -a/r = -a'/r', \text{ etc.} \qquad \text{and} \qquad -b/r = -b'/r', \quad a/r = a'/r', \text{ etc.}$$

2.7   Determine the signs of the functions sine, cosine, and tangent in each of the quadrants.

sin $\theta = y/r$. Since y is positive in quadrants I and II and negative in quadrants III and IV and r is always positive, sin θ is positive in quadrants I and II and negative in quadrants III and IV.

cos $\theta = x/r$. Since x is positive in quadrants I and IV and negative in II and III, cos θ is positive in quadrants I and IV and negative in quadrants II and III.

tan $\theta = y/x$. Since x and y have the same signs in quadrants I and III and opposite signs in quadrants II and IV, tan θ is positive in quadrants I and III and negative in quadrants II and IV.

(NOTE: The reciprocal of a trigonometric function has the same sign in each quadrant as the function.)

2.8   Determine the values of the trigonometric functions of angle θ (smallest positive angle in standard position) if P is a point on the terminal side of θ and the coordinates of P are

(a) $P(3, 4)$,   (b) $P(-3, 4)$,   (c) $P(-1, -3)$

(a) $r = \sqrt{3^2 + 4^2} = 5$ (b) $r = \sqrt{(-3)^2 + 4^2} = 5$ (c) $r = \sqrt{(-1)^2 + (-3)^2} = \sqrt{10}$
[See Fig. 2-12(a).]   [See Fig. 2-12(b).]   [See Fig. 2-12(c).]

$\sin \theta = y/r = 4/5$ $\sin \theta = 4/5$ $\sin \theta = -3/\sqrt{10} = -3\sqrt{10}/10$

$\cos \theta = x/r = 3/5$ $\cos \theta = -3/5$ $\cos \theta = -1/\sqrt{10} = -\sqrt{10}/10$

$\tan \theta = y/x = 4/3$ $\tan \theta = 4/(-3) = -4/3$ $\tan \theta = -3/(-1) = 3$

$\cot \theta = x/y = 3/4$ $\cot \theta = -3/4$ $\cot \theta = -1/(-3) = 1/3$

$\sec \theta = r/x = 5/3$ $\sec \theta = 5/(-3) = -5/3$ $\sec \theta = \sqrt{10}/(-1) = -\sqrt{10}$

$\csc \theta = r/y = 5/4$ $\csc \theta = 5/4$ $\csc \theta = \sqrt{10}/(-3) = -\sqrt{10}/3$

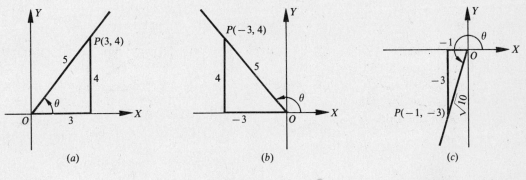

Fig. 2-12

Note the reciprocal relationships. For example, in (b)

$$\sin \theta = \frac{1}{\csc \theta} = \frac{4}{5} \qquad \cos \theta = \frac{1}{\sec \theta} = \frac{-3}{5} \qquad \tan \theta = \frac{1}{\cot \theta} = \frac{-4}{3} \qquad \text{etc.}$$

Note in (c) the rationalizing of the denominators:

$$\sin \theta = -\frac{3}{\sqrt{10}} = -\frac{3}{\sqrt{10}} \cdot \frac{\sqrt{10}}{\sqrt{10}} = -\frac{3\sqrt{10}}{10}$$

and

$$\cos \theta = -\frac{1}{\sqrt{10}} = -\frac{1}{\sqrt{10}} \cdot \frac{\sqrt{10}}{\sqrt{10}} = -\frac{\sqrt{10}}{10}$$

Whenever the denominator of a fraction is an irrational number, an equivalent fraction with a rational denominator will be given as well.

2.9   In what quadrant will θ terminate, if

(a) $\sin \theta$ and $\cos \theta$ are both negative?   (c) $\sin \theta$ is positive and secant θ is negative?

(b) $\sin \theta$ and $\tan \theta$ are both positive?   (d) $\sec \theta$ is negative and $\tan \theta$ is negative?

(a)   Since $\sin \theta = y/r$ and $\cos \theta = x/r$, both x and y are negative. (Recall that r is always positive.) Thus, θ is a third-quadrant angle.

(b)   Since $\sin \theta$ is positive, y is positive; since $\tan \theta = y/x$ is positive, x is also positive. Thus, θ is a first-quadrant angle.

(c)   Since $\sin \theta$ is positive, y is positive; since $\sec \theta$ is negative, x is negative. Thus, θ is a second-quadrant angle.

(d)   Since $\sec \theta$ is negative, x is negative; since $\tan \theta$ is negative, y is then positive. Thus, θ is a second-quadrant angle.

2.10   In what quadrants may θ terminate, if
(a) $\sin \theta$ is positive?   (b) $\cos \theta$ is negative?   (c) $\tan \theta$ is negative?   (d) $\sec \theta$ is positive?

(a)   Since $\sin \theta$ is positive, y is positive. Then x may be positive or negative and θ is a first- or second-quadrant angle.

(b)   Since $\cos \theta$ is negative, x is negative. Then y may be positive or negative and θ is a second- or third-quadrant angle.

(c)   Since $\tan \theta$ is negative, either y is positive and x is negative or y is negative and x is positive. Thus, θ may be a second- or fourth-quadrant angle.

(d)   Since $\sec \theta$ is positive, x is positive. Thus, θ may be a first- or fourth-quadrant angle.

2.11   Find the values of $\cos \theta$ and $\tan \theta$, given $\sin \theta = \frac{8}{17}$ and θ in quandrant I.

Let P be a point on the terminal line of θ. Since $\sin \theta = y/r = 8/17$, we take $y = 8$ and $r = 17$. Since θ is in quadrant I, x is positive; thus

$$x = \sqrt{r^2 - y^2} = \sqrt{(17)^2 - (8)^2} = 15$$

To draw the figure, locate the point $P(15, 8)$, join it to the origin, and indicate the angle θ. (See Fig. 2-13.)

$$\cos \theta = \frac{x}{r} = \frac{15}{17} \qquad \text{and} \qquad \tan \theta = \frac{y}{x} = \frac{8}{15}$$

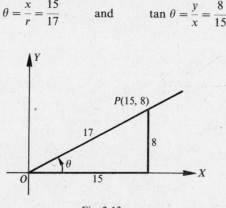

Fig. 2-13

The choice of $y = 8$ and $r = 17$ is one of convenience. Note that $8/17 = 16/34$ and we might have taken $y = 16$, $r = 34$. Then $x = 30$, $\cos \theta = 30/34 = 15/17$ and $\tan \theta = 16/30 = 8/15$.

(See Prob. 2.6.)

2.12   Find the values of $\sin \theta$ and $\tan \theta$, given $\cos \theta = \frac{5}{6}$.

Since $\cos \theta$ is positive, θ is in quadrant I or IV.

Since $\cos \theta = x/r = 5/6$, we take $x = 5$ and $r = 6$; $y = \pm\sqrt{(6)^2 - (5)^2} = \pm\sqrt{11}$.

(a)   For θ in quadrant I [Fig. 2-14(a)] we have $x = 5$, $y = \sqrt{11}$, and $r = 6$; then

$$\sin \theta = \frac{y}{r} = \frac{\sqrt{11}}{6} \qquad \text{and} \qquad \tan \theta = \frac{y}{x} = \frac{\sqrt{11}}{5}$$

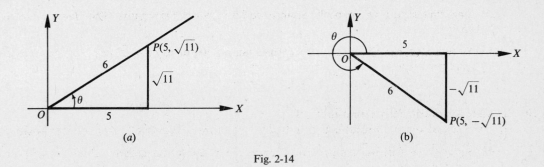

Fig. 2-14

(b)   For θ in quadrant IV [Fig. 2-14(b)] we have $x = 5$, $y = -\sqrt{11}$, and $r = 6$; then

$$\sin \theta = \frac{y}{r} = \frac{-\sqrt{11}}{6} \qquad \text{and} \qquad \tan \theta = \frac{y}{x} = \frac{-\sqrt{11}}{5}$$

2.13   Find the values of $\sin \theta$ and $\cos \theta$, given $\tan \theta = -\frac{3}{4}$.

Since $\tan \theta = y/x$ is negative, θ is in quadrant II (take $x = -4$ and $y = 3$) or in quadrant IV (take $x = 4$ and $y = -3$). In either case $r = \sqrt{16 + 9} = 5$.

(a)   For θ in quadrant II [Fig. 2-15(a)], $\sin \theta = y/r = 3/5$ and $\cos \theta = x/r = -4/5$.

(b)   For θ in quadrant IV [Fig. 2-15(b)], $\sin \theta = y/r = -3/5$ and $\cos \theta = x/r = 4/5$.

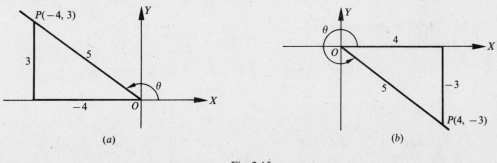

Fig. 2-15

2.14   Find $\sin \theta$, given $\cos \theta = -\frac{4}{5}$ and that $\tan \theta$ is positive.

Since $\cos \theta = x/r$ is negative, x is negative. Since also $\tan \theta = y/x$ is positive, y must be negative. Then θ is in quadrant III. (See Fig. 2-16.)

Take $x = -4$ and $r = 5$; then $y = -\sqrt{5^2 - (-4)^2} = -3$. Thus, $\sin \theta = y/r = -3/5$.

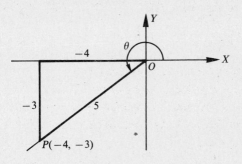

Fig. 2-16

2.15 Find the values of the remaining functions of θ, given $\sin \theta = \sqrt{3}/2$ and $\cos \theta = -1/2$.

Since $\sin \theta = y/r$ is positive, y is positive. Since $\cos \theta = x/r$ is negative, x is negative. Thus, θ is in quadrant II. (See Fig. 2-17.)

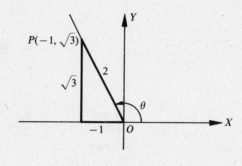

Fig. 2-17

Taking $x = -1$, $y = \sqrt{3}$, and $r = \sqrt{(-1)^2 + (\sqrt{3})^2} = 2$, we have

$$\tan \theta = \frac{y}{x} = \frac{\sqrt{3}}{-1} = -\sqrt{3} \qquad \cot \theta = \frac{1}{\tan \theta} = \frac{-1}{\sqrt{3}} = \frac{-\sqrt{3}}{3}$$

$$\sec \theta = \frac{1}{\cos \theta} = -2 \qquad \csc \theta = \frac{1}{\sin \theta} = \frac{2}{\sqrt{3}} = \frac{2\sqrt{3}}{3}$$

2.16 Determine the values of $\cos \theta$ and $\tan \theta$ if $\sin \theta = m/n$, a negative fraction.

Since $\sin \theta$ is negative, θ is in quadrant III or IV.

(a) In quadrant III: Take $y = m$, $r = n$, $x = -\sqrt{n^2 - m^2}$; then

$$\cos \theta = \frac{x}{r} = \frac{-\sqrt{n^2 - m^2}}{n} \qquad \text{and} \qquad \tan \theta = \frac{y}{x} = \frac{-m}{\sqrt{n^2 - m^2}} = \frac{-m\sqrt{n^2 - m^2}}{n^2 - m^2}$$

(b) In quadrant IV: Take $y = m$, $r = n$, $x = +\sqrt{n^2 - m^2}$; then

$$\cos \theta = \frac{x}{r} = \frac{\sqrt{n^2 - m^2}}{n} \qquad \text{and} \qquad \tan \theta = \frac{y}{x} = \frac{m}{\sqrt{n^2 - m^2}} = \frac{m\sqrt{n^2 - m^2}}{n^2 - m^2}$$

2.17 Determine the values of the trigonometric functions of
(a) $0°$, (b) $90°$, (c) $180°$, (d) $270°$

Let P be any point (not 0) on the terminal side of θ. When $\theta = 0°$, $x = r$ and $y = 0$; when $\theta = 90°$, $x = 0$ and $y = r$; when $\theta = 180°$, $x = -r$ and $y = 0$; and when $\theta = 270°$, $x = 0$ and $y = -r$.

(a) $\theta = 0°$; $x = r$, $y = 0$ (c) $\theta = 180°$; $x = -r$, $y = 0$
[See Fig. 2-18(a).]   [See Fig. 2-18(c).]

$\sin 0° = y/r = 0/r = 0$ $\sin 180° = y/r = 0/r = 0$

$\cos 0° = x/r = r/r = 1$ $\cos 180° = x/r = -r/r = -1$

$\tan 0° = y/x = 0/r = 0$ $\tan 180° = y/x = 0/(-r) = 0$

$\cot 0° = x/y = $ undefined $\cot 180° = x/y = $ undefined

$\sec 0° = r/x = r/r = 1$ $\sec 180° = r/x = r/(-r) = -1$

$\csc 0° = r/y = $ undefined $\csc 180° = r/y = $ undefined

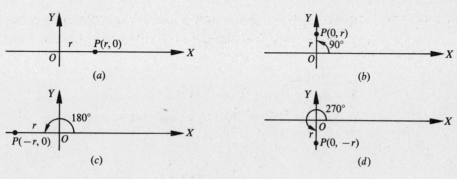

Fig. 2-18

(b) $\theta = 90°;\ x = 0,\ y = r$
    (See Fig. 2-18(b).)

$\sin 90° = y/r = r/r = 1$

$\cos 90° = x/r = 0/r = 0$

$\tan 90° = y/x = \text{undefined}$

$\cot 90° = x/y = 0/r = 0$

$\sec 90° = r/x = \text{undefined}$

$\csc 90° = r/y = r/r = 1$

(d) $\theta = 270°;\ x = 0,\ y = -r$
    (See Fig. 2-18(d).)

$\sin 270° = y/r = -r/r = -1$

$\cos 270° = x/r = 0/r = 0$

$\tan 270° = y/x = \text{undefined}$

$\cot 270° = x/y = 0/(-r) = 0$

$\sec 270° = r/x = \text{undefined}$

$\csc 270° = r/y = r/(-r) = -1$

2.18 Evaluate: (a) $\sin 0° + 2\cos 0° + 3\sin 90° + 4\cos 90° + 5\sec 0° + 6\csc 90°$

(b) $\sin 180° + 2\cos 180° + 3\sin 270° + 4\cos 270° - 5\sec 180° - 6\csc 270°$

(a) $0 + 2(1) + 3(1) + 4(0) + 5(1) + 6(1) = 16$

(b) $0 + 2(-1) + 3(-1) + 4(0) - 5(-1) - 6(-1) = 6$

2.19 Using a protractor, construct an angle of 20° in standard position. With O as center describe an arc of radius 10 units meeting the terminal side in P. From P drop a perpendicular to the x axis, meeting it in A. By actual measurement, $OA = 9.4$, $AP = 3.4$, and P has coordinates (9.4, 3.4). Then find the trigonometric functions of 20° (see Fig. 2-19).

$\sin 20° = 3.4/10 = 0.34$ $\qquad \cot 20° = 9.4/3.4 = 2.8$

$\cos 20° = 9.4/10 = 0.94$ $\qquad \sec 20° = 10/9.4 = 1.1$

$\tan 20° = 3.4/9.4 = 0.36$ $\qquad \csc 20° = 10/3.4 = 2.9$

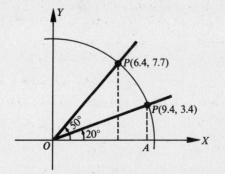

Fig. 2-19

2.20 Obtain the trigonometric functions of 50° as in Prob. 2.19. Refer to Fig. 2.19.

By actual measurement P, on the terminal side at a distance 10 units from the origin, has coordinates (6.4, 7.7). Then

$$\sin 50° = 7.7/10 = 0.77 \qquad \cot 50° = 6.4/7.7 = 0.83$$

$$\cos 50° = 6.4/10 = 0.64 \qquad \sec 50° = 10/6.4 = 1.6$$

$$\tan 50° = 7.7/6.4 = 1.2 \qquad \csc 50° = 10/7.7 = 1.3$$

Supplementary Problems

2.21 State the quadrant in which each angle terminates and the signs of the sine, cosine, and tangent of each angle.
(a) 125°, (b) 75°, (c) 320°, (d) 212°, (e) 460°, (f) 750°, (g) −250°, (h) −1000°

Ans.   (a) II; +,−,−;   (b) I; +,+,+;   (c) IV; −,+,−;   (d) III; −,−,+;
(e) II;+,−,−;   (f) I; +,+,+;   (g) II; +,−,−;   (h) I; +,+,+

2.22 In what quadrant will θ terminate if

    (a) $\sin \theta$ and $\cos \theta$ are both positive?

    (b) $\cos \theta$ and $\tan \theta$ are both positive?

    (c) $\sin \theta$ and $\sec \theta$ are both negative?

    (d) $\cos \theta$ and $\cot \theta$ are both negative?

    (e) $\tan \theta$ is positive and $\sec \theta$ is negative?

    (f) $\tan \theta$ is negative and $\sec \theta$ is positive?

    (g) $\sin \theta$ is positive and $\cos \theta$ is negative?

    (h) $\sec \theta$ is positive and $\csc \theta$ is negative?

Ans.   (a) I, (b) I, (c) III, (d) II, (e) III, (f) IV, (g) II, (h) IV

2.23 Denote by θ the smallest positive angle whose terminal side passes through the given point and find the trigonometric functions of θ:
(a) $P(-5, 12)$, (b) $P(7, -24)$, (c) $P(2, 3)$, (d) $P(-3, -5)$

Ans.   Answers listed in the order $\sin \theta$, $\cos \theta$, $\tan \theta$, $\cot \theta$, $\sec \theta$, $\csc \theta$

    (a) $12/13, -5/13, -12/5, -5/12, -13/5, 13/12$

    (b) $-24/25, 7/25, -24/7, -7/24, 25/7, -25/24$

    (c) $3/\sqrt{13} = 3\sqrt{13}/13, 2/\sqrt{13} = 2\sqrt{13}/13, 3/2, 2/3, \sqrt{13}/2, \sqrt{13}/3$

    (d) $-5/\sqrt{34} = -5\sqrt{34}/34, -3/\sqrt{34} = -3\sqrt{34}/34, 5/3, 3/5, -\sqrt{34}/3, -\sqrt{34}/5$

2.24 Find the values of the trigonometric functions of θ, given:

    (a) $\sin \theta = 7/25$ (d) $\cot \theta = 24/7$ (g) $\tan \theta = 3/5$ (j) $\csc \theta = -2/\sqrt{3} = -2\sqrt{3}/3$

    (b) $\cos \theta = -4/5$ (e) $\sin \theta = -2/3$ (h) $\cot \theta = \sqrt{6}/2$

    (c) $\tan \theta = -5/12$ (f) $\cos \theta = 5/6$ (i) $\sec \theta = -\sqrt{5}$

Ans.   Answers listed in the order $\sin \theta$, $\cos \theta$, $\tan \theta$, $\cot \theta$, $\sec \theta$, $\csc \theta$
    (a)   I: $7/25, 24/25, 7/24, 24/7, 25/24, 25/7$
        II: $7/25, -24/25, -7/24, -24/7, -25/24, 25/7$

    (b)   II: $3/5, -4/5, -3/4, -4/3, -5/4, 5/3$;
        III: $-3/5, -4/5, 3/4, 4/3, -5/4, -5/3$

(c) II: 5/13, −12/13, −5/12, −12/5, −13/12, 13/5
     IV: −5/13, 12/13, −5/12, −12/5, 13/12, −13/5

(d) I: 7/25, 24/25, 7/24, 24/7, 25/24, 25/7
     III: −7/25, −24/25, 7/24, 24/7, −25/24, −25/7

(e) III: −2/3, −$\sqrt{5}$/3, 2/$\sqrt{5}$ = 2$\sqrt{5}$/5, $\sqrt{5}$/2, −3/$\sqrt{5}$ = −3$\sqrt{5}$/5, −3/2
     IV: −2/3, $\sqrt{5}$/3, −2/$\sqrt{5}$ = −2$\sqrt{5}$/5, −$\sqrt{5}$/2, 3/$\sqrt{5}$ = 3$\sqrt{5}$/5, −3/2

(f) I: $\sqrt{11}$/6, 5/6, $\sqrt{11}$/5, 5/$\sqrt{11}$ = 5$\sqrt{11}$/11, 6/5, 6/$\sqrt{11}$ = 6$\sqrt{11}$/11
     IV: −$\sqrt{11}$/6, 5/6, −$\sqrt{11}$/5, −5/$\sqrt{11}$ = −5$\sqrt{11}$/11, 6/5, −6/$\sqrt{11}$ = −6$\sqrt{11}$/11

(g) I: 3/$\sqrt{34}$ = 3$\sqrt{34}$/34, 5/$\sqrt{34}$ = 5$\sqrt{34}$/34, 3/5, 5/3, $\sqrt{34}$/5, $\sqrt{34}$/3
     III: −3/$\sqrt{34}$ = −3$\sqrt{34}$/34, −5/$\sqrt{34}$ = −5$\sqrt{34}$/34, 3/5, 5/3, −$\sqrt{34}$/5, −$\sqrt{34}$/3

(h) I: 2/$\sqrt{10}$ = $\sqrt{10}$/5, $\sqrt{3}$/$\sqrt{5}$ = $\sqrt{15}$/5, 2/$\sqrt{6}$ = $\sqrt{6}$/3, $\sqrt{6}$/2, $\sqrt{5}$/$\sqrt{3}$ = $\sqrt{15}$/3, $\sqrt{10}$/2
     III: −2/$\sqrt{10}$ = −$\sqrt{10}$/5, −$\sqrt{3}$/$\sqrt{5}$ = −$\sqrt{15}$/5, 2/$\sqrt{6}$ = $\sqrt{6}$/3, $\sqrt{6}$/2, −$\sqrt{5}$/$\sqrt{3}$ = −$\sqrt{15}$/3, −$\sqrt{10}$/2

(i) II: 2/$\sqrt{5}$ = 2$\sqrt{5}$/5, −1/$\sqrt{5}$ = −$\sqrt{5}$/5, −2, −1/2, −$\sqrt{5}$, $\sqrt{5}$/2
     III: −2/$\sqrt{5}$ = −2$\sqrt{5}$/5, −1/$\sqrt{5}$ = −$\sqrt{5}$/5, 2, 1/2, −$\sqrt{5}$, −$\sqrt{5}$/2

(j) III: −$\sqrt{3}$/2, −1/2, $\sqrt{3}$, 1/$\sqrt{3}$ = $\sqrt{3}$/3, −2, −2/$\sqrt{3}$ = −2$\sqrt{3}$/3
     IV: −$\sqrt{3}$/2, 1/2, −$\sqrt{3}$, −1/$\sqrt{3}$ = −$\sqrt{3}$/3, 2, −2/$\sqrt{3}$ = −2$\sqrt{3}$/3

2.25 Evaluate each of the following:

(a) $\tan 180° - 2 \cos 180° + 3 \csc 270° + \sin 90°$

(b) $\sin 0° + 3 \cot 90° + 5 \sec 180° - 4 \cos 270°$.

(c) $3 \sin \pi + 4 \cos 0 - 3 \cos \pi + \sin \pi/2$

(d) $4 \cos \pi/2 - 5 \sin 3\pi/2 - 2 \sin \pi/2 + \sin 0$

Ans.   (a) 0,   (b) −5,   (c) 6,   (d) 3

2.26 State the quadrant in which each angle, in radian measure, terminates:
(a) $\pi/4$, (b) $5\pi/6$, (c) $11\pi/3$, (d) $-3\pi/4$, (e) $8\pi/3$, (f) $17\pi/6$, (g) $23\pi/6$

Ans.   (a) I,   (b) II,   (c) IV,   (d) III,   (e) II,   (f) II,   (g) IV

2.27 State the point on the unit circle that corresponds to each real number:
(a) 17π, (b) $-13\pi/2$, (c) $7\pi/2$, (d) 28π

Ans.   (a) $W(17\pi) = W(\pi) = (\cos \pi, \sin \pi) = (-1, 0)$

(b) $W(-13\pi/2) = W(\pi/2) = (\cos \pi/2, \sin \pi/2) = (0, 1)$

(c) $W(7\pi/2) = W(3\pi/2) = (\cos 3\pi/2, \sin 3\pi/2) = (0, -1)$

(d) $W(28\pi) = W(0) = (\cos 0, \sin 0) = (1, 0)$

Chapter 3

Trigonometric Functions of an Acute Angle

3.1 TRIGONOMETRIC FUNCTIONS OF AN ACUTE ANGLE

In dealing with any right triangle, it will be convenient (see Fig. 3-1) to denote the vertices as A, B, and C with C the vertex of the right angle, to denote the angles of the triangle as A, B, and C with $C = 90°$, and to denote the sides opposite the angles as a, b, and c respectively. With respect to angle A, a will be called the *opposite side* and b will be called the *adjacent side*; with respect to angle B, b will be called a the *opposite side* and the *adjacent side*. Side c will always be called the *hypotenuse*.

If now the right triangle is placed in a coordinate system (Fig. 3-2) so that angle A is in standard position, the point B on the terminal side of angle A has coordinates (b, a), and the distance $c = \sqrt{a^2 + b^2}$, then the trigonometric functions of angle A may be defined in terms of the sides of the right triangle, as follows:

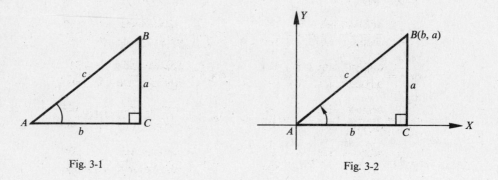

Fig. 3-1   Fig. 3-2

$$\sin A = \frac{a}{c} = \frac{\text{opposite side}}{\text{hypotenuse}} \qquad \cot A = \frac{b}{a} = \frac{\text{adjacent side}}{\text{opposite side}}$$

$$\cos A = \frac{b}{c} = \frac{\text{adjacent side}}{\text{hypotenuse}} \qquad \sec A = \frac{c}{b} = \frac{\text{hypotenuse}}{\text{adjacent side}}$$

$$\tan A = \frac{a}{b} = \frac{\text{opposite side}}{\text{adjacent side}} \qquad \csc A = \frac{c}{a} = \frac{\text{hypotenuse}}{\text{opposite side}}$$

3.2 TRIGONOMETRIC FUNCTIONS OF COMPLEMENTARY ANGLES

The acute angles A and B of the right triangle ABC are complementary; that is, $A + B = 90°$. From Fig. 3-1, we have

$$\sin B = b/c = \cos A \qquad \cot B = a/b = \tan A$$

$$\cos B = a/c = \sin A \qquad \sec B = c/a = \csc A$$

$$\tan B = b/a = \cot A \qquad \csc B = c/b = \sec A$$

These relations associate the functions in pairs—sine and cosine, tangent and cotangent, secant and cosecant—each function of a pair being called the *cofunction* of the other. Thus, any function of an acute angle is equal to the corresponding cofunction of the complementary angle.

EXAMPLE 3.1 Find the values of the trigonometric functions of the angles of the right triangle ABC in Fig. 3-3.

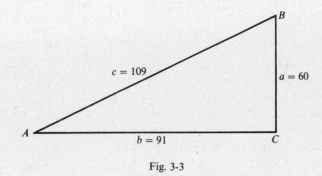

Fig. 3-3

$$\sin A = \frac{\text{opposite side}}{\text{hypotenuse}} = \frac{a}{c} = \frac{60}{109} \qquad \csc A = \frac{\text{hypotenuse}}{\text{opposite side}} = \frac{c}{a} = \frac{109}{60}$$

$$\cos A = \frac{\text{adjacent side}}{\text{hypotenuse}} = \frac{b}{c} = \frac{91}{109} \qquad \sec A = \frac{\text{hypotenuse}}{\text{adjacent side}} = \frac{c}{b} = \frac{109}{91}$$

$$\tan A = \frac{\text{opposite side}}{\text{adjacent side}} = \frac{a}{b} = \frac{60}{91} \qquad \cot A = \frac{\text{adjacent side}}{\text{opposite side}} = \frac{b}{a} = \frac{91}{60}$$

$$\sin B = \frac{\text{opposite side}}{\text{hypotenuse}} = \frac{b}{c} = \frac{91}{109} \qquad \csc B = \frac{\text{hypotenuse}}{\text{opposite side}} = \frac{c}{b} = \frac{109}{91}$$

$$\cos B = \frac{\text{adjacent side}}{\text{hypotenuse}} = \frac{a}{c} = \frac{60}{109} \qquad \sec B = \frac{\text{hypotenuse}}{\text{adjacent side}} = \frac{c}{a} = \frac{109}{60}$$

$$\tan B = \frac{\text{opposite side}}{\text{adjacent side}} = \frac{b}{a} = \frac{91}{60} \qquad \cot B = \frac{\text{adjacent side}}{\text{opposite side}} = \frac{a}{b} = \frac{60}{91}$$

3.3 TRIGONOMETRIC FUNCTIONS OF 30°, 45°, AND 60°

The special acute angles 30°, 45°, and 60° (see Appendix 1 Geometry) have trigonometric function values that can be computed exactly. The following results are obtained in Probs. 3.8 and 3.9. For each fraction that had an irrational number denominator, only the equivalent fraction with a rational number denominator is stated in the table.

Angle θ	$\sin\theta$	$\cos\theta$	$\tan\theta$	$\cot\theta$	$\sec\theta$	$\csc\theta$
30°	$\frac{1}{2}$	$\frac{1}{2}\sqrt{3}$	$\frac{1}{3}\sqrt{3}$	$\sqrt{3}$	$\frac{2}{3}\sqrt{3}$	2
45°	$\frac{1}{2}\sqrt{2}$	$\frac{1}{2}\sqrt{2}$	1	1	$\sqrt{2}$	$\sqrt{2}$
60°	$\frac{1}{2}\sqrt{3}$	$\frac{1}{2}$	$\sqrt{3}$	$\frac{1}{3}\sqrt{3}$	2	$\frac{2}{3}\sqrt{3}$

3.4 TRIGONOMETRIC FUNCTION VALUES

For many application problems, values of trigonometric functions are needed for angles that are not special angles. These values may be found in tables of trigonometric functions or by using a scientific calculator. Problems 3.10 to 3.15 illustrate a number of simple applications of trigonometric functions. For these problems, a two-decimal-place table is included below.

Angle θ	$\sin \theta$	$\cos \theta$	$\tan \theta$	$\cot \theta$	$\sec \theta$	$\csc \theta$
15°	0.26	0.97	0.27	3.73	1.04	3.86
20°	0.34	0.94	0.36	2.75	1.06	2.92
30°	0.50	0.87	0.58	1.73	1.15	2.00
40°	0.64	0.77	0.84	1.19	1.31	1.56
45°	0.71	0.71	1.00	1.00	1.41	1.41
50°	0.77	0.64	1.19	0.84	1.56	1.31
60°	0.87	0.50	1.73	0.58	2.00	1.15
70°	0.94	0.34	2.75	0.36	2.92	1.06
75°	0.97	0.26	3.73	0.27	3.86	1.04

When using a calculator to find values for trigonometric functions, be sure to follow the procedure indicated in the instruction manual for your calculator. In general the procedure is (1) make sure the calculator is in degree mode, (2) enter the number of degrees in the angle, (3) press the key for the trigonometric function wanted, and (4) read the function value from the display.

EXAMPLE 3.2   Find tan 15° using a calculator. With the calculator in degree mode, enter 15 and press the (tan) key. The number 0.267949 will appear on the display; thus tan 15° = 0.267949. The number of digits that are displayed depends on the calculator used, but most scientific calculators show at least six digits. In this book if the value displayed on a calculator is not exact, it will be rounded to six digits when stated in a problem or example. Rounding procedures for final results will be introduced as needed.

Using a calculator to find an acute angle when the value of a trigonometric function is given requires the use of the inverse (inv) key or the second function (2nd) key. The value of the function is entered, the (inv) key is pressed, and then the trigonometric function key is pressed. The degree mode is used to get answers in degree measure.

EXAMPLE 3.3   Find acute angle A when sin $A = 0.2651$. With the calculator in degree mode, enter .2651 and press the (inv) key and the (sin) key. The number 15.3729 on the display is the degree measure of acute angle A. Thus to the nearest degree $A = 15°$.

3.5   ACCURACY OF RESULTS USING APPROXIMATIONS

When using approximate numbers, the results need to be rounded. In this chapter, we will report angles to the nearest degree and lengths to the nearest unit. If a problem has intermediate values to be computed, wait to round numbers until the final result is found. Each intermediate value should have at least one more digit than the final result is to have so that each rounding does not directly involve the unit of accuracy.

3.6   SELECTING THE FUNCTION IN PROBLEM SOLVING

In finding a side of a right triangle when an angle and a side are known, there are two trigonometric functions which can be used, a function and its reciprocal. When manually solving the problem, the choice is usually made so the unknown side is in the numerator of the fraction. This is done so the operation needed to solve the equation will be multiplication rather than division. When a calculator is used, the function selected is sine, cosine, or tangent since these functions are represented by keys on the calculator.

EXAMPLE 3.4 A support wire is anchored 12 m up from the base of a flagpole and the wire makes a 15° angle with the ground. How long is the wire?

From Fig. 3-4, it can be seen that both sin 15° and csc 15° involve the known length 12 m and the requested length x. Either function can be used to solve the problem. The manual solution, that is using tables not a calculator, is

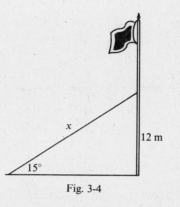

Fig. 3-4

easier using csc 15° but not all trigonometric tables list values for secant and cosecant. The calculator solution will use sin 15° since there is no function key for cosecant.

	Manual Solution		Calculator Solution
$\csc 15° = \dfrac{x}{12}$	or	$\sin 15° = \dfrac{12}{x}$	$\sin 15° = \dfrac{12}{x}$
$x = 12 \csc 15°$		$x = \dfrac{12}{\sin 15°}$	$x = \dfrac{12}{\sin 15°}$
$x = 12(3.86)$		$x = \dfrac{12}{0.26}$	$x = \dfrac{12}{0.258819}$
$x = 46.32$		$x = 46.15$	$x = 46.3644$
$x = 46$ m		$x = 46$ m	$x = 46$ m

The wire is 46 m long.

In each solution, the result to the nearest meter is the same but the results of the computations are different because of the rounding used in determining the value of the function used. Rounding to a few decimal places, as in the table provided in this section, often leads to different computational results. Using the four-decimal-place tables in Appendix 2 will result in very few situations where the choice of functions affects the results of the computation. Also, when these tables are used, the results will more frequently agree with those found using a calculator.

For the problems in this chapter, a manual solution and a calculator solution will be shown and an answer for each procedure will be indicated. In later chapters, an answer for each method will be indicated only when the two procedures produce different results.

The decision to use or not to use a calculator is a personal one for you to make. If you will not be able to use a calculator when you apply the procedures studied, then do not practice them using a calculator. Occasionally there will be procedures discussed that are used only with tables and others that apply to calculator solutions only. These will be clearly indicated and can be omitted if you are not using that solution method.

3.7 ANGLES OF DEPRESSION AND ELEVATION

An angle of depression is the angle from the horizontal down to the line of sight from the observer to an object below. The angle of elevation is the angle from the horizontal up to the line of sight from the observer to an object above.

In Fig. 3-5, the angle of depression from point A to point B is α and the angle of elevation from point B to point A is β. Since both angles are measured from horizontal lines, which are parallel, the line of sight AB is a transversal, and since alternate interior angles for parallel lines are equal, $\alpha = \beta$. (See Appendix 1 Geometry.)

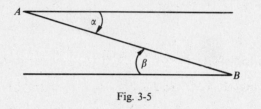

Fig. 3-5

Solved Problems

3.1 Find the trigonometric functions of the acute angles of the right triangle ABC, Fig. 3-6, given $b = 24$ and $c = 25$.

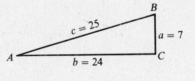

Fig. 3-6

Since $a^2 = c^2 - b^2 = (25)^2 - (24)^2 = 49$, $a = 7$. Then

$$\sin A = \frac{\text{opposite side}}{\text{hypotenuse}} = \frac{7}{25} \qquad \cot A = \frac{\text{adjacent side}}{\text{opposite side}} = \frac{24}{7}$$

$$\cos A = \frac{\text{adjacent side}}{\text{hypotenuse}} = \frac{24}{25} \qquad \sec A = \frac{\text{hypotenuse}}{\text{adjacent side}} = \frac{25}{24}$$

$$\tan A = \frac{\text{opposite side}}{\text{adjacent side}} = \frac{7}{24} \qquad \csc A = \frac{\text{hypotenuse}}{\text{opposite side}} = \frac{25}{7}$$

and
$$\sin B = 24/25 \qquad \cot B = 7/24$$

$$\cos B = 7/25 \qquad \sec B = 25/7$$

$$\tan B = 24/7 \qquad \csc B = 25/24$$

3.2 Find the values of the trigonometric functions of the acute angles of the right triangle ABC, Fig. 3-7, given $a = 2$ and $c = 2\sqrt{5}$.

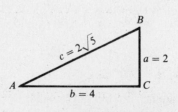

Fig. 3-7

Since $b^2 = c^2 - a^2 = (2\sqrt{5})^2 - (2)^2 = 20 - 4 = 16$, $b = 4$. Then

$$\sin A = 2/2\sqrt{5} = \sqrt{5}/5 = \cos B \qquad \cot A = 4/2 = 2 = \tan B$$

$$\cos A = 4/2\sqrt{5} = 2\sqrt{5}/5 = \sin B \qquad \sec A = 2\sqrt{5}/4 = \sqrt{5}/2 = \csc B$$

$$\tan A = 2/4 = 1/2 = \cot B \qquad \csc A = 2\sqrt{5}/2 = \sqrt{5} = \sec B$$

3.3 Find the values of the trigonometric functions of the acute angle A, given $\sin A = 3/7$.

Construct the right triangle ABC, Fig. 3-8, with $a = 3$, $c = 7$, and $b = \sqrt{7^2 - 3^2} = 2\sqrt{10}$ units. Then

$$\sin A = 3/7 \qquad\qquad \cot A = 2\sqrt{10}/3$$

$$\cos A = 2\sqrt{10}/7 \qquad\qquad \sec A = 7/2\sqrt{10} = 7\sqrt{10}/20$$

$$\tan A = 3/2\sqrt{10} = 3\sqrt{10}/20 \qquad \csc A = 7/3$$

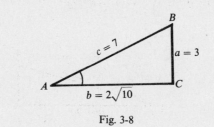

Fig. 3-8

3.4 Find the values of the trigonometric functions of the acute angle B, given $\tan B = 1.5$.

Refer to Fig. 3-9. Construct the right triangle ABC having $b = 15$ and $a = 10$ units. (Note that $1.5 = \frac{3}{2}$, and therefore a right triangle with $b = 3$ and $a = 2$ will serve equally well.)

Then $c = \sqrt{a^2 + b^2} = \sqrt{10^2 + 15^2} = 5\sqrt{13}$ and

$$\sin B = 15/5\sqrt{13} = 3\sqrt{13}/13 \qquad \cot B = 2/3$$

$$\cos B = 10/5\sqrt{13} = 2\sqrt{13}/13 \qquad \sec B = 5\sqrt{13}/10 = \sqrt{13}/2$$

$$\tan B = 15/10 = 3/2 \qquad\qquad \csc B = 5\sqrt{13}/15 = \sqrt{13}/3$$

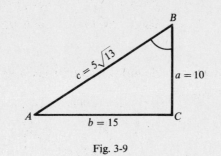

Fig. 3-9

3.5 If A is acute and $\sin A = 2x/3$, determine the values of the remaining functions.

Construct the right triangle ABC having $a = 2x < 3$ and $c = 3$, as in Fig. 3-10.

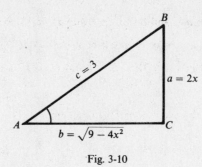

Fig. 3-10

Then $b = \sqrt{c^2 - a^2} = \sqrt{9 - 4x^2}$ and

$$\sin A = \frac{2x}{3} \qquad \cos A = \frac{\sqrt{9 - 4x^2}}{3} \qquad \tan A = \frac{2x}{\sqrt{9 - 4x^2}} = \frac{2x\sqrt{9 - 4x^2}}{9 - 4x^2}$$

$$\cot A = \frac{\sqrt{9 - 4x^2}}{2x} \qquad \sec A = \frac{3}{\sqrt{9 - 4x^2}} = \frac{3\sqrt{9 - 4x^2}}{9 - 4x^2} \qquad \csc A = \frac{3}{2x}$$

3.6 If A is acute and $\tan A = x = x/1$, determine the values of the remaining functions.

Construct the right triangle ABC having $a = x$ and $b = 1$, as in Fig. 3-11. Then $c = \sqrt{x^2 + 1}$ and

$$\sin A = \frac{x}{\sqrt{x^2 + 1}} = \frac{x\sqrt{x^2 + 1}}{x^2 + 1} \qquad \cos A = \frac{1}{\sqrt{x^2 + 1}} = \frac{\sqrt{x^2 + 1}}{x^2 + 1} \qquad \tan A = x$$

$$\cot A = \frac{1}{x} \qquad \sec A = \sqrt{x^2 + 1} \qquad \csc A = \frac{\sqrt{x^2 + 1}}{x}$$

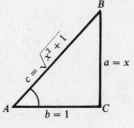

Fig. 3-11

3.7 If A is an acute angle:

(a) Why is $\sin A < 1$?   (d) Why is $\sin A < \tan A$?

(b) When is $\sin A = \cos A$?   (e) When is $\sin A < \cos A$?

(c) Why is $\sin A < \csc A$?   (f) When is $\tan A > 1$?

In the right triangle ABC:

(a) Side $a <$ side c; therefore $\sin A = a/c < 1$.

(b) $\sin A = \cos A$ when $a/c = b/c$; then $a = b$, $A = B$, and $A = 45°$.

(c) $\sin A < 1$ (above) and $\csc A = 1/\sin A > 1$.

(d) Sin $A = a/c$, tan $A = a/b$, and $b < c$; therefore $a/c < a/b$ or sin $A <$ tan A.

(e) Sin $A <$ cos A when $a < b$; then $A < B$ pr $A < 90° - A$. and $A < 45°$.

(f) Tan $A = a/b > 1$ when $a > b$; then $A > B$ and $A > 45°$.

3.8 Find the exact values of the trigonometric functions of 45°. (See Fig. 3-12.)

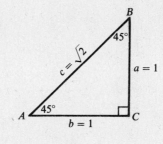

Fig. 3-12

In any isosceles right triangle ABC, $A = B = 45°$ and $a = b$. Let $a = b = 1$; then $c = \sqrt{1 + 1} = \sqrt{2}$ and

$$\sin 45° = 1/\sqrt{2} = \tfrac{1}{2}\sqrt{2} \qquad \cot 45° = 1$$

$$\cos 45° = 1/\sqrt{2} = \tfrac{1}{2}\sqrt{2} \qquad \sec 45° = \sqrt{2}$$

$$\tan 45° = 1/1 = 1 \qquad \csc 45° = \sqrt{2}$$

3.9 Find the exact values of the trigonometric functions of 30° and 60°. (See Fig. 3-13.)

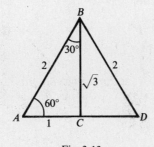

Fig. 3-13

In any equilateral triangle ABD, each angle is 60°. The bisector of any angle, like B, is the perpendicular bisector of the opposite side. Let the sides of the equilateral triangle be of length 2 units. Then in the right triangle ABC, $AB = 2$, $AC = 1$, and $BC = \sqrt{2^2 - 1^2} = \sqrt{3}$.

$$\sin 30° = 1/2 = \cos 60° \qquad\qquad \cot 30° = \sqrt{3} = \tan 60°$$

$$\cos 30° = \sqrt{3}/2 = \sin 60° \qquad\qquad \sec 30° = 2/\sqrt{3} = 2\sqrt{3}/3 = \csc 60°$$

$$\tan 30° = 1/\sqrt{3} = \sqrt{3}/3 = \cot 60° \qquad \csc 30° = 2 = \sec 60°$$

(NOTE: In Probs. 3.10 to 3.15 two solution procedures are shown, one for manual solution and one for calculator solution, whenever the two are different. Which one you use depends upon your access to a calculator during your problem-solving work. If your access to a calculator is restricted, then focus only on the manual solutions. In the calculator solutions, steps are shown to illustrate the procedures rather than as a guide to work steps that need to be shown. The steps shown in each solution are to allow you to see all the details of the procedure used.)

3.10   When the sun is 20° above the horizon, how long is the shadow cast by a building 50 m high?

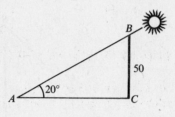

Fig. 3-14

In Fig. 3-14, $A = 20°$, $CB = 50$, and AC is to be found.

Manual Solution	Calculator Solution
$\cot A = \dfrac{AC}{CB}$	$\tan A = \dfrac{CB}{AC}$
$AC = CB \cot A$	$AC = \dfrac{CB}{\tan A}$
$AC = 50 \cot 20°$	$AC = \dfrac{50}{\tan 20°}$
$AC = 50(2.75)$	$AC = \dfrac{50}{0.363970}$
$AC = 137.5$	$AC = 137.374$
$AC = 138$ m	$AC = 137$ m

(NOTE: The difference in the answers for the two procedures is because cot 20° was rounded to two decimal places in the table. Each answer is the correct one for that procedure.)

3.11   A tree 100 ft tall casts a shadow 120 ft long. Find the angle of elevation of the sun.

In Fig. 3-15, $CB = 100$, $AC = 120$, and we want to find A.

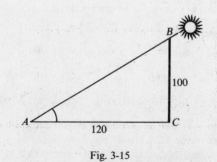

Fig. 3-15

Manual Solution	Calculator Solution
$\tan A = \dfrac{CB}{AC}$	$\tan A = \dfrac{CB}{AC}$
$\tan A = \dfrac{100}{120}$	$\tan A = \dfrac{100}{120}$
$\tan A = 0.83$	$\tan A = 0.833333$
$A = 40°$	$A = 39.8056°$
	$A = 40°$

(Since $\tan 40°$ has the closest value to 0.83, we used $A = 40°$.)

3.12 A ladder leans against the side of a building with its foot 12 ft from the building. How far from the ground is the top of the ladder and how long is the ladder if it makes an angle of 70° with the ground?

From Fig. 3-16, $\tan A = CB/AC$; then $CB = AC \tan A = 12 \tan 70° = 12(2.75) = 33$. The top of the ladder is 33 ft above the ground. The calculator solution procedure is the same. Manual: $\sec A = AB/AC$; then $AB = AC \sec A = 12 \sec 70° = 12(2.92) = 35.04$.

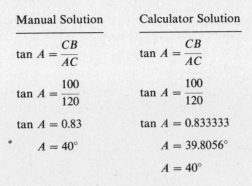

Fig. 3-16

Calculator: $\cos A = AC/AB$; then $AB = AC/(\cos A) = 12/(\cos 70°) = 12/0.342020 = 35.0857$.

The ladder is 35 ft long.

3.13 From the top of a lighthouse, 120 m above the sea, the angle of depression of a boat is 15°. How far is the boat from the lighthouse?

In Fig. 3-17, the right triangle ABC has $A = 15°$ and $CB = 120$. Then solutions follow.

Manual: $\cot A = AC/CB$ and $AC = CB \cot A = 120 \cot 15° = 120(3.73) = 447.6$.

Calculator: $\tan A = CB/AC$ and $AC = CB/(\tan A) = 120/(\tan 15°) = 120/0.267949 = 447.846$.

The boat is 448 m from the lighthouse.

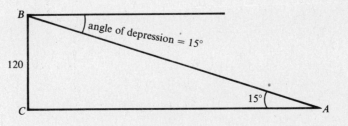

Fig. 3-17

3.14 Find the length of the chord of a circle of radius 20 cm subtended by a central angle of 150°.

In Fig. 3-18, OC bisects $\angle AOB$. Then $BC = AC$ and OAC is a right triangle.

Manual: In $\triangle OAC$, $\sin \angle COA = AC/OA$ and $AC = OA \sin \angle COA = 20 \sin 75° = 20(0.97) = 19.4$; $BA = 2(19.4) = 38.8$.

Calculator: $AC = OA \sin \angle COA = 20 \sin 75° = 20(0.965926) = 19.3185$; $BA = 2(19.3185) = 38.6370$.

The length of the chord is 39 cm.

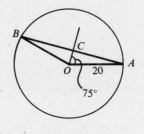

Fig. 3-18

3.15 Find the height of a tree if the angle of elevation of its top changes from 20° to 40° as the observer advances 75 ft toward its base. See Fig. 3-19.

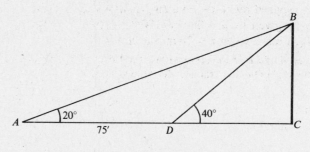

Fig. 3-19

In the right triangle ABC, $\cot A = AC/CB$; then $AC = CB \cot A$ or $DC + 75 = CB \cot 20°$.
In the right triangle DBC, $\cot D = DC/CB$; then $DC = CB \cot 40°$.

Manual:
$$DC = CB \cot 20° - 75 = CB \cot 40°$$
$$CB(\cot 20° - \cot 40°) = 75$$
$$CB(2.75 - 1.19) = 75$$

and
$$CB = 75/1.56 = 48.08$$

Calculator:
$$\cot 20° = 1/\tan 20° = 1/0.363970 = 2.74748$$
$$\cot 40° = 1/\tan 40° = 1/0.839100 = 1.19175$$
$$CB(\cot 20° - \cot 40°) = 75$$
$$CB(2.74748 - 1.19175) = 75$$
$$CB(1.55573) = 75$$
$$CB = 75/1.55573 = 48.2089$$

The tree is 48 ft tall.

3.16 A tower standing on level ground is due north of point A and due west of point B, a distance c ft from A. If the angles of elevation of the top of the tower as measured from A and B are α and β respectively, find the height h of the tower.

In the right triangle ACD of Fig. 3-20, $\cot \alpha = AC/h$; and in the right triangle BCD, $\cot \beta = BC/h$. Then $AC = h \cot \alpha$ and $BC = h \cot \beta$.

Since ABC is a right triangle, $(AC)^2 + (BC)^2 = c^2 = h^2(\cot \alpha)^2 + h^2(\cot \beta)^2$ and

$$h = \frac{c}{\sqrt{(\cot \alpha)^2 + (\cot \beta)^2}}$$

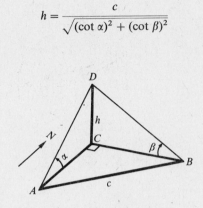

Fig. 3-20

3.17 If holes are to be spaced regularly on a circle, show that the distance d between the centers of two successive holes is given by $d = 2r \sin (180°/n)$, where $r =$ the radius of the circle and $n =$ the number of holes. Find d when $r = 20$ in and $n = 4$.

In Fig. 3-21, let A and B be the centers of two consecutive holes on the circle of radius r and center O. Let the bisector of the angle O of the triangle AOB meet AB at C. In right triangle AOC,

$$\sin \angle AOC = \frac{AC}{r} = \frac{\frac{1}{2}d}{r} = \frac{d}{2r}$$

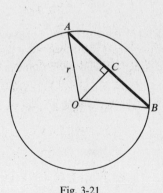

Fig. 3-21

Then
$$d = 2r \sin \angle AOC = 2r \sin \tfrac{1}{2} \angle AOB$$
$$= 2r \sin \frac{1}{2}\left(\frac{360°}{n}\right) = 2r \sin \frac{180°}{n}$$

When $r = 20$ and $n = 4$, $d = 2 \cdot 20 \sin 45° = 2 \cdot 20(\sqrt{2}/2) = 20\sqrt{2}$ in.

Supplementary Problems

3.18 Find the exact values of the trigonometric functions of the acute angles of the right triangle ABC, given:
(a) $a = 3, b = 1$;   (b) $a = 2, c = 5$;   (c) $b = \sqrt{7}, c = 4$

Ans.   Answers are in the order sine, cosine, tangent, cotangent, secant, and cosecant.

(a) $A: 3/\sqrt{10} = 3\sqrt{10}/10, 1/\sqrt{10} = \sqrt{10}/10, 3, 1/3, \sqrt{10}, \sqrt{10}/3$;
$B: 1/\sqrt{10} = \sqrt{10}/10, 3/\sqrt{10} = 3\sqrt{10}/10, 1/3, 3, \sqrt{10}/3, \sqrt{10}$

(b) $A: 2/5, \sqrt{21}/5, 2/\sqrt{21} = 2\sqrt{21}/21, \sqrt{21}/2, 5/\sqrt{21} = 5\sqrt{21}/21, 5/2$;
$B: \sqrt{21}/5, 2/5, \sqrt{21}/2, 2/\sqrt{21} = 2\sqrt{21}/21, 5/2, 5/\sqrt{21} = 5\sqrt{21}/21$

(c) $A: 3/4, \sqrt{7}/4, 3/\sqrt{7} = 3\sqrt{7}/7, \sqrt{7}/3, 4/\sqrt{7} = 4\sqrt{7}/7, 4/3$;
$B: \sqrt{7}/4, 3/4, \sqrt{7}/3, 3/\sqrt{7} = 3\sqrt{7}/7, 4/3, 4/\sqrt{7} = 4\sqrt{7}/7$

3.19 Which is the greater and why:

(a) $\sin 55°$ or $\cos 55°$ (c) $\tan 15°$ or $\cot 15°$

(b) $\sin 40°$ or $\cos 40°$ (d) $\sec 55°$ or $\csc 55°$

Ans.   (a) $\sin 55°$,   (b) $\cos 40°$,   (c) $\cot 15°$,   (d) $\sec 55°$

3.20 Find the exact value of each of the following.

(a) $\sin 30° + \tan 45°$

(b) $\cot 45° + \cos 60°$

(c) $\sin 30° \cos 60° + \cos 30° \sin 60°$

(d) $\cos 30° \cos 60° - \sin 30° \sin 60°$

(e) $\dfrac{\tan 60° - \tan 30°}{1 + \tan 60° \tan 30°}$

(f) $\dfrac{\csc 30° + \csc 60° + \csc 90°}{\sec 0° + \sec 30° + \sec 60°}$

Ans.   (a) 3/2,   (b) 3/2,   (c) 1,   (d) 0,   (e) $1/\sqrt{3} = \sqrt{3}/3$,   (f) 1

3.21 A man drives 500 m along a road which is inclined 20° to the horizontal. How high above his starting point is he?

Ans.   Manual: 170 m; calculator: 171 m (manual answer differs owing to rounding of table values).

3.22 A tree broken over by the wind forms a right triangle with the ground. If the broken part makes an angle of 50° with the ground and the top of the tree is now 20 ft from its base, how tall was the tree?

Ans.   55 ft

3.23 Two straight roads intersect to form an angle of 75°. Find the shortest distance from one road to a gas station on the other road 1000 m from the junction.

Ans.   Manual: 3730 m; calculator; 3732 (manual answer differs owing to rounding of table values).

3.24 Two buildings with flat roofs are 60 m apart. From the roof of the shorter building, 40 m in height, the angle of elevation to the edge of the roof of the taller building is 40°. How high is the taller building?

Ans.   90 m

3.25 A ladder, with its foot in the street, makes an angle of 30° with the street when its top rests on a building on one side of the street and makes an angle of 40° with the street when its top rests on a building on the other side of the street. If the ladder is 50 ft long, how wide is the street?

Ans. 82 ft

3.26 Find the perimeter of an isosceles triangle whose base is 40 cm and whose base angle is 70°.

Ans. 157 cm

Chapter 4

Solution of Right Triangles

4.1 INTRODUCTION

The solution of right triangles depends on having approximate values for trigonometric functions of acute angles. An important part of the solution is determining the appropriate value to use for a trigonometric function. This part of the solution is different when you are using tables as in Secs. 4.2 to 4.4 from when you are using a scientific calculator as in Secs. 4.5 and 4.6.

In general, the procedure will be to use the given data to write an equation using a trigonometric function and then to solve for the unknown value in the equation. The given data will consist either of two sides of a right triangle or of one side and an acute angle. Once one value has been found, a second acute angle and the remaining side can be found. The second acute angle is found using the fact that the acute angles of a right triangle are complementary (i.e., add up to 90°). The third side is found by using a definition of a second trigonometric function or by using the pythagorean theorem (see Appendix 1 Geometry).

4.2 FOUR-PLACE TABLES OF TRIGONOMETRIC FUNCTIONS

Appendix 2 Tables has three different four-decimal-place tables of values for trigonometric functions, with Table 1 giving angles in 10' intervals, Table 2 giving angles in 0.1° intervals, and Table 3 giving angles in 0.01-rad intervals. Tables published in texts differ in several ways, such as in the number of digits listed, the number of decimal places in each value, whether or not secant and cosecant values are listed, and the measurement unit of the angles.

The angles in Tables 1 and 2 are listed in the left- and right-hand columns. Angles less than 45° are located in the left-hand column and the function is read from the top of the page. Angles greater than 45° are located in the right-hand column and the function is read from the bottom of the page. In each row, the sum of the angles in the left- and right-hand columns is 90°, and the tables are based on the fact that cofunctions of complementary angles are equal.

In Table 3, the angles in radians are listed in the left-hand column only and the function is read from the top of the page.

4.3 TABLES OF VALUES FOR TRIGONOMETRIC FUNCTIONS

In this chapter, Table 1 or Table 2 will be used to find values of trigonometric functions whenever a manual solution is used. If the angle contains a number of degrees only or a number of degrees and a multiple of 10', the value of the function is read directly from the table.

EXAMPLE 4.1 Find sin 24°40'.

Opposite 24°40' (<45°) in the left-hand column read the entry 0.4173 in the column labeled sin A at the top.

EXAMPLE 4.2 Find cos 72°.

Opposite 72° (>45°) in the right-hand column read the entry 0.3090 in the column labeled cos A at the bottom.

EXAMPLE 4.3

(a) tan 55°20' = 1.4460. Read *up* the page since 55°20' > 45°.

(b) cot 41°50' = 1.1171. Read down the page since 41°50' < 45°.

If the number of minutes in the given angle is not a multiple of 10, as in 24°43′, interpolate between the values of the functions of the two nearest angles (24°40′ and 24°50′) using the method of proportional parts.

EXAMPLE 4.4   Find sin 24°43′.

We find

$$\sin 24°40' = 0.4173$$
$$\sin 24°50' = 0.4200$$

Difference for 10′ = 0.0027 = tabular difference

Correction = difference for 3′ = 0.3(0.0027) = 0.00081 or 0.0008 when rounded off to four decimal places.
As the angle increases, the sine of the angle increases; thus,

$$\sin 24°43' = 0.4173 + 0.0008 = 0.4181$$

If a five-place table is available, the value 0.41813 can be read directly from the table and then rounded off to 0.4181.

EXAMPLE 4.5   Find cos 64°26′.

We find

$$\cos 64°20' = 0.4331$$
$$\cos 64°30' = 0.4305$$

Tabular difference = 0.0026

Correction = 0.6(0.0026) = 0.00156 or 0.0016 to four decimal places.
As the angle increases, the cosine of the angle decreases. Thus

$$\cos 64°26' = 0.4331 - 0.0016 = 0.4315$$

To save time, we should proceed as follows in Example 4.4:

(a)   Locate sin 24°40′ = 0.4173. For the moment, disregard the decimal point and use only the sequence 4173.

(b)   Find (mentally) the tabular difference 27, that is, the difference between the sequence 4173 corresponding to 24°40′ and the sequence 4200 corresponding to 24°50′.

(c)   Find 0.3(27) = 8.1 and round off to the nearest integer. This is the correction.

(d)   Add (since sine) the correction to 4173 and obtain 4181. Then

$$\sin 24°43' = 0.4181$$

When, as in Example 4.4, we interpolate from the smaller angle to the larger: (1) The correction is added in finding the sine, tangent, and secant. (2) The correction is subtracted in finding the cosine, cotangent, and cosecant.

EXAMPLE 4.6   Find cos 27.23°.

We find

$$\cos 27.20° = 0.8894$$
$$\cos 27.30° = 0.8886$$

Tabular difference = 0.0008

Correction = 0.3(0.0008) = 0.00024 or 0.0002 to four decimal places.
As the angle increases, the cosine decreases and thus

$$\cos 27.23° = 0.8894 - 0.0002 = 0.8892$$

EXAMPLE 4.7   Find sec 57.08°.

We find

$$\sec 57.00° = 1.8361$$
$$\sec 57.10° = 1.8410$$

Tabular difference = 0.0049

Correction = 0.8(0.0049) = 0.00392 or 0.0039 to four decimal places.
As the angle increases, the secant increases and thus

$$\sec 57.08° = 1.8361 + 0.0039 = 1.8400$$

(See Probs. 4.1 and 4.2.)

4.4   USING TABLES TO FIND AN ANGLE GIVEN A FUNCTION VALUE

The process is a reversal of that given above.

EXAMPLE 4.8   Reading directly from Table 1, we find

$$0.2924 = \sin 17° \qquad 2.7725 = \tan 70°10'$$

EXAMPLE 4.9   Find A, given $\sin A = 0.4234$. (Use Table 1.)
The given value is not an entry in the table. We find, however,

$$
\begin{array}{ll}
0.4226 = \sin 25°0' & \qquad 0.4226 = \sin 25°0' \\
0.4253 = \sin 25°10' & \qquad 0.4234 = \sin A \\
\overline{0.0027} = \text{tabular difference} & \qquad \overline{0.0008} = \text{partial difference}
\end{array}
$$

Correction $= \dfrac{0.0008}{0.0027}(10') = \dfrac{8}{27}(10') = 3'$, to the nearest minute.

Adding (since sine) the correction, we have $25°0' + 3' = 25°3' = A$.

EXAMPLE 4.10   Find A, given $\cot A = 0.6345$. (Use Table 1.)

We find

$$
\begin{array}{ll}
0.6330 = \cot 57°40' & \qquad 0.6330 = \cot 57°40' \\
0.6371 = \cot 57°30' & \qquad 0.6345 = \cot A \\
\overline{0.0041} = \text{tabular diff.} & \qquad \overline{0.0015} = \text{partial difference}
\end{array}
$$

Correction $= \dfrac{0.0015}{0.0041}(10') = \dfrac{15}{41}(10') = 4'$, to the nearest minute.

Subtracting (since cot) the correction, we have $57°40' - 4' = 57°36' = A$.

To save time, we should proceed as follows in Example 4.9:

(*a*)   Locate the next smaller entry, $0.4226 = \sin 25°0'$. For the moment use only the sequence 4226.

(*b*)   Find the tabular difference, 27.

(*c*)   Find the partial difference, 8, between 4226 and the given sequence 4234.

(*d*)   Find $\frac{8}{27}(10') = 3'$ and add to $25°0'$.

EXAMPLE 4.11   Find A, given $\sin A = 0.4234$. (Use Table 2.)
The given value is not an entry in the table. We find

$$
\begin{array}{ll}
0.4226 = \sin 25.00° & \qquad 0.4226 = \sin 25.00° \\
0.4242 = \sin 25.10° & \qquad 0.4234 = \sin A \\
\overline{0.0016} = \text{tabular difference} & \qquad \overline{0.0008} = \text{partial difference}
\end{array}
$$

Correction $= \dfrac{0.0008}{0.0016}(0.1) = 0.05$, to the nearest hundredth.

Adding (since sine) the correction, we have $A = 25.00° + 0.05° = 25.05°$.

EXAMPLE 4.12   Find A, given $\cot A = 0.6345$. (Use Table 2.)

We find

$$
\begin{array}{ll}
0.6322 = \cot 57.60° & \qquad 0.6322 = \cot 57.60° \\
0.6346 = \cot 57.50° & \qquad 0.6345 = \cot A \\
\overline{0.0024} = \text{tabular difference} & \qquad \overline{0.0023} = \text{partial difference}
\end{array}
$$

Correction $= \dfrac{0.0023}{0.0024}(0.1) = 0.10$, to the nearest hundredth.

Subtracting (since cotangent) the correction, we have $A = 57.60° - 0.10° = 57.50°$.

(See Prob. 4.4.)

4.5 CALCULATOR VALUES OF TRIGONOMETRIC FUNCTIONS

Calculators give values of trigonometric functions based on the number of digits that can be displayed, usually 8, 10, or 12. The number of decimal places shown varies with the size of the number but is usually at least four. When a calculator is used in this book, all trigonometric function values shown will be rounded to six digits unless the value is exact using fewer digits.

EXAMPLE 4.13   Find sin 24°40′.

(a)   Put the calculator in degree mode.

(b)   Enter 24, press (+) key, enter 40, press (÷) key, enter 60, and press (=) key.

(c)   Press (sin) key.

(d)   sin 24°40′ = 0.417338 rounded to six digits.

EXAMPLE 4.14   Find tan 48°23′.

(a)   Put the calculator in degree mode.

(b)   Enter 48, press (+) key, enter 23, press (÷) key, enter 60, and press (=) key.

(c)   Press (tan) key.

(d)   tan 48°23′ = 1.12567 rounded to six digits.

EXAMPLE 4.15   Find cos 53.28°.

(a)   Put the calculator in degree mode.

(b)   Enter 53.28.

(c)   Press (cos) key.

(d)   cos 53.28° = 0.597905 rounded to six digits.

For values of cotangent, secant, and cosecant, the reciprocal of the value of the reciprocal function is used. (See Sec. 2.4.)

EXAMPLE 4.16   Find cot 37°20′.

(a)   Put the calculator in degree mode.

(b)   Enter 37, press (+) key, enter 20, press (÷) key, enter 60, and press (=) key.

(c)   Press (tan) key.

(d)   Press (1/x) key or divide 1 by the value of tan 37°20′ from (c).

(e)   cot 37°20′ = 1.31110 rounded to six digits.

If the calculator being used has parentheses keys, "(" and ")", they may be used to simplify problems by performing continuous operations.

Using parentheses in Example 4.14 yields:

(a)   Put the calculator in degree mode.

(b)   Press (( ) key, enter 48, press (+) key, enter 23, press (÷) key, enter 60, press ( )) key, and press (tan) key.

(c)   tan 48°23′ = 1.12567 rounded to six digits.

This procedure will be indicated in calculator solutions by showing $\tan 48°23′ = \tan (48 + \frac{23}{60})° = 1.12567$.

(See Prob. 4.3.)

4.6 FIND AN ANGLE GIVEN A FUNCTION VALUE USING A CALCULATOR

The values of angles can easily be found as a number of degrees plus a decimal. If angles are wanted in minutes, then the decimal part of the angle measure is multiplied by 60' and this result is rounded to the nearest 10', 1', or 0.1', as desired.

EXAMPLE 4.17   Find A, when $\sin A = 0.4234$.

(a) Put the calculator in degree mode.

(b) Enter 0.4234, press (inv) key, and press (sin) key.

(c) $A = 25.05°$ to the nearest hundredth degree OR

(d) Record the whole number of degrees, 25°.

(e) Press (−) key, enter 25, press (=) key, press (×) key, enter 60, and press (=) key.

(f) To the nearest minute, the displayed value is 3'.

(g) $A = 25°3'$ to the nearest minute.

EXAMPLE 4.18   Find A, when $\cos A = 0.8163$.

(a) Put the calculator in degree mode.

(b) Enter 0.8163, press (inv) key, and press (cos) key.

(c) $A = 35.28°$ to the nearest hundredth degree OR

(d) Record the whole number of degrees, 35°.

(e) Press (−) key, enter 35, press (=) key, press (×) key, enter 60, and press (=) key.

(f) To the nearest minute, the displayed value is 17'.

(g) $A = 35°17'$ to the nearest minute.

When values of cotangent, secant, or cosecant are given, the reciprocal of the given function value is found and then the reciprocal function is used.

EXAMPLE 4.19   Find A, when $\sec A = 3.4172$.

(a) Put the calculator in degree mode.

(b) Enter 3.4172 and press (1/x) key or enter 1, press (÷) key, enter 3.4172, and press (=) key.

(c) Press (inv) key and press (cos) key.

(d) $A = 72.98°$ to the nearest hundredth degree OR

(e) Record the whole number of degrees, 72°.

(f) Press (−) key, enter 72, press (=) key, press (×) key, enter 60, and press (=) key.

(g) To the nearest minute, the displayed value is 59'.

(h) $A = 72°59'$ to the nearest minute.

4.7 ACCURACY IN COMPUTED RESULTS

Errors in computed results arise from:

(a) Errors in the given data. These errors are always present in data resulting from measurements.

(b) The use of values of trigonometric functions, whether from a table or a calculator, that are usually approximations of never-ending decimals.

A measurement recorded as 35 m means that the result is correct to the nearest meter; that is, the true length is between 34.5 and 35.5 m. Similarly, a recorded length of 35.0 m means that the true length is between 34.95 and 35.05 m; a recorded length of 35.8 m means that the true length is between 35.75 and 35.85 m; a recorded length of 35.80 m means that the true length is between 35.795 and 35.805 m; and so on.

In the number 35 there are two significant digits, 3 and 5. They are also the significant digits in 3.5, 0.35, 0.035, 0.0035 but not in 35.0, 3.50, 0.350, 0.0350. In the numbers 35.0, 3.50, 0.350, 0.0350 there are three significant digits, 3, 5, and 0. This is another way of saying that 35 and 35.0 are not the same measurement.

It is impossible to determine the significant figures in a measurement recorded as 350, 3500, 35000, ... For example, 350 may mean that the true result is between 345 and 355 or between 349.5 and 350.5. One way to indicate that a whole number ending in a zero has units as its digit of accuracy is to insert a decimal point; thus 3500. has four significant digits. Zeroes included between nonzero significant digits are significant digits.

A computed result should not show more decimal places than that shown in the least accurate of the measured data. Of importance here are the following relations giving comparable degrees of accuracy in lengths and angles:

(a) Distances expressed to 2 significant digits and angles expressed to the nearest degree.

(b) Distances expressed to 3 significant digits and angles expressed to the nearest 10′ or to the nearest 0.1°.

(c) Distances expressed to 4 significant digits and angles expressed to the nearest 1′ or to the nearest 0.01°.

(d) Distances expressed to 5 significant digits and angles expressed to the nearest 0.1′ or to the nearest 0.001°.

(NOTE: If several approximations are used when finding an answer, each intermediate step should use at least one more significant digit than is required for the accuracy of the final result.)

Solved Problems

4.1 Find the function value using tables.

(a) $\sin 56°34' = 0.8345$; $8339 + 0.4(16) = 8339 + 6$

(b) $\cos 19°45' = 0.9412$; $9417 - 0.5(10) = 9417 - 5$

(c) $\tan 77°12' = 4.4016$; $43897 + 0.2(597) = 43897 + 119$

(d) $\cot 40°36' = 1.1667$; $11708 - 0.6(68) = 11708 - 41$

(e) $\sec 23°47' = 1.0928$; $10918 + 0.7(14) = 10918 + 10$

(f) $\csc 60°4' = 1.1539$; $11547 - 0.4(19) = 11547 - 8$

(g) $\sin 46.35° = 0.7236$; $7230 + 0.5(12) = 7230 + 6$

(h) $\cos 18.29° = 0.9495$; $9500 - 0.9(6) = 9500 - 5$

(i) $\tan 82.19° = 7.2908$; $72066 + 0.9(936) = 72066 + 842$

(j) $\cot 13.84° = 4.0591$; $40713 - 0.4(305) = 40713 - 122$

(k) $\sec 29.71° = 1.1513$; $11512 + 0.1(12) = 11512 + 1$

(l) $\csc 11.08° = 5.2035$; $52408 - 0.8(466) = 52408 - 373$

4.2   In the manual solution, if the correction is 6.5, 13.5, 10.5, etc., we shall round off so that the *final* result is even.

    (*a*) $\sin 28°37' = 0.4790$; $4772 + 0.7(25) = 4772 + 17.5$

    (*b*) $\cot 65°53' = 0.4476$; $4487 - 0.3(35) = 4487 - 10.5$

    (*c*) $\cos 35°25' = 0.8150$; $8158 - 0.5(17) = 8158 - 8.5$

    (*d*) $\sec 39°35' = 1.2976$; $12960 + 0.5(31) = 12960 + 15.5$

4.3   Find the function value using a calculator.

    (*a*) $\sin 56°34' = 0.834527$; $\sin(56 + 34/60)°$

    (*b*) $\cos 19°45' = 0.941176$; $\cos(19 + 45/60)°$

    (*c*) $\tan 77°12' = 4.40152$; $\tan(77 + 12/60)°$

    (*d*) $\cot 40°36' = 1.16672$; $1/\tan 40°36' = 1/\tan(40 + 36/60)°$

    (*e*) $\sec 23°47' = 1.09280$; $1/\cos 23°47' = 1/\cos(23 + 47/60)°$

    (*f*) $\csc 60°4' = 1.15393$; $1/\sin 60°4' = 1/\sin(60 + 4/60)°$

    (*g*) $\sin 46.35° = 0.723570$

    (*h*) $\cos 18.29° = 0.949480$

    (*i*) $\tan 82.19° = 7.29071$

    (*j*) $\cot 13.84° = 4.05904$

    (*k*) $\sec 29.71° = 1.15135$

    (*l*) $\csc 11.08° = 5.20347$

4.4   Find A to the nearest minute and to the nearest hundredth of a degree.

    (*a*) $\sin A = 0.6826$, $A = 43°3'$; $43°0' + \dfrac{6}{21}(10') = 43°0' + 3'$: $A = 43.05°$;

$$43.00° + \frac{6}{13}(0.1°) = 43.00° + 0.05°$$

    (*b*) $\cos A = 0.5957$, $A = 53°26'$; $53°30' - \dfrac{9}{24}(10') = 53°30' - 4'$: $A = 53.44°$;

$$53.50° - \frac{9}{14}(0.1°) = 53.50° - 0.06°$$

    (*c*) $\tan A = 0.9470$, $A = 43°26'$; $43°20' + \dfrac{35}{55}(10') = 43°20' + 6'$: $A = 43.44°$;

$$43.40° + \frac{13}{33}(0.1°) = 43.40° + 0.04°$$

    (*d*) $\cot A = 1.7580$, $A = 29°38'$; $29°40' - \dfrac{24}{119}(10') = 29°40' - 2'$: $A = 29.63°$;

$$29.70° - \frac{48}{71}(0.1°) = 29.70° - 0.07°$$

    (*e*) $\sec\ A = 2.3198$, $A = 64°28'$; $64°20' + \dfrac{110}{140}(10') = 64°20' + 8'$: $A = 64.46°$; $64.40° + \dfrac{54}{84}(0.1°) =$ $64.40° + 0.06°$

    (*f*) $\csc A = 1.5651$, $A = 39°43'$; $39°50' - \dfrac{40}{55}(10') = 39°50' - 7$: $A = 39.71°$;

$$39.80° - \frac{29}{33}(0.1°) = 39.80° - 0.09°$$

4.5   Solve the right triangle in which $A = 35°10'$ and $c = 72.5$.

$B = 90° - 35°10' = 54°50'$. (See Fig. 4-1.)

$$a/c = \sin A \qquad a = c \sin A = 72.5(0.5760) = 41.8$$

$$b/c = \cos A \qquad b = c \cos A = 72.5(0.8175 = 59.3$$

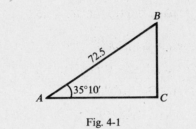

Fig. 4-1

4.6   Solve the right triangle in which $a = 24.36$ and $A = 58°53'$.

$B = 90° - 58°53' = 31°7'$. (See Fig. 4-2.)

$$b/a = \cot A \qquad b = a \cot A = 24.36(0.6036) = 14.70$$

or $\qquad a/b = \tan A \qquad b = a/\tan A = 24.36/1.6567 = 14.70$

$$c/a = \csc A \qquad c = a \csc A = 24.36(1.1681) = 28.45$$

or $\qquad a/c = \sin A \qquad c = a/\sin A = 24.36/0.8562 = 28.45$

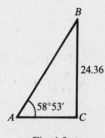

Fig. 4-2

4.7   Solve the right triangle ABC in which $a = 43.9$ and $b = 24.3$. (See Fig. 4-3.)

$\tan A = \dfrac{43.9}{24.3} = 1.8066$; $A = 61°0'$ and $B = 90° - A = 29°0'$, or $A = 61.0°$ and $B = 90° - A = 29.0°$.

$$c/a = \csc A \qquad c = a \csc A = 43.9(1.1434) = 50.2$$

or $\qquad a/c = \sin A, \qquad c = a/\sin A = 43.9/0.8746 = 50.2$

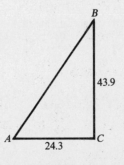

Fig. 4-3

4.8   Solve the right triangle ABC in which $b = 15.25$ and $c = 32.68$. (See Fig. 4-4.)

$$\sin B = \frac{15.25}{32.68} = 0.4666; \ B = 27°49' \text{ and } A = 90° - B = 62°11', \text{ or } B = 27.82° \text{ and } A = 90° - B = 62.18°.$$

	$a/b = \cot B$	$a = b \cot B = 15.25(1.8953) = 28.90$
or	$b/a = \tan B$	$a = b/\tan B = 15.25/0.5276 = 28.90$

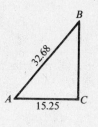

Fig. 4-4

(NOTE: See Appendix 1 Geometry for properties used in Probs. 4.9 to 4.11.)

4.9   The base of an isosceles triangle is 20.4 and the base angles are 48°40′. Find the equal sides and the altitude of the triangle.

In Fig. 4-5, BD is perpendicular to AC and bisects it.
In the right triangle ABD,

	$AB/AD = \sec A$	$AB = 10.2(1.5141) = 15.4$
or	$AD/AB = \cos A$	$AB = 10.2/0.6604 = 15.4$
	$DB/AD = \tan A$	$DB = 10.2(1.1369) = 11.6$

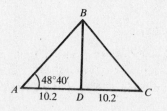

Fig. 4-5

4.10   Considering the earth as a sphere of radius 3960 mi, find the radius r of the 40th parallel of latitude. Refer to Fig. 4-6.

In the right triangle OCB, $\angle OBC = 40°$ and $OB = 3960$.

Then $\cos \angle OBC = CB/OB$ and $r = CB = 3960 \cos 40°$

Manual: $r = 3960 \cos 40° = 3960(0.7660) = 3033$.

Calculator: $r = 3960 \cos 40° = 3960(0.766044) = 3033.53$.

To three significant digits, the radius r of the 40th parallel is 3030 mi.

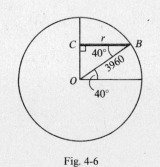

Fig. 4-6

4.11   Find the perimeter of a regular octagon inscribed in a circle of radius 150 cm.

In Fig. 4-7, two consecutive vertices A and B of the octagon are joined to the center O of the circle. The triangle OAB is isosceles with equal sides 150 and $\angle AOB = 360°/8 = 45°$. As in Prob. 4.9, we bisect $\angle AOB$ to form the right triangle MOB.

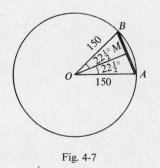

Fig. 4-7

Then $MB = OB \sin \angle MOB = 150 \sin 22°30' = 150(0.3827) = 57.4$, and the perimeter of the octagon is $16MB = 16(57.4) = 918$ cm.

4.12   To find the width of a river, a surveyor set up his transit at C on one bank and sighted across to a point B on the opposite bank; then turning through an angle of 90°, he laid off a distance $CA = 225$ m. Finally, setting the transit at A, he measured $\angle CAB$ as 48°20′. Find the width of the river.

See Fig. 4-8. In the right triangle ACB,

$$CB = AC \tan \angle CAB = 225 \tan 48°20' = 225(1.1237) = 253 \text{ m}$$

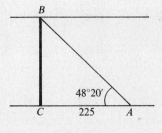

Fig. 4-8

4.13   In Fig. 4-9, the line AD crosses a swamp. In order to locate a point on this line, a surveyor turned through an angle 51°16′ at A and measured 1585 feet to a point C. He then turned through an angle of 90° at C and ran a line CB. If B is on AD, how far must he measure from C to reach B?

$$CB = AC \tan 51°16′$$

$$= 1585(1.2467) = 1976 \text{ ft}$$

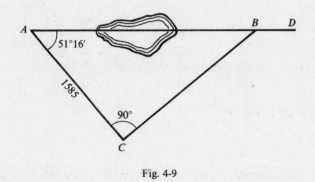

Fig. 4-9

4.14   From a point A on level ground, the angles of elevation of the top D and bottom B of a flagpole situated on the top of a hill are measured as 47°54′ and 39°45′. Find the height of the hill if the height of the flagpole is 115.5 ft. See Fig. 4.10.

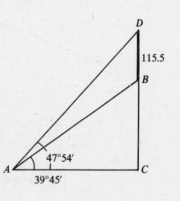

Fig. 4-10

Manual: Let the line of the pole meet the horizontal through A in C.
           In the right triangle ACD, $AC = DC \cot 47°54′ = (115.5 + BC)(0.9036)$.
           In the right triangle ACB, $AC = BC \cot 39°45′ = BC(1.2024)$.

Then $(115.5 + BC)(0.9036) = BC(1.2024)$

and $BC = \dfrac{115.5(0.9036)}{1.2024 - 0.9036} = 349.283$

Calculator: In the right triangle ACD, $AC = DC/\tan 47°54′ = (DB + BC)/\tan 47°54′$.
             In the right triangle ACB, $AC = BC/\tan 39°45′$.

Then
$$\frac{BC}{\tan 39°45'} = \frac{DB + BC}{\tan 47°54'}$$

$$BC \tan 47°54' = DB \tan 39°45' + BC \tan 39°45'$$

$$BC \tan 47°54' - BC \tan 39°45' = DB \tan 39°45'$$

$$(\tan 47°54' - \tan 39°45')BC = DB \tan 39°45'$$

$$BC = \frac{DB \tan 39°45'}{\tan 47°54' - \tan 39°45'}$$

$$= \frac{115.5 \tan (39 + 45/60)°}{\tan (47 + 54/60)° - \tan (39 + 45/60)°}$$

$$= 349.271$$

The height of the hill is 349.3 ft.

4.15 From the top of a lighthouse, 175 ft above the water, the angle of depression of a boat due south is 18°50'. Calculate the speed of the boat if, after it moves due west for 2 min, the angle of depression is 14°20'.

In Fig. 4-11, AD is the lighthouse, C is the position of the boat when due south of the lighthouse, and B is the position 2 min later.

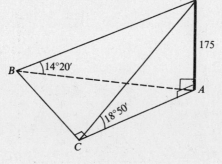

Fig. 4-11

Manual: In the right triangle CAD, $AC = AD \cot \angle ACD = 175 \cot 18°50' = 175(2.9319) = 513$.
In the right triangle BAD, $AB = AD \cot \angle ABD = 175 \cot 14°20' = 175(3.9136) = 685$.
In the right triangle ABC, $BC = \sqrt{(AB)^2 - (AC)^2} = \sqrt{(685)^2 - (513)^2} = 453.6$.

Calculator: In the right triangle CAD, $AC = 175/\tan 18°50'$.
In the right triangle BAD, $AB = 175/\tan 14°20'$.
In the right triangle ABC, $BC = \sqrt{(AB)^2 - (AC)^2}$.

$$BC = \sqrt{[175/\tan (14 + 20/60)°]^2 - [175/\tan (18 + 50/60)°]^2}$$

$$= 453.673$$

The boat travels 454 ft in 2 min; its speed is 227 ft/min.

Supplementary Problems

4.16 Find, to four decimal places, the values of the six trigonometric functions of each of the following angles:
(a) 18°47′, (b) 32°13′, (c) 58°24′, (d) 79°45′

Ans.		sine	cosine	tangent	cotangent	secant	cosecant
(a)	18°47′	0.3220	0.9468	0.3401	2.9403	1.0563	3.1057
(b)	32°13′	0.5331	0.8460	0.6301	1.5869	1.1820	1.8757
(c)	58°24′	0.8517	0.5240	1.6255	0.6152	1.9084	1.1741
(d)	79°45′	0.9840	0.1780	5.5304	0.1808	5.6201	1.0162

[NOTE: With a calculator, the values are the same except for (a) cos 18°47′ = 0.9467, (d) cos 79°45′ = 0.1779, (d) tan 79°45′ = 5.5301, and (d) sec 79°45′ = 5.6198.]

4.17 Find, to four decimal places, the values of the six trigonometric functions of each of the following angles:
(a) 29.43°, (b) 73.67°, (c) 61.72°, (d) 12.08°

Ans.		sine	cosine	tangent	cotangent	secant	cosecant
(a)	29.43°	0.4914	0.8710	0.5642	1.7725	1.1482	2.0352
(b)	73.67°	0.9596	0.2812	3.4131	0.2930	3.5566	1.0420
(c)	61.72°	0.8807	0.4738	1.8588	0.5380	2.1107	1.1355
(d)	12.08°	0.2093	0.9779	0.2140	4.6726	1.0226	4.7784

[NOTE: With a calculator, the values are the same except for (b) sin 73.67° = 0.9597, (c) sin 61.72° = 0.8806, and (d) cot 12.08° = 4.6725.]

4.18 Find (acute) angle A, given:

(a) sin A = 0.5741   *Ans.*   A = 35° 2′ or 35.04°   (e) cos A = 0.9382   *Ans.*   A = 20°15′ or 20.25°

(b) sin A = 0.9468   A = 71°13′ or 71.23°   (f) cos A = 0.6200   A = 51°41′ or 51.68°

(c) sin A = 0.3510   A = 20°33′ or 20.55°   (g) cos A = 0.7120   A = 44°36′ or 44.60°

(d) sin A = 0.8900   A = 62°52′ or 62.88°   (h) cos A = 0.4651   A = 62°17′ or 62.28°

(i) tan A = 0.2725   A = 15°15′ or 15.24°   (m) cot A = 0.2315   A = 76°58′ or 76.97°

(j) tan A = 1.1652   A = 49°22′ or 49.38°   (n) cot A = 2.9715   A = 18°36′ or 18.60°

(k) tan A = 0.5200   A = 27°28′ or 27.47°   (o) cot A = 0.7148   A = 54°27′ or 54.44°

(l) tan A = 2.7775   A = 70°12′ or 70.20°   (p) cot A = 1.7040   A = 30°24′ or 30.41°

(q) sec A = 1.1161   A = 26°22′ or 26.37°   (u) csc A = 3.6882   A = 15°44′ or 15.73°

(r) sec A = 1.4382   A = 45°57′ or 45.95°   (v) csc A = 1.0547   A = 71°28′ or 71.47°

(s) sec A = 1.2618   A = 37°35′ or 37.58°   (w) csc A = 1.7631   A = 34°33′ or 34.55°

(t) sec A = 2.1584   A = 62°24′ or 62.40°   (x) csc A = 1.3436   A = 48° 6′ or 48.10°

[NOTE: Calculator answers are the same except for (b) 71°14′, (d) 62.87°, and (j) 49.36°.]

4.19　Solve each of the right triangles ABC, given:

　　(a)　$A = 35°20'$, $c = 112$　　　　*Ans.*　$B = 54°40'$, $a = 64.8$, $b = 91.4$

　　(b)　$B = 48°40'$, $c = 225$　　　　　　　　$A = 41°20'$, $a = 149$, $b = 169$

　　(c)　$A = 23°18'$, $c = 346.4$　　　　　　　$B = 66°42'$. $a = 137.0$, $b = 318.1$

　　(d)　$B = 54°12'$, $c = 182.5$　　　　　　　$A = 35°48'$, $a = 106.7$, $b = 148.0$

　　(e)　$A = 32°10'$, $a = 75.4$　　　　　　　$B = 57°50'$, $b = 120$, $c = 142$

　　(f)　$A = 58°40'$, $b = 38.6$　　　　　　　$B = 31°20'$, $a = 63.4$, $c = 74.2$

　　(g)　$B = 49°14'$, $b = 222.2$　　　　　　　$A = 40°46'$, $a = 191.6$, $c = 293.4$

　　(h)　$A = 66°36'$, $a = 112.6$　　　　　　　$B = 23°24'$, $b = 48.73$, $c = 122.7$

　　(i)　$A = 29°48'$, $b = 458.2$　　　　　　　$B = 60°12'$, $a = 262.4$, $c = 528.0$

　　(j)　$a = 25.4$, $b = 38.2$　　　　　　　$A = 33°40'$, $B = 56°20'$, $c = 45.9$ or $A = 33.6°$, $B = 56.4°$

　　(k)　$a = 45.6$, $b = 84.8$　　　　　　　$A = 28°20'$, $B = 61°40'$, $c = 96.3$ or $A = 28.3°$, $B = 61.7°$

　　(l)　$a = 38.64$, $b = 48.74$　　　　　$A = 38°24'$, $B = 51°36'$, $c = 62.20$ or $A = 38.41°$, $B = 51.59°$

　　(m)　$a = 506.2$, $c = 984.8$　　　　$A = 30°56'$, $B = 59°4'$, $b = 844.7$ or $A = 30.93°$, $B = 59.07°$

　　(n)　$b = 672.9$, $c = 888.1$　　　　$A = 40°44'$, $B = 49°16'$, $a = 579.6$ or $A = 40.74°$, $B = 49.26°$

　　(o)　$A = 23.2°$, $c = 117$　　　　　　　$B = 66.8°$, $a = 46.1$, $b = 108$

　　(p)　$A = 58.61°$, $b = 87.24$　　　　　$B = 31.39°$, $a = 143.0$, $c = 167.5$

[NOTE:　With a calculator, the values are the same except for (d) $a = 106.8$.]

4.20　Find the base and altitude of an isosceles triangle whose vertical angle is $65°$ and whose equal sides are 415 cm.

　　Ans.　Base = 446 cm, altitude = 350 cm

4.21　The base of an isosceles triangle is 15.90 in and the base angles are $54°28'$. Find the equal sides and the altitude.

　　Ans.　Side = 13.68 in, altitude = 11.13 in

4.22　The radius of a circle is 21.4 m. Find (a) the length of the chord subtended by a central angle of $110°40'$ and (b) the distance between two parallel chords on the same side of the center subtended by central angles $118°40'$ and $52°20'$.

　　Ans.　(a) 35.2 m, (b) 8.29 m

4.23　Show that the base b of an isosceles triangle whose equal sides are a and whose vertical angle is θ is given by $b = 2a \sin \frac{1}{2}\theta$.

4.24　Show that the perimeter P of a regular polygon of n sides inscribed in a circle of radius r is given by $P = 2nr \sin (180°/n)$.

4.25　A wheel, 5 ft in diameter, rolls up an incline of $18°20'$. What is the height of the center of the wheel above the base of the incline when the wheel has rolled 5 ft up the incline?

　　Ans.　3.95 ft

4.26　A wall is 15 ft high and 10 ft from a house. Find the length of the shortest ladder which will just touch the top of the wall and reach a window 20.5 ft above the ground.

　　Ans.　42.5 ft

Chapter 5

Practical Applications

5.1 BEARING

The bearing of a point B from a point A, in a horizontal plane, is usually defined as the angle (always acute) made by the ray drawn from A through B with the north-south line through A. The bearing is then read from the north or south line toward the east or west. The angle used in expressing a bearing is usually stated in degrees and minutes. For example, see Fig. 5-1.

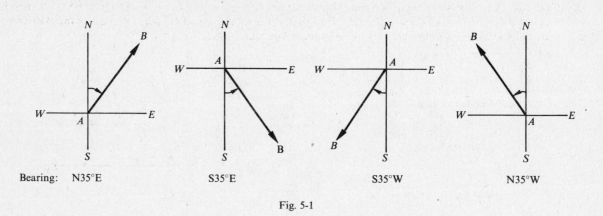

| Bearing: N35°E | S35°E | S35°W | N35°W |

Fig. 5-1

In aeronautics the bearing of B from A is more often given as the angle made by the ray AB with the north line through A, measured clockwise from the north (i.e., from the north around through the east). For example, see Fig. 5-2.

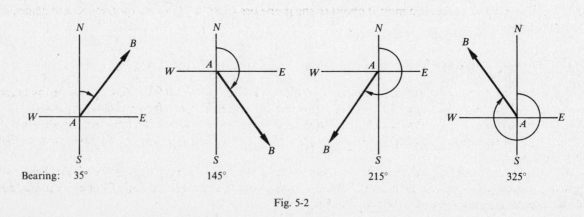

| Bearing: 35° | 145° | 215° | 325° |

Fig. 5-2

5.2 VECTORS

Any physical quantity, like force or velocity, which has both magnitude and direction is called a *vector quantity*. A vector quantity may be represented by a directed line segment (arrow) called a *vector*. The *direction* of the vector is that of the given quantity and the *length* of the vector is proportional to the magnitude of the quantity.

EXAMPLE 5.1 An airplane is traveling N40°E at 200 mi/h. Its velocity is represented by the vector **AB** in Fig. 5-3.

54

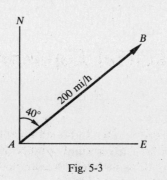

Fig. 5-3

EXAMPLE 5.2 A motor boat having the speed 12 mi/h in still water is headed directly across a river whose current is 4 mi/h. In Fig. 5-4, the vector **CD** represents the velocity of the current and the vector **AB** represents, to the same scale, the velocity of the boat in still water. Thus, vector **AB** is three times as long as vector **CD**.

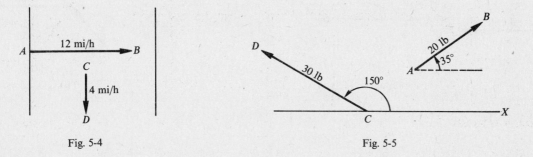

Fig. 5-4   Fig. 5-5

EXAMPLE 5.3 In Fig. 5-5, vector **AB** represents a force of 20 lb making an angle of 35° with the positive direction on the x axis and vector **CD** represents a force of 30 lb at 150° with the positive direction on the x axis. Both vectors are drawn to the same scale.

Two vectors are said to be equal if they have the same magnitude and direction. A vector has no fixed position in a plane and may be moved about in the plane provided only that its magnitude and direction are not changed.

5.3 VECTOR ADDITION

The *resultant* or *vector sum* of a number of vectors, all in the same plane, is that vector in the plane which would produce the same effect as that produced by all the original vectors acting together.

If two vectors **α** and **β** have the same direction, their resultant is a vector **R** whose magnitude is equal to the sum of the magnitudes of the two vectors and whose direction is that of the two vectors. See Fig. 5-6(a).

If two vectors have opposite directions, their resultant is a vector **R** whose magnitude is the difference (greater magnitude − smaller magnitude) of the magnitudes of the two vectors and whose direction is that of the vector of greater magnitude. See Fig. 5-6(b).

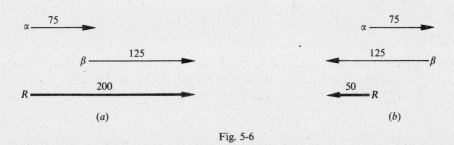

(a)   (b)

Fig. 5-6

In all other cases, the magnitude and direction of the resultant of two vectors is obtained by either of the following two methods.

(1) *Parallelogram Method.* Place the tail ends of both vectors at any point O in their plane and complete the parallelogram having these vectors as adjacent sides. The directed diagonal issuing from O is the resultant or vector sum of the two given vectors. Thus, in Fig. 5-7(*b*), the vector **R** is the resultant of the vectors α and β of Fig. 5-7(*a*).

(2) *Triangle Method.* Choose one of the vectors and label its tail end O. Place the tail end of the other vector at the arrow end of the first. The resultant is then the line segment closing the triangle and directed from O. Thus, in Figs. 5-7(*c*) and 5-7(*d*), **R** is the resultant of the vectors α and β.

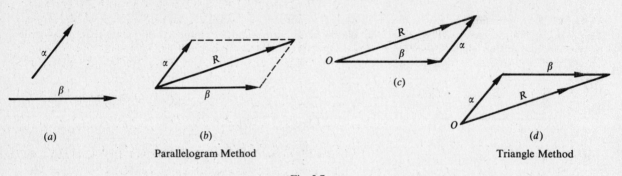

     (*a*)   (*b*)   (*d*)

             Parallelogram Method   Triangle Method

Fig. 5-7

EXAMPLE 5.4 The resultant **R** of the two vectors of Example 5.2 represents the speed and direction in which the boat travels. Figure 5-8(*a*) illustrates the parallelogram method; Fig. 5-8(*b*) and (*c*) illustrates the triangle method.

The magnitude of $\mathbf{R} = \sqrt{(12)^2 + 4^2} = 13$ mi/h rounded.

From Fig. 5-8(*a*) or (*b*), $\tan \theta = \frac{4}{12} = 0.3333$ and $\theta = 18°$.

Thus, the boat moves downstream in a line making an angle $\theta = 18°$ with the direction in which it is headed or making an angle $90° - \theta = 72°$ with the bank of the river. (See Sec. 4.7 for rounding procedures.)

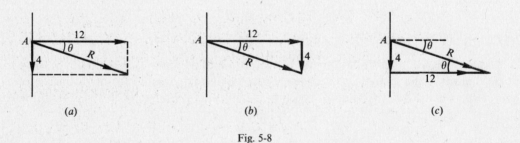

         (*a*)   (*b*)   (*c*)

Fig. 5-8

5.4 COMPONENTS OF A VECTOR

The component of a vector α along a line L is the perpendicular projection of the vector α on L. It is often very useful to resolve a vector into two components along a pair of perpendicular lines.

EXAMPLE 5.5 In Fig. 5-8(*a*), (*b*), and (*c*) the components of **R** are (1) 4 mi/h in the direction of the current and (2) 12 mi/h in the direction perpendicular to the current.

EXAMPLE 5.6 In Fig. 5-9, the force **F** has horizontal component $\mathbf{F}_h = \mathbf{F} \cos 30°$ and vertical component $\mathbf{F}_v = \mathbf{F} \sin 30°$. Note that **F** is the vector sum or resultant of $\mathbf{F}_h$ and $\mathbf{F}_v$.

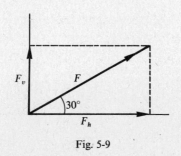

Fig. 5-9

5.5 AIR NAVIGATION

The *heading* of an airplane is the direction (determined from a compass reading) in which the airplane is pointed. The heading is measured clockwise from the north and expressed in degrees and minutes.

The *airspeed* (determined from a reading of the airspeed indicator) is the speed of the airplane in still air.

The *course* (or *track*) of an airplane is the direction in which it moves relative to the ground. The course is measured clockwise from the north.

The *groundspeed* is the speed of the airplane relative to the ground.

The *drift angle* (or wind-correction angle) is the difference (positive) between the heading and the course.

In Fig. 5-10: ON is the true north line through O

$\angle NOA$ is the heading

$OA =$ the airspeed

AN is the true north line through A

$\angle NAW$ is the wind angle, measured clockwise from north line

$AB =$ the windspeed

$\angle NOB$ is the course

$OB =$ the groundspeed

$\angle AOB$ is the drift angle

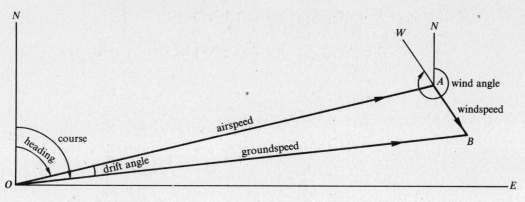

Fig. 5-10

Note that there are three vectors involved: **OA** representing the airspeed and heading, **AB** representing the direction and speed of the wind, and **OB** representing the groundspeed and course. The groundspeed vector is the resultant of the airspeed vector and the wind vector.

EXAMPLE 5.7   Figure 5-11 illustrates an airplane flying at 240 mi/h on a heading of 60° when the wind is 30 mi/h from 330°.

In constructing the figure put in the airspeed vector at O, then follow through (note the directions of the arrows) with the wind vector, and close the triangle. Note further that the groundspeed vector does not follow through from the wind vector.

In the resulting triangle: Groundspeed $= \sqrt{(240)^2 + (30)^2} = 242$ mi/h
$$\tan \theta = 30/240 = 0.1250 \text{ and } \theta = 7°10'$$
$$\text{Course} = 60° + \theta = 67°10'$$

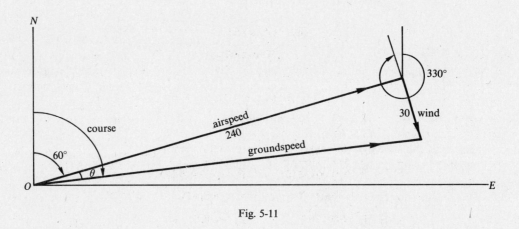

Fig. 5-11

5.6   INCLINED PLANE

An object with weight W on an inclined plane which has an angle of inclination α exerts a force $\mathbf{F}_a$ against the inclined plane and a force $\mathbf{F}_d$ down the inclined plane. The forces $\mathbf{F}_a$ and $\mathbf{F}_d$ are the component vectors for the weight W. See Fig. 5-12(a).

The angle θ formed by the force $\mathbf{F}_a$ against the inclined plane and the weight W is equal to the angle of inclination α. Since $\theta = \alpha$, $\mathbf{F}_a = W \cos \alpha$ and $\mathbf{F}_d = W \sin \alpha$. See Fig. 5-12($b$).

The minimum force needed to keep an object from sliding down an inclined plane (ignoring friction) has the same magnitude but opposite direction from $\mathbf{F}_d$.

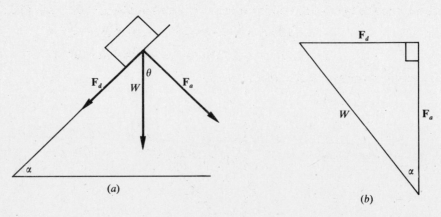

Fig. 5-12

EXAMPLE 5.8   A 500-lb barrel rests on an 11.2° inclined plane. What is the minimum force (ignoring friction) needed to keep the barrel from rolling down the incline and what is the force the barrel exerts against the surface of the inclined plane? (See Fig. 5-13.)

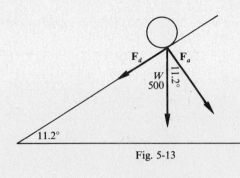

Fig. 5-13

$$\mathbf{F}_d = 500 \sin 11.2°$$

$$= 500(0.1942)$$

$$= 97.1$$

$$\mathbf{F}_d = 97.1 \text{ lb}$$

$$\mathbf{F}_a = 500 \cos 11.2°$$

Manual	Calculator
$\mathbf{F}_a = 500(0.9810)$	$\mathbf{F}_a = 500(0.980955)$
$= 490.5$	$= 490.478$
$\mathbf{F}_a = 491 \text{ lb}$	$\mathbf{F}_a = 490 \text{ lb}$

The minimum force needed to keep the barrel from rolling down the incline is 97.1 lb and the force against the inclined plane is 491 lb (or 490 lb using a calculator).

Solved Problems

Use rounding procedures stated in Sec. 4.7.

5.1 A motor boat moves in the direction N40°E for 3 h at 20 mi/h. How far north and how far east does it travel?

Suppose the boat leaves A. Using the north-south line through A, draw the ray AD so that the bearing of D from A is N40°E. On $\overrightarrow{AD}$ locate B such that $AB = 3(20) = 60$ mi. Through B pass a line perpendicular to the line NAS, meeting it in C. In the right triangle ABC, see Fig. 5-14,

$$AC = AB \cos A = 60 \cos 40° = 60(0.7660) = 45.96$$

and $$CB = AB \sin A = 60 \sin 40° = 60(0.6428) = 38.57$$

The boat travels 46 mi north and 39 mi east.

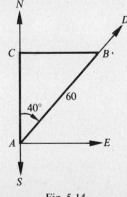

Fig. 5-14

5.2 Three ships are situated as follows: A is 225 mi due north of C, and B is 375 mi due east of C. What is the bearing (a) of B from A and (b) of A from B?

In the right triangle ABC, see Fig. 5-15,

$$\tan \angle CAB = 375/225 = 1.6667 \qquad \text{and} \qquad \angle CAB = 59°0'$$

(a) The bearing of B from A (angle SAB) is S59°0'E.
(b) The bearing of A from B (angle $N'BA$) is N59°0'W.

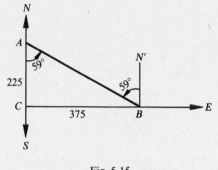

Fig. 5-15

5.3 Three ships are situated as follows: A is 225 miles west of C while B, due south of C, bears S25°10'E from A. (a) How far is B from A? (b) How far is B from C? (c) What is the bearing of A from B?

From Fig. 5-16, $\angle SAB = 25°10'$ and $\angle BAC = 64°50'$. Then

$$AB = AC \sec \angle BAC = 225 \sec 64°50' = 225(2.3515) = 529.1$$

or $\qquad\qquad AB = AC/\cos \angle BAC = 225/\cos 64°50' = 225/0.4253 = 529.0$

and $\qquad\qquad CB = AC \tan \angle BAC = 225 \tan 64°50' = 225(2.1283) = 478.9$

(a) B is 529 miles from A.   (b) B is 479 miles from C.

(c) Since $\angle CBA = 25°10'$, the bearing of A from B is N25°10'W.

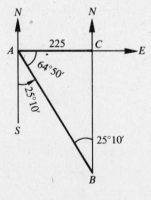

Fig. 5-16

5.4 From a boat sailing due north at 16.5 km/h, a wrecked ship K and an observation tower T are observed in a line due east. One hour later the wrecked ship and the tower have bearings S34°40'E and S65°10'E. Find the distance between the wrecked ship and the tower.

In Fig. 5-17, C, K, and T represent respectively the boat, the wrecked ship, and the tower when in a line. One hour later the boat is at A, 16.5 km due north of C. In the right triangle ACK,

$$CK = 16.5 \tan 34°40' = 16.5(0.6916)$$

In the right triangle ACT,

$$CT = 16.5 \tan 65°10' = 16.5(2.1609)$$

Then

$$KT = CT - CK = 16.5(2.1609 - 0.6916) = 24.2 \text{ km}$$

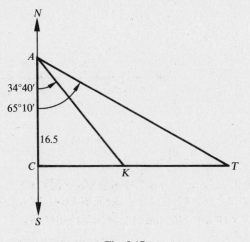

Fig. 5-17

5.5 A ship is sailing due east when a light is observed bearing N62°10′E. After the ship has traveled 2250 m, the light bears N48°25′E. If the course is continued, how close will the ship approach the light?

In Fig. 5-18, L is the position of the light, A is the first position of the ship, B is the second position, and C is the position when nearest L.

In the right triangle ACL, $AC = CL \cot \angle CAL = CL \cot 27°50' = 1.8940CL$.

In the right triangle BCL, $BC = CL \cot \angle CBL = CL \cot 41°35' = 1.1270CL$.

Since $AC = BC + 2250$, $1.8940CL = 1.1270CL + 2250$ and $CL = \dfrac{2250}{1.8940 - 1.1270} = 2934$ m.

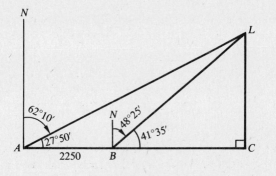

Fig. 5-18

5.6   Refer to Fig. 5-19. A body at O is being acted upon by two forces, one of 150 lb due north and the
other of 200 lb due east. Find the magnitude and direction of the resultant.

In the right triangle OBC, $OC = \sqrt{(OB)^2 + (BC)^2} = \sqrt{(200)^2 + (150)^2} = 250$ lb,

$$\tan \angle BOC = 150/200 = 0.7500, \text{ and } \angle BOC = 36°50'.$$

The magnitude of the resultant force is 250 lb and its direction is N53°10′E.

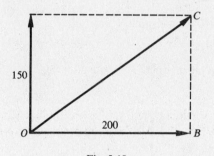

Fig. 5-19

5.7   An airplane is moving horizontally at 240 mi/h when a bullet is shot with speed 2750 ft/s at right
angles to the path of the airplane. Find the resultant speed and direction of the bullet.

The speed of the airplane is $240 \text{ mi/h} = \dfrac{240(5280)}{60(60)} \text{ ft/s} = 352$ ft/s.

In Fig. 5-20, the vector **AB** represents the velocity of the airplane, the vector **AC** represents the initial
velocity of the bullet, and the vector **AD** represents the resultant velocity of the bullet.

In the right triangle ACD, $AD = \sqrt{(352)^2 + (2750)^2} = 2770$ ft/s,

$$\tan \angle CAD = 352/2750 = 0.1280, \text{ and } \angle CAD = 7°20' \text{ or } 7.3°.$$

Thus, the bullet travels at 2770 ft/s along a path making an angle of 82°40′ or 82.7° with the path of the
airplane.

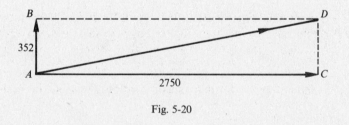

Fig. 5-20

5.8   A river flows due south at 125 ft/min. A motor boat, moving at 475 ft/min in still water, is headed
due east across the river. (*a*) Find the direction in which the boat moves and its speed. (*b*) In what
direction must the boat be headed in order that it move due east and what is its speed in that
direction?

(*a*)   Refer to Fig. 5-21. In right triangle OAB, $OB = \sqrt{(475)^2 + (125)^2} = 491$,

$$\tan \theta = 125/475 = 0.2632, \text{ and } \theta = 14°40'.$$

Thus the boat moves at 491 ft/min in the direction S75°20′E.

(*b*)   Refer to Fig. 5-22. In right triangle OAB, $\sin \theta = 125/475 = 0.2632$ and $\theta = 15°20'$.

Thus the boat must be headed N74°40′E and its speed in that direction is

$$OB = \sqrt{(475)^2 - (125)^2} = 458 \text{ ft/min}$$

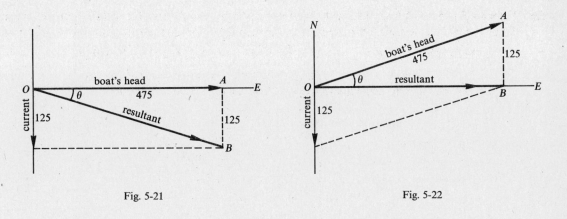

Fig. 5-21   Fig. 5-22

5.9 A telegraph pole is kept vertical by a guy wire which makes an angle of 25° with the pole and which exerts a pull of $\mathbf{F} = 300$ lb on the top. Find the horizontal and vertical components $\mathbf{F}_h$ and $\mathbf{F}_v$ of the pull $\mathbf{F}$. See Fig. 5-23.

$$\mathbf{F}_h = 300 \sin 25° = 300(0.4226) = 127 \text{ lb}$$

$$\mathbf{F}_v = 300 \cos 25° = 300(0.9063) = 272 \text{ lb}$$

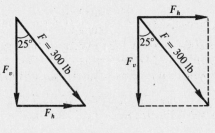

Fig. 5-23

5.10 A man pulls a rope attached to a sled with a force of 100 lb. The rope makes an angle of 27° with the ground. (*a*) Find the effective pull tending to move the sled along the ground and the effective pull tending to lift the sled vertically. (*b*) Find the force which the man must exert in order that the effective force tending to move the sled along the ground is 100 lb.

(*a*)   In Figs. 5-24 and 5-25, the 100 lb pull in the rope is resolved into horizontal and vertical components, $\mathbf{F}_h$ and $\mathbf{F}_v$, respectively. Then $\mathbf{F}_h$ is the force tending to move the sled along the ground and $\mathbf{F}_v$ is the force tending to lift the sled.

$$\mathbf{F}_h = 100 \cos 27° = 100(0.8910) = 89 \text{ lb} \qquad \mathbf{F}_v = 100 \sin 27° = 100(0.4540) = 45 \text{ lb}$$

(*b*)   In Fig. 5-26, the horizontal component of the required force $\mathbf{F}$ is $\mathbf{F}_h = 100$ lb. Then

$$\mathbf{F} = 100/\cos 27° = 100/0.8910 = 112 \text{ lb}$$

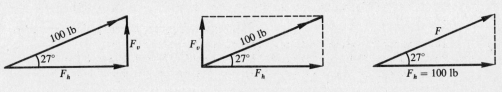

Fig. 5-24   Fig. 5-25   Fig. 5-26

5.11 A block weighing $W = 500$ lb rests on a ramp inclined $29°$ with the horizontal. (a) Find the force tending to move the block down the ramp and the force of the block on the ramp. (b) What minimum force must be applied to keep the block from sliding down the ramp? Neglect friction.

(a) Refer to Fig. 5-27. Resolve the weight W of the block into components F_1 and F_2, respectively parallel and perpendicular to the ramp. F_1 is the force tending to move the block down the ramp and F_2 is the force of the block on the ramp.

$$F_1 = W \sin 29° = 500(0.4848) = 242 \text{ lb} \qquad F_2 = W \cos 29° = 500(0.8746) = 437 \text{ lb}$$

(b) 242 lb up the ramp

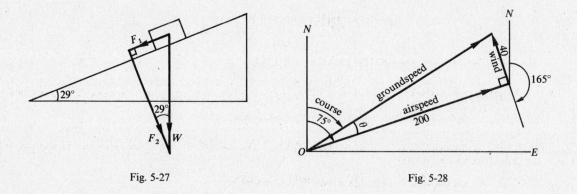

Fig. 5-27   Fig. 5-28

5.12 The heading of an airplane is $75°$ and the airspeed is 200 mi/h. Find the groundspeed and course if there is a wind of 40 mi/h from $165°$. Refer to Fig. 5-28.

Construction: Put in the airspeed vector from O, follow through with the wind vector, and close the triangle.

Solution: Groundspeed $= \sqrt{(200)^2 + (40)^2} = 204$ mi/h, $\tan \theta = 40/200 = 0.2000$ and $\theta = 11°20'$, and course $= 75° - \theta = 63°40'$.

5.13 The airspeed of an airplane is 200 km/h. There is a wind of 30 km/h from $270°$. Find the heading and groundspeed in order to track $0°$. Refer to Fig. 5-29.

Construction: The groundspeed vector is along ON. Lay off the wind vector from O, follow through with the airspeed vector (200 units from the head of the wind vector to a point on ON), and close the triangle.

Solution: Groundspeed $= \sqrt{(200)^2 - (30)^2} = 198$ km/h, $\sin \theta = 30/200 = 0.1500$ and $\theta = 8°40'$, and heading $= 360° - \theta = 351°20'$.

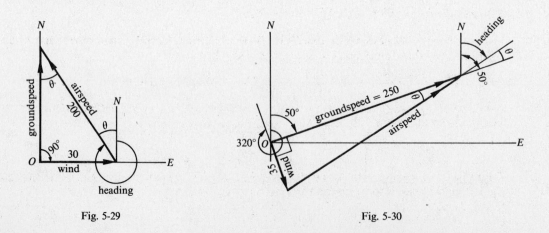

Fig. 5-29   Fig. 5-30

5.14 There is a wind of 35 mi/h from 320°. Find the airspeed and heading in order that the groundspeed and course be 250 mi/h and 50° respectively. Refer to Fig. 5-30.

Construction: Lay off the groundspeed vector from O, put in the wind vector at O so that it does not follow through to the groundspeed vector, and close the triangle.

Solution: Airspeed = $\sqrt{(250)^2 + (35)^2} = 252$ mi/h, $\tan \theta = 35/250 = 0.1400$ and $\theta = 8°$, and heading = $50° - 8° = 42°$.

Supplementary Problems

Use rounding procedures stated in Section 4.7.

5.15 An airplane flies 100 km in the direction S38°10′E. How far south and how far east of the starting point is it?

Ans.   78.6 km south, 61.8 km east

5.16 A plane is headed due east with airspeed 240 km/h. If a wind at 40 km/h from the north is blowing, find the groundspeed and course.

Ans.   Groundspeed, 243 km/h; course, 99°30′ or S80°30′E

5.17 A body is acted upon by a force of 75 lb, due west, and a force of 125 lb, due north. Find the magnitude and direction of the resultant force.

Ans.   146 lb, N31°0′W

5.18 Find the rectangular components of a force of 525 lb in a direction 38.4° with the horizontal.

Ans.   411 lb, 326 lb

5.19 An aviator heads his airplane due west. He finds that because of a wind from the south, the course makes an angle of 20° with the heading. If his airspeed is 100 mi/h, what is his groundspeed and what is the speed of the wind?

Ans.   Groundspeed, 106 mi/h, wind, 36 mi/h

5.20 An airplane is headed west while a 40 mi/h wind is blowing from the south. What is the necessary airspeed to follow a course N72°W and what is the groundspeed?

Ans.   Airspeed, 123 mi/h; groundspeed, 129 mi/h

5.21 A barge is being towed north at the rate 18 mi/h. A man walks across the deck from west to east at the rate 6 ft/s. Find the magnitude and direction of the actual velocity.

Ans.   27 ft/s, N12°50′E

5.22 A ship at A is to sail to C, 56 km north and 258 km east of A. After sailing N25°10′E for 120 mi to P, the ship is headed toward C. Find the distance of P from C and the required course to reach C.

Ans.   214 km, S75°40′E

5.23 A guy wire 78 ft long runs from the top of a telephone pole 56 ft high to the ground and pulls on the pole with a force of 290 lb. What is the horizontal pull on the top of the pole?

Ans.   201 lb

5.24 A weight of 200 lb is placed on a smooth plane inclined at an angle of 37.6° with the horizontal and held in place by a rope parallel to the surface and fastened to a peg in the plane. Find the pull on the string.

Ans. 122 lb

5.25 A man wishes to raise a 300-lb weight to the top of a wall 20 m high by dragging it up an incline. What is the length of the shortest inclined plane he can use if his pulling strength is 140 lb?

Ans. 43 m

5.26 A 150-lb shell is dragged up a runway inclined 40° to the horizontal. Find (*a*) the force of the shell against the runway and (*b*) the force required to drag the shell.

Ans. (*a*) 115 lb, (*b*) 96 lb

Chapter 6

Applying Logarithms to Trigonometry

6.1 INTRODUCTION

Logarithms can be used to simplify some of the computation needed to solve trigonometry problems. When manually solving problems, logarithms offer another way to do multiplying, dividing, raising to a power, and finding a root. If a calculator is being used, logarithms are not needed since procedures for multiplication, division, powers, and roots are already available.

Since logarithms are not used to perform addition or subtraction, the decision to use or not use logarithms depends in part on the problem-solving procedure being used. If additions and subtractions can be done at the beginning or at the end of the procedure, then logarithms can be easily used to do the other operations. Appendix 3 Logarithms reviews the rules for logarithms and provides examples using logarithms to perform a variety of computations.

6.2 LOGARITHMS OF TRIGONOMETRIC FUNCTIONS

In Appendix 2, Table 4 is a four-place table of common logarithms and can be used to find the logarithm of a number with three or fewer significant digits directly. If the number has four significant digits, we interpolate, using the method of proportional parts to find the logarithm. If the number has more than four significant digits, we round the number to four significant digits and then interpolate to find the logarithm.

Other frequently used logarithm tables are a five-place table that allows the logarithm of a number with four significant digits to be found directly and tables of logarithms of trigonometric functions that combine a trigonometric function table with a logarithm table. We will find the value of the trigonometric function using Table 1, round the result to four significant digits if necessary, and find the logarithm of the rounded value using Table 4.

EXAMPLE 6.1 Find the logarithm.

(a) $\log \sin 22°30' = \log (0.3827) = 9.5829 - 10$

(b) $\log \tan 23°50' = \log (2.2460) = 0.3514$

(c) $\log \csc 3°40' = \log (15.6368) = \log (15.64) = 1.1942$

(d) $\log \cos 38°21' = \log (0.7842) = 9.8944 - 10$

(e) $\log \cot 87°34' = \log (0.0425) = 8.6284 - 10$

(f) $\log \sec 67°28' = \log (2.6095) = \log (2.610) = 0.4166$

EXAMPLE 6.2 Find angle A to the nearest minute.

(a) $\log \sin A = 9.3975 - 10$ (d) $\log \cot A = 0.4471$

(b) $\log \cos A = 9.5964 - 10$ (e) $\log \sec A = 0.3354$

(c) $\log \tan A = 9.9862 - 10$ (f) $\log \csc A = 0.1983$

(a) $\sin A = \text{antilog} (9.3975 - 10) = 0.2498; \ A = 14°28'$

(b) $\cos A = \text{antilog} (9.5964 - 10) = 0.3948; \ A = 66°45'$

(c) $\tan A = \text{antilog} (9.9862 - 10) = 0.9688; \ A = 44°6'$

(d) $\cot A = \text{antilog} (0.4471) = 2.799; \ A = 19°40'$

(e) $\sec A = \text{antilog} (0.3354) = 2.164; \ A = 62°29'$

(f) $\csc A = \text{antilog} (0.1983) = 1.579; \ A = 39°18'$

6.3 SOLVING RIGHT TRIANGLES

Any right triangle may be solved using the trigonometric functions of one of the acute angles, the angle relation $A + B = 90°$, and the Pythagorean theorem $a^2 + b^2 = c^2$. (See Fig. 6-1.)

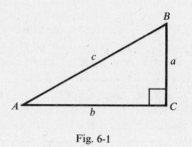

Fig. 6-1

When an acute angle and a side are known, the angle relation can be used to solve for the second acute angle, and trigonometric functions of the known angle can be used to find the two unknown sides.

EXAMPLE 6.3 Solve triangle ABC, given $A = 24°18'$ and $a = 291.1$ cm. (See Fig. 6-1.)
To find angle B, use $B = 90° - A = 90° - 24°18' = 65°42'$.
To find side b, use $b = a \cot A = 291.1 \cot 24°18' = 291.1(2.2148)$.

$$\log b = \log [291.1(2.215)] = \log 291.1 + \log 2.215 = 2.4640 + 0.3454 = 2.8094$$

$$b = \text{antilog } 2.8094 = 644.7 \text{ cm}$$

To find side c, use $c = a \csc A = 291.1 \csc 24°18' = 291.1(2.4300)$.

$$\log c = \log [291.1(2.430)] = \log 291.1 + \log 2.430 = 2.4640 + 0.3856 = 2.8496$$

$$c = \text{antilog } 2.8496 = 707.3 \text{ cm}$$

When given two sides of a right triangle, the third side can be found by using the pythagorean theorem. An acute angle is found by calculating the ratio of the two known sides and looking up the corresponding trigonometric function having this value. The second acute angle is found by using the angle relation.

EXAMPLE 6.4 Solve the right triangle ABC, given $a = 48.62$ m and $b = 37.64$ m. (See Fig. 6-1.)
To find side c, use $c = \sqrt{a^2 + b^2} = \sqrt{(48.62)^2 + (37.64)^2}$.

$$\log (48.62)^2 = 2 \log 48.62 = 2(1.6868) = 3.3736$$

$$(48.62)^2 = \text{antilog } 3.3736 = 2364$$

$$\log (37.64)^2 = 2 \log 37.64 = 2(1.5756) = 3.1512$$

$$(37.64)^2 = \text{antilog } 3.1512 = 1416$$

$$c = \sqrt{(48.62)^2 + (37.64)^2} = \sqrt{2364 + 1416} = \sqrt{3780}$$

$$\log c = \log \sqrt{3780} = \tfrac{1}{2} \log 3780 = \tfrac{1}{2}(3.5775) = 1.7888$$

$$c = \text{antilog } 1.7888 = 61.49 \text{ m}$$

To find angle A, use $\tan A = a/b = 48.62/37.64$.

$$\log \tan A = \log (48.62/37.64) = \log 48.62 - \log 37.64$$
$$= 1.6868 - 1.5756 = 0.1112$$
$$\tan A = \text{antilog } 0.1112 = 1.292$$
$$A = 52°16'$$

To find angle B, use $B = 90° - A = 90° - 52°16' = 37°44'$.

Solved Problems

6.1 Verify each of the following.

(a) $\log \sin 14°28' = \log 0.2476 = 9.3938 - 10$

(b) $\log \cos 66°45' = \log 0.3948 = 9.5964 - 10$

(c) $\log \tan 31°26' = \log 0.6112 = 9.7862 - 10$

(d) $\log \cot 45°55' = \log 0.9685 = 9.9861 - 10$

(e) $\log \sec 72°14' = \log 3.2772 = \log 3.277 = 0.5155$

(f) $\log \csc 32°37' = \log 1.8552 = \log 1.855 = 0.2684$

(g) $\log \tan 70°21' = \log 2.8006 = \log 2.801 = 0.4474$

(h) $\log \cot 11°17' = \log 5.0123 = \log 5.012 = 0.7000$

6.2 Verify each of the following.

(a) If $\log \sin A = 9.9002 - 10$, $\sin A = 0.7947$ and $A = 52°38'$.

(b) If $\log \cos A = 9.9360 - 10$, $\cos A = 0.8630$ and $A = 30°21'$.

(c) If $\log \tan A = 9.8715 - 10$, $\tan A = 0.7438$ and $A = 36°38'$.

(d) If $\log \cot A = 9.1015 - 10$, $\cot A = 0.1263$ and $A = 82°48'$.

(e) If $\log \sec A = 0.4598$, $\sec A = 2.883$ and $A = 69°42'$.

(f) If $\log \csc A = 0.1993$, $\csc A = 1.582$ and $A = 39°12'$.

(g) If $\log \tan A = 1.2261$, $\tan A = 16.83$ and $A = 86°36'$.

(h) If $\log \cot A = 0.0125$, $\cot A = 1.029$ and $A = 44°11'$.

6.3 Solve right triangle ABC, given $a = 562.8$ cm and $A = 64°24'$. (See Fig. 6-2.)

$$B = 90° - A = 90° - 64°24' = 25°36'$$

$$b = a \cot A = 562.8 \cot 64°24' = 562.8(0.4792)$$

$$\log b = \log [562.8(0.4792)] = \log 562.8 + \log 0.4792$$

$$= 2.7503 + 9.6805 - 10 = 2.4308$$

$$b = \text{antilog } 2.4308 = 269.6 \text{ cm}$$

$$c = a \csc A = 562.8 \csc 64°24' = 562.8(1.1089)$$

$$\log c = \log [562.8(1.109)] = \log 562.8 + \log 1.109$$

$$= 2.7053 + 0.0449 = 2.7952$$

$$c = \text{antilog } 2.7952 = 624.0 \text{ cm}$$

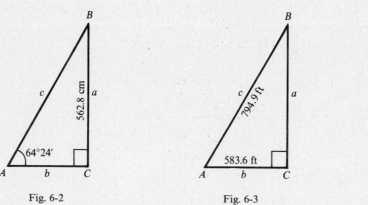

Fig. 6-2   Fig. 6-3   Fig. 6-4

6.4 Solve right triangle ABC, given $b = 583.6$ ft and $c = 794.9$ ft. (See Fig. 6-3.)

$$a = \sqrt{c^2 - b^2} = \sqrt{(c - b)(c + b)}$$

$$= \sqrt{(794.9 - 583.6)(794.9 + 583.6)} = \sqrt{211.3(1378.5)}$$

$$\log a = \log \sqrt{211.3(1379)} = \tfrac{1}{2}[\log 211.3 + \log 1379]$$

$$= \tfrac{1}{2}(2.3249 + 3.1396) = \tfrac{1}{2}(5.4645) = 2.7322$$

$$a = \text{antilog } 2.7322 = 539.8 \text{ ft}$$

$$\cos A = \frac{b}{c} = \frac{583.6}{794.9}$$

$$\log \cos A = \log (583.6/794.9) = \log 583.6 - \log 794.9$$

$$= 2.7661 - 2.9003 = (12.7661 - 10) - 2.9003$$

$$= (12.7661 - 2.9003) - 10 = 9.8658 - 10$$

$$\cos A = \text{antilog } (9.8658 - 10) = 0.7342$$

$$A = 42°46'$$

$$B = 90° - A = 90° - 42°46' = 47°14'$$

6.5 Solve right triangle ABC, given $c = 84.72$ in and $B = 41°41'$.
(See Fig. 6-4.)

$$A = 90° - B = 90° - 41°41' = 48°19'$$

$$a = c \cos B = 84.72 \cos 41°41' = 84.72(0.7468)$$

$$\log a = \log [84.72(0.7468)] = \log 84.72 + \log 0.7468$$

$$= 1.9280 + 9.8732 - 10 = 11.8012 - 10 = 1.8012$$

$$a = \text{antilog } 1.8012 = 63.27 \text{ in}$$

$$b = c \sin B = 84.72 \sin 41°41' = 84.72(0.6650)$$

$$\log b = \log [84.72(0.6650)] = \log 84.72 + \log 0.6650$$

$$= 1.9280 + 9.8228 - 10 = 11.7508 - 10 = 1.7508$$

$$b = \text{antilog } 1.7508 = 56.34 \text{ in}$$

6.6 At a height of 23,240 ft a pilot of an airplane measures the angle of depression of a light at an airport as $28°45'$. How far is he from the light?

In the Fig. 6-5, A is the position of the light, B is the position of the pilot, and $c = AB$ is the required distance. Then

$$c = a/\sin A$$

$$\log a = 4.3663$$
$$(-) \log \sin A = 9.6821 - 10$$
$$\overline{\log c = 4.6842}$$

$$c = 48,330$$

The required distance is 48,330 ft.

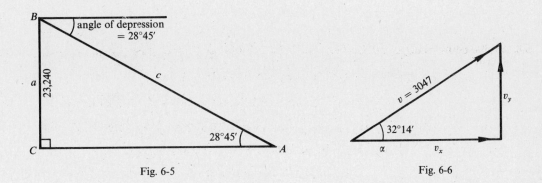

Fig. 6-5   Fig. 6-6

6.7 A shell is fired at an angle of elevation $32°14'$ with initial velocity 3047 ft/s. Find the initial horizontal and vertical velocities.

From Fig. 6-6, $v = 3047$, $\alpha = 32°14'$, and

$$v_x = v \cos \alpha \qquad\qquad\qquad v_y = v \sin \alpha$$

$$\begin{array}{rl} \log v = 3.4839 & \qquad \log v = 3.4839 \\ (+) \log \cos \alpha = 9.9273 - 10 & \qquad (+) \log \sin \alpha = 9.7270 - 10 \\ \hline \log v_x = 3.4112 & \qquad \log v_y = 3.2109 \end{array}$$

$$v_x = 2578 \text{ ft/s} \qquad\qquad\qquad v_y = 1625 \text{ ft/s}$$

6.8   Two forces of 151.7 lb and 225.8 lb act at right angles. Find the magnitude of the resultant and the angle that it makes with the larger force. (See Fig. 6-7.)

Using the right triangle ABC,

$$\tan A = CB/AC \qquad\qquad AB = CB/\sin A$$

$\log CB = 2.1810$	$\log CB = 2.1810$
$(-)\log AC = 2.3537$	$(-)\log \sin A = 9.7461 - 10$
$\log \tan A = 9.8273 - 10$	$\log AB = 2.4349$
$A = 33°52'$	$AB = 272.2$

The magnitude of the resultant force is 272.2 lb, and it makes an angle of 33°52′ with the larger force.

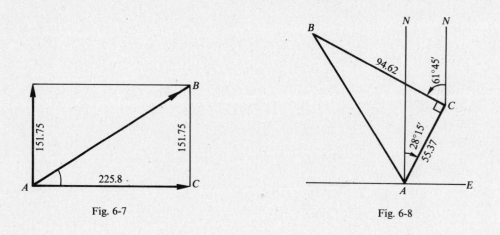

Fig. 6-7   Fig. 6-8

6.9   A boat travels N28°15′E for 55.37 mi and then N61°45′W for 94.62 mi. What is its distance and bearing from the starting point?

In Fig. 6-8, the boat starts at A, travels to C, and then to B. In the right triangle ABC,

$$\tan \angle CAB = BC/AC \qquad\qquad AB = BC/\sin \angle CAB$$

$\log BC = 1.9760$	$\log BC = 1.9760$
$(-)\log AC = 1.7433$	$(-)\log \sin \angle CAB = 9.9360 - 10$
$\log \tan \angle CAB = 0.2327$	$\log AB = 2.0400$
$\angle CAB = 59°40'$	$AB = 109.6$

The boat is then 109.6 mi from the starting point. Since $\angle NAB = \angle CAB - \angle CAN = 59°40' - 28°15' = 31°25'$, the required bearing is N31°25′W.

Supplementary Problems

Solve each of the following right triangles ABC, given:

6.10 $a = 25.72$, $A = 36°20'$ *Ans.* $B = 53°40'$, $b \doteq 34.97$, $c = 43.41$

6.11 $a = 342.9$, $A = 55°33'$ *Ans.* $B = 34°27'$, $b = 235.2$, $c = 416.0$

6.12 $a = 574.2$, $B = 56°21'$ *Ans.* $A = 33°39'$, $b = 862.6$, $c = 1037$

6.13 $c = 44.26, A = 56°14'$ *Ans.* $B = 33°46', a = 36.80, b = 24.60$

6.14 $c = 287.7, A = 38°10'$ *Ans.* $B = 51°50', a = 177.8, b = 226.2$

6.15 $c = 67.55, B = 47°26'$ *Ans.* $A = 42°34', a = 45.69, b = 49.75$

6.16 $a = 42.42, b = 58.48$ *Ans.* $A = 35°58', B = 54°2', c = 72.25$

6.17 $a = 384.7, b = 254.9$ *Ans.* $A = 56°28', B = 33°32', c = 461.6$

6.18 A straight road is to be constructed joining two towns A and B. If B is located 133.8 mi to the east and 256.8 mi to the north of A, find the length and direction of the road from A.

   Ans.   289.6 mi, N27°31'E

6.19 Two forces of 281.7 lb and 323.5 lb act at right angles. Find the magnitude of the resultant force and the angle that it makes with the larger force.

   Ans.   428.9 lb, 41°3'

6.20 Find the base of an isosceles triangle whose vertex angle is 48°28' and whose equal legs are 168.1.

   Ans.   138.0

Chapter 7

Reduction to Functions of Positive Acute Angles

7.1 COTERMINAL ANGLES

Let θ be any angle; then

$$\sin (\theta + n360°) = \sin \theta \qquad \cot (\theta + n360°) = \cot \theta$$

$$\cos (\theta + n360°) = \cos \theta \qquad \sec (\theta + n360°) = \sec \theta$$

$$\tan (\theta + n360°) = \tan \theta \qquad \csc (\theta + n360°) = \csc \theta$$

where n is any positive or negative integer or zero.

EXAMPLE 7.1 (a) $\sin 400° = \sin (40° + 360°) = \sin 40°$

(b) $\cos 850° = \cos (130° + 2 \cdot 360°) = \cos 130°$

(c) $\tan (-1000°) = \tan (80° - 3 \cdot 360°) = \tan 80°$

If x is an angle in radian measure, then

$$\sin (x + 2n\pi) = \sin x \qquad \cot (x + 2n\pi) = \cot x$$

$$\cos (x + 2n\pi) = \cos x \qquad \sec (x + 2n\pi) = \sec x$$

$$\tan (x + 2n\pi) = \tan x \qquad \csc (x + 2n\pi) = \csc x$$

where n is any integer.

EXAMPLE 7.2 (a) $\sin 11\pi/5 = \sin (\pi/5 + 2\pi) = \sin \pi/5$

(b) $\cos (-27\pi/11) = \cos [17\pi/11 - 2(2\pi)] = \cos 17\pi/11$

(c) $\tan 137\pi = \tan [\pi + 68(2\pi)] = \tan \pi$

7.2 FUNCTIONS OF A NEGATIVE ANGLE

Let θ be any angle; then

$$\sin (-\theta) = -\sin \theta \qquad \cot (-\theta) = -\cot \theta$$

$$\cos (-\theta) = \cos \theta \qquad \sec (-\theta) = \sec \theta$$

$$\tan (-\theta) = -\tan \theta \qquad \csc (-\theta) = -\csc \theta$$

EXAMPLE 7.3 $\sin (-50°) = -\sin 50°$, $\cos (-30°) = \cos 30°$, $\tan (-200°) = -\tan 200°$. For a proof of these relations, see Prob. 7.1.

7.3 REFERENCE ANGLES

If θ is a quadrantal angle, then the function values are the same as in Sec. 2.6 and a reference angle is not needed. Since any angle A can be written as $\theta + n360°$, where n is an integer and $0° \leq \theta < 360°$, reference angles will be found for angles from $0°$ to $360°$.

A reference angle R for an angle θ in standard position is the positive acute angle between the x axis and the terminal side of angle θ. The values of the six trigonometric functions of the reference angle for θ, R, agree with the function values for θ except possibly in sign. When the signs of the functions of R are determined by the quadrant of angle θ, as in Sec. 2.5, then any function of θ can be expressed as a function of the acute angle R. Thus, our tables can be used to find the value of a trigonometric function of any angle.

Quadrant for θ	Relationship	Function Signs
I	$R = \theta$	All functions are positive.
II	$R = 180° - \theta$	Only sin R and csc R positive.
III	$R = \theta - 180°$	Only tan R and cot R positive.
IV	$R = 360° - \theta$	Only cos R and sec R positive.

See Prob. 7.2 for a verification of the equality of the values of the trigonometric functions of θ and signed values of its reference angle R.

EXAMPLE 7.4 Express each as a function of an acute angle.

(a) sin 232°, (b) cos 312°, (c) tan 912°, (d) sec (−227°)

(a) sin 232° = −sin (232° − 180°) = −sin 52°
232° is in quadrant III, so the sine is negative and $R = \theta - 180°$.

(b) cos 312° = +cos (360° − 312°) = cos 48°
312° is in quadrant IV, so the cosine is positive and $R = 360° - \theta$.

(c) tan 912° = tan [192° + 2(360°)] = tan 192°
= +tan (192° − 180°) = tan 12°
Since 912° ≥ 360°, we find the coterminal angle first.
192° is in quadrant III, so the tangent is positive and $R = \theta - 180°$.

(d) sec (−227°) = sec (133° − 360°) = sec 133°
= −sec (180° − 133°) = −sec 47°
Since −227° < 0°, we find a coterminal angle first.
133° is in quadrant II, so the secant is negative and $R = 180° - \theta$.

When finding the value of a trigonometric function by using a calculator, a reference angle is unnecessary. The function value is found as indicated in Sec. 4.5. However, when an angle having a given function value is to be found and that angle is to be in a specific quadrant, a reference angle is usually needed even when using a calculator.

7.4 ANGLES WITH A GIVEN FUNCTION VALUE

Since coterminal angles have the same value for a function, there are an unlimited number of angles that have the same value for a trigonometric function. Even when we restrict the angles to the interval of 0° to 360°, there are usually two angles that have the same function value. All the angles that have the same function value also have the same reference angle. The quadrants for the angle are determined by the sign of the function value. The relationships from Sec. 7.3 are used to find the angle θ, once the reference angle is found from a table (see Sec. 4.4) or a calculator (see Sec. 4.6).

EXAMPLE 7.5 Find all angles θ between 0° and 360° when:

(a) sin θ = 0.6293, (b) cos θ = −0.3256, (c) tan θ = −1.2799

(a) Since sin θ = 0.6293 is positive, solutions for θ are in quadrants I and II because the sine is positive in these quadrants.
sin R = 0.6293; thus R = 39°.
In quadrant I, $R = \theta$, so θ = 39°.
In quadrant II, $R = 180° - \theta$, so $\theta = 180° - R = 180° - 39° = 141°$.
θ = 39° and 141°.

(b) Since $\cos \theta = -0.3256$ is negative, solutions for θ are in quadrants II and III because the cosine is negative in these quadrants.

$\cos R = 0.3256$; thus $R = 71°$.

In quadrant II, $R = 180° - \theta$, so $\theta = 180° - R = 180° - 71° = 109°$.
In quadrant III, $R = \theta - 180°$, so $\theta = 180° + R = 180° + 71° = 251°$.
$\theta = 109°$ and $251°$.

(c) Since $\tan \theta = -1.2799$ is negative, solutions for θ are in quadrants II and IV because the tangent is negative in these quadrants.

$\tan R = 1.2799$; thus $R = 52°$.

In quadrant II, $R = 180° - \theta$, so $\theta = 180° - R = 180° - 52° = 128°$.
In quadrant IV, $R = 360° - \theta$, so $\theta = 360° - R = 360° - 52° = 308°$.
$\theta = 128°$ and $308°$.

EXAMPLE 7.6 Find all angles θ when (a) $\sin \theta = -0.2079$ and (b) $\tan \theta = 0.5543$.

(a) $\sin R = 0.2079$; thus $R = 12°$. The sine is negative in quadrants III and IV.
In quadrant III, $\theta = 180° + R = 180° + 12° = 192°$.
In quadrant IV, $\theta = 360° - R = 360° - 12° = 348°$.
All angles coterminal with these values of θ are needed, so $\theta = 192° + n360°$ and $348° + n360°$ where n is any integer.

(b) $\tan R = 0.5543$; thus $R = 29°$. The tangent is positive in quadrants I and III.
In quadrant I, $\theta = R = 29°$.
In quadrant III, $\theta = 180° + R = 180° + 29° = 209°$.
All angles coterminal with these values of θ are needed, so $\theta = 29° + n360°$ and $209° + n360°$ where n is any integer.

Solved Problems

7.1 Derive formulas for the functions of $-\theta$ in terms of the functions of θ.

In Fig. 7-1, θ and $-\theta$ are constructed in standard position and are numerically equal. On their respective terminal sides the points $P(x, y)$ and $P_1(x_1, y_1)$ are located so that $OP = OP_1$. In each of the figures the two triangles are congruent and $r_1 = r$, $x_1 = x$, and $y_1 = -y$. Then

$$\sin(-\theta) = \frac{y_1}{r_1} = \frac{-y}{r} = -\frac{y}{r} = -\sin\theta \qquad \cot(-\theta) = \frac{x_1}{y_1} = \frac{x}{-y} = -\frac{x}{y} = -\cot\theta$$

$$\cos(-\theta) = \frac{x_1}{r_1} = \frac{x}{r} = \cos\theta \qquad\qquad \sec(-\theta) = \frac{r_1}{x_1} = \frac{r}{x} = \sec\theta$$

$$\tan(-\theta) = \frac{y_1}{x_1} = \frac{-y}{x} = -\frac{y}{x} = -\tan\theta \qquad \csc(-\theta) = \frac{r_1}{y_1} = \frac{r}{-y} = -\frac{r}{y} = -\csc\theta$$

Except for those cases in which a function is not defined, the above relations are also valid when θ is a quadrantal angle. This may be verified by making use of the fact that $-0°$ and $0°$, $-90°$ and $270°$, $-180°$ and $180°$, and $-270°$ and $90°$ are coterminal.

For example, $\sin(-0°) = \sin 0° = 0 = -\sin 0°$, $\sin(-90°) = \sin 270° = -1 = -\sin 90°$, $\cos(-180°) = \cos 180°$, and $\cot(-270°) = \cot 90° = 0 = -\cot 270°$.

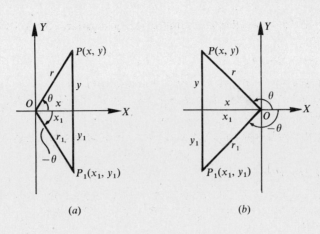

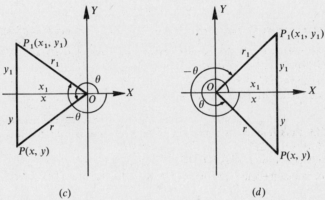

Fig. 7-1

7.2 Verify the equality of the trigonometric functions for θ and its reference angle R where $x > 0$, $y > 0$, and $r = \sqrt{x^2 + y^2}$.

(a) θ is in quadrant I. See Fig. 7-2(a).

$$\sin \theta = \frac{y}{r} = \sin R \qquad\qquad \cot \theta = \frac{x}{y} = \cot R$$

$$\cos \theta = \frac{x}{r} = \cos R \qquad\qquad \sec \theta = \frac{r}{x} = \sec R$$

$$\tan \theta = \frac{y}{x} = \tan R \qquad\qquad \csc \theta = \frac{r}{y} = \csc R$$

(b) θ is in quadrant II. See Fig. 7-2(b).

$$\sin \theta = \frac{y}{r} = \sin R \qquad\qquad \cot \theta = \frac{-x}{y} = -\left(\frac{x}{y}\right) = -\cot R$$

$$\cos \theta = \frac{-x}{r} = -\left(\frac{x}{r}\right) = -\cos R \qquad\qquad \sec \theta = \frac{r}{-x} = -\left(\frac{r}{x}\right) = -\sec R$$

$$\tan \theta = \frac{y}{-x} = -\left(\frac{y}{x}\right) = -\tan R \qquad\qquad \csc \theta = \frac{r}{y} = \csc R$$

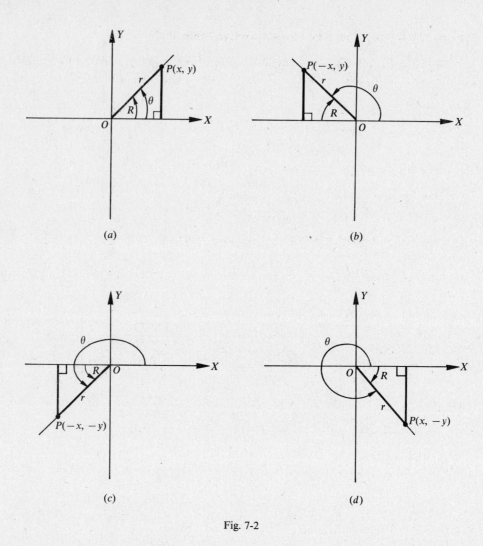

Fig. 7-2

(c) θ is in quadrant III. See Fig. 7-2(c).

$$\sin \theta = \frac{-y}{r} = -\left(\frac{y}{r}\right) = -\sin R \qquad \cot \theta = \frac{-x}{-y} = \frac{x}{y} = \cot R$$

$$\cos \theta = \frac{-x}{r} = -\left(\frac{x}{r}\right) = -\cos R \qquad \sec \theta = \frac{r}{-x} = -\left(\frac{r}{x}\right) = -\sec R$$

$$\tan \theta = \frac{-y}{-x} = \frac{y}{x} = \tan R \qquad \csc \theta = \frac{r}{-y} = -\left(\frac{r}{y}\right) = -\csc R$$

(d) θ is in quadrant IV. See Fig. 7-2(d).

$$\sin \theta = \frac{-y}{r} = -\left(\frac{y}{r}\right) = -\sin R \qquad \cot \theta = \frac{x}{-y} = -\left(\frac{x}{y}\right) = -\cot R$$

$$\cos \theta = \frac{x}{r} = \cos R \qquad \sec \theta = \frac{r}{x} = \sec R$$

$$\tan \theta = \frac{-y}{x} = -\left(\frac{y}{x}\right) = -\tan R \qquad \csc \theta = \frac{r}{-y} = -\left(\frac{r}{y}\right) = -\csc R$$

7.3   Express the following as functions of positive acute angles.

(a) $\sin 130°$, (b) $\tan 325°$, (c) $\sin 200°$, (d) $\cos 370°$, (e) $\tan 165°$, (f) $\sec 250°$, (g) $\sin 670°$, (h) $\cot 930°$, (i) $\csc 865°$, (j) $\sin(-100°)$, (k) $\cos(-680°)$, (l) $\tan(-290°)$

   (a) $\sin 130° = +\sin(180° - 130°) = \sin 50°$

   (b) $\tan 325° = -\tan(360° - 325°) = -\tan 35°$

   (c) $\sin 200° = -\sin(200° - 180°) = -\sin 20°$

   (d) $\cos 370° = \cos(10° + 360°) = \cos 10°$

   (e) $\tan 165° = -\tan(180° - 165°) = -\tan 15°$

   (f) $\sec 250° = -\sec(250° - 180°) = -\sec 70°$

   (g) $\sin 670° = \sin(310° + 360°) = \sin 310° = -\sin(360° - 310°)$
          $= -\sin 50°$

   (h) $\cot 930° = \cot[210° + 2(360°)] = \cot 210° = +\cot(210° - 180°)$
          $= \cot 30°$

   (i) $\csc 865° = \csc[145° + 2(360°)] = \csc 145° = +\csc(180° - 145°)$
          $= \csc 35°$

   (j) $\sin(-100°) = -\sin 100° = -[+\sin(180° - 100°)] = -\sin 80°$
          or $\sin(-100°) = \sin(260° - 360°) = \sin 260° = -\sin(260° - 180°)$
          $= -\sin 80°$

   (k) $\cos(-680°) = +\cos 680° = \cos(320° + 360°) = \cos 320°$
          $= +\cos(360° - 320°) = \cos 40°$
          or $\cos(-680°) = \cos[40° - 2(360°)] = \cos 40°$

   (l) $\tan(-290°) = -\tan 290° = -[-\tan(360° - 290°)] = +\tan 70°$
          or $\tan(-290°) = \tan(70° - 360°) = \tan 70°$

7.4   Find the exact value of the sine, cosine, and tangent of
   (a) $120°$, (b) $210°$, (c) $315°$, (d) $-135°$, (e) $-240°$, (f) $-330°$

   (a) $120°$ is in quadrant II; reference angle $= 180° - 120° = 60°$.

   $\sin 120° = \sin 60° = \dfrac{\sqrt{3}}{2}$ $\cos 120° = -\cos 60° = -\dfrac{1}{2}$ $\tan 120° = -\tan 60° = -\sqrt{3}$

   (b) $210°$ is in quadrant III; reference angle $= 210° - 180° = 30°$.

   $\sin 210° = -\sin 30° = -\dfrac{1}{2}$ $\cos 210° = -\cos 30° = -\dfrac{\sqrt{3}}{2}$ $\tan 210° = \tan 30° = \dfrac{\sqrt{3}}{3}$

   (c) $315°$ is in quadrant IV; reference angle $= 360° - 315° = 45°$.

   $\sin 315° = -\sin 45° = -\dfrac{\sqrt{2}}{2}$ $\cos 315° = \cos 45° = \dfrac{\sqrt{2}}{2}$ $\tan 315° = -\tan 45° = -1$

   (d) $-135°$ is coterminal with $-135° + 360° = 225°$; $225°$ is in quadrant III; reference angle $= 225° - 180° = 45°$.

   $$\sin(-135°) = -\sin 45° = -\frac{\sqrt{2}}{2} \qquad \cos(-135°) = -\cos 45° = -\frac{\sqrt{2}}{2}$$

   $$\tan(-135°) = \tan 45° = 1$$

(e) $-240°$ is coterminal with $-240° + 360° = 120°$; $120°$ is in quadrant II; reference angle $= 180° - 120° = 60°$.

$$\sin(-240°) = \sin 60° = \frac{\sqrt{3}}{2} \qquad \cos(-240°) = -\cos 60° = -\frac{1}{2}$$

$$\tan(-240°) = -\tan 60° = -\sqrt{3}$$

(f) $-330°$ is coterminal with $-330° + 360° = 30°$; $30°$ is in quadrant I; reference angle $= 30°$.

$$\sin(-330°) = \sin 30° = \frac{1}{2} \qquad \cos(-330°) = \cos 30° = \frac{\sqrt{3}}{2} \qquad \tan(-330°) = \tan 30° = \frac{\sqrt{3}}{3}$$

7.5 Use Table 1 to find:

(a) $\sin 125°14' = +\sin(180° - 125°14') = \sin 54°46' = 0.8168$

(b) $\cos 169°40' = -\cos(180° - 169°40') = -\cos 10°20' = -0.9838$

(c) $\tan 200°23' = +\tan(200°23' - 180°) = \tan 20°23' = 0.3716$

(d) $\cot 250°44' = +\cot(250°44' - 180°) = \cot 70°44' = 0.3495$

(e) $\cos 313°18' = +\cos(360° - 313°18') = \cos 46°42' = 0.6858$

(f) $\sin 341°52' = -\sin(360° - 341°52') = -\sin 18°8' = -0.3112$

7.6 Use Table 2 to find:

(a) $\tan 97.2° = -\tan(180° - 97.2°) = -\tan 82.8° = -7.9158$

(b) $\cos 147.8° = -\cos(180° - 147.8°) = -\cos 32.2° = -0.8462$

(c) $\cot 241.28° = +\cot(241.28° - 180°) = \cot 61.28° = 0.5480$

(d) $\sin 194.37° = -\sin(194.37° - 180°) = -\sin 14.37° = -0.2482$

(e) $\cos 273.1° = +\cos(360° - 273.1°) = \cos 86.9° = 0.0541$

(f) $\tan 321.61° = -\tan(360° - 321.61°) = -\tan 38.39° = -0.7923$

7.7 Use a calculator to find:

(a) $\sin 158°38' = \sin(158 + 38/60)° = 0.364355$

(b) $\cos 264°21' = \cos(264 + 21/60)° = -0.098451$

(c) $\tan 288°14' = \tan(288 + 14/60)° = -3.03556$

(d) $\tan 112.68° = -2.39292$

(e) $\sin 223.27° = -0.685437$

(f) $\cos 314.59° = 0.702029$

7.8 Show that $\sin \theta$ and $\tan \frac{1}{2}\theta$ have the same sign.

(a) Suppose $\theta = n \cdot 180°$. If n is even (including zero), say $2m$, then $\sin(2m \cdot 180°) = \tan(m \cdot 180°) = 0$. The case when n is odd is excluded since then $\tan \frac{1}{2}\theta$ is not defined.

(b) Suppose $\theta = n \cdot 180° + \phi$, where $0° < \phi < 180°$. If n is even, including zero, θ is in quadrant I or quadrant II and $\sin \theta$ is positive while $\frac{1}{2}\theta$ is in quadrant I or quadrant III and $\tan \frac{1}{2}\theta$ is positive. If n is odd, θ is in quadrant III or IV and $\sin \theta$ is negative while $\frac{1}{2}\theta$ is in quadrant II or IV and $\tan \frac{1}{2}\theta$ is negative.

7.9 Find all positive values of θ less than $360°$ for which $\sin \theta = -\frac{1}{2}$.

There will be two angles (see Chap. 2), one in the third-quadrant and one in the fourth-quadrant. The reference angle of each has its sine equal to $+\frac{1}{2}$ and is $30°$. Thus the required angles are $\theta = 180° + 30° = 210°$ and $\theta = 360° - 30° = 330°$.

(NOTE: To obtain *all* values of θ for which $\sin \theta = -\frac{1}{2}$, add $n \cdot 360°$ to each of the above solutions; thus $\theta = 210° + n \cdot 360°$ and $\theta = 330° + n \cdot 360°$, where n is any integer.)

7.10 Find all positive values of θ less than $360°$ for which $\cos \theta = 0.9063$.

There are two solutions, $\theta = 25°$ in the first-quadrant and $\theta = 360° - 25° = 335°$ in the fourth-quadrant.

7.11 Find all positive values of $\frac{1}{4}\theta$ less than $360°$, given $\sin \theta = 0.6428$.

The two positive angles less than $360°$ for which $\sin \theta = 0.6428$ are $\theta = 40°$ and $\theta = 180° - 40° = 140°$. But if $\frac{1}{4}\theta$ is to include all values less than $360°$, θ must include all values less than $4 \cdot 360° = 1440°$. Hence, for θ we take the two angles above and all coterminal angles less than $1440°$; that is,

$$\theta = 40°, 400°, 760°, 1120°; 140°, 500°, 860°, 1220°$$

and $$\tfrac{1}{4}\theta = 10°, 100°, 190°, 280°; 35°, 125°, 215°, 305°$$

Supplementary Problems

7.12 Express each of the following in terms of functions of a positive acute angle.

(a) $\sin 145°$ (d) $\cot 155°$ (g) $\sin(-200°)$ (j) $\cot 610°$

(b) $\cos 215°$ (e) $\sec 325°$ (h) $\cos(-760°)$ (k) $\sec 455°$

(c) $\tan 440°$ (f) $\csc 190°$ (i) $\tan(-1385°)$ (l) $\csc 825°$

Ans.   (a) $\sin 35°$ (g) $\sin 20°$

(b) $-\cos 35°$ (h) $\cos 40°$

(c) $\tan 80°$ (i) $\tan 55°$

(d) $-\cot 25°$ (j) $\cot 70°$

(e) $\sec 35°$ (k) $-\sec 85°$

(f) $-\csc 10°$ (l) $\csc 75°$

7.13 Find the exact values of the sine, cosine, and tangent of
(a) $150°$, (b) $225°$, (c) $300°$, (d) $-120°$, (e) $-210°$, (f) $-315°$

Ans.   (a) $1/2, -\sqrt{3}/2, -1/\sqrt{3} = -\sqrt{3}/3$ (d) $-\sqrt{3}/2, -1/2, \sqrt{3}$

(b) $-\sqrt{2}/2, -\sqrt{2}/2, 1$ (e) $1/2, -\sqrt{3}/2, -1/\sqrt{3} = -\sqrt{3}/3$

(c) $-\sqrt{3}/2, 1/2, -\sqrt{3}$ (f) $\sqrt{2}/2, \sqrt{2}/2, 1$

7.14 Using appropriate tables, verify:

    Ans.   (*a*) $\sin 155°13' = 0.4192$ (*f*) $\tan 129.48° = -1.2140$

             (*b*) $\cos 104°38' = -0.2526$ (*g*) $\sin 110.32° = 0.9378$

             (*c*) $\tan 305°24' = -1.4071$ (*h*) $\cos 262.35° = -0.1332$

             (*d*) $\sin 114°18' = 0.9114$ (*i*) $\tan 211.84° = 0.6210$

             (*e*) $\cos 166°51' = -0.9738$ (*j*) $\cos 314.92° = 0.7061$

7.15 Find all angles, $0° \leq \theta < 360°$, for which:

    (*a*) $\sin \theta = \sqrt{2}/2$, (*b*) $\cos \theta = -1$, (*c*) $\sin \theta = -0.6180$, (*d*) $\cos \theta = 0.5125$, (*e*) $\tan \theta = -1.5301$

    Ans.   (*a*) $45°, 135°$ (*c*) $218°10', 321°50'$ or $218.17°, 321.83°$

             (*b*) $180°$ (*d*) $59°10', 300°50'$ or $59.17°, 300.83°$

             (*e*) $123°10', 303°10'$ or $123.17°, 303.17°$

Chapter 8

Variations and Graphs of the Trigonometric Functions

8.1 LINE REPRESENTATIONS OF TRIGONOMETRIC FUNCTIONS

Let θ be any given angle in standard position. (See Fig. 8-1 for θ in each of the quadrants.) With the vertex O as center, describe a circle of radius one unit cutting the initial side OX of θ at A, the positive y axis at B, and the terminal side of θ at P. Draw MP perpendicular to OX; draw also the tangents to the circle at A and B meeting the terminal side of θ or its extension through O in the points Q and R, respectively.

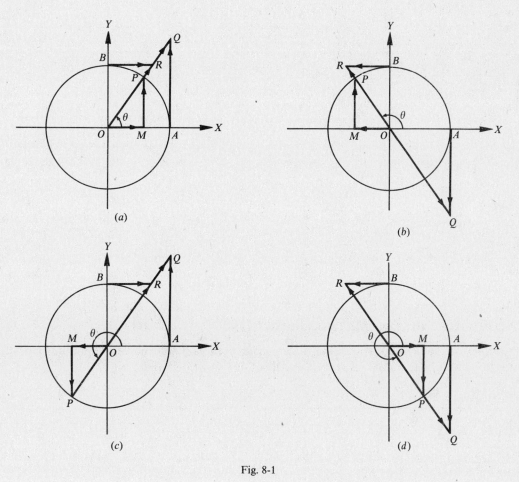

Fig. 8-1

In each of the parts of Fig. 8-1, the right triangles OMP, OAQ, and OBR are similar, and

$$\sin\theta = \frac{MP}{OP} = MP \qquad\qquad \cot\theta = \frac{OM}{MP} = \frac{BR}{OB} = BR$$

$$\cos\theta = \frac{OM}{OP} = OM \qquad\qquad \sec\theta = \frac{OP}{OM} = \frac{OQ}{OA} = OQ$$

$$\tan\theta = \frac{MP}{OM} = \frac{AQ}{OA} = AQ \qquad\qquad \csc\theta = \frac{OP}{MP} = \frac{OR}{OB} = OR$$

83

The segments MP, OM, AQ, etc., are directed line segments. The magnitude of a function is given by the length of the corresponding segment and the sign is given by the indicated direction. The directed segments OQ and OR are to be considered positive when measured on the terminal side of the angle and negative when measured on the terminal side extended.

8.2 VARIATIONS OF TRIGONOMETRIC FUNCTIONS

Let P move counterclockwise about the unit circle, starting at A, so that $\theta = \angle AOP$ varies continuously from $0°$ to $360°$. Using Fig. 8-1, see how the trigonometric functions vary ($I.$ = increases, $D.$ = decreases):

As θ increases from	0° to 90°	90° to 180°	180° to 270°	270° to 360°
sin θ	I. from 0 to 1	D. from 1 to 0	D. from 0 to −1	I. from −1 to 0
cos θ	D. from 1 to 0	D. from 0 to −1	I. from −1 to 0	I. from 0 to 1
tan θ	I. from 0 without limit (0 to +∞)	I. from large negative values to 0 (−∞ to 0)	I. from 0 without limit (0 to +∞)	I. from large negative values to 0 (−∞ to 0)
cot θ	D. from large positive values to 0 (+∞ to 0)	D. from 0 without limit (0 to −∞)	D. from large positive values to 0 (+∞ to 0)	D. from 0 without limit (0 to −∞)
sec θ	I. from 1 without limit (1 to +∞)	I. from large negative values to −1 (−∞ to −1)	D. from −1 without limit (−1 to −∞)	D. from large positive values to 1 (+∞ to 1)
csc θ	D. from large positive values to 1 (+∞ to 1)	I. from 1 without limit (1 to +∞)	I. from large negative values to −1 (−∞ to −1)	D. from −1 without limit (−1 to −∞)

8.3 GRAPHS OF TRIGONOMETRIC FUNCTIONS

In the following table, values of the angle x are given in radians. Whenever a trigonometric function is undefined for the value of x, $\pm \infty$ is recorded instead of a function value.

x	$y = \sin x$	$y = \cos x$	$y = \tan x$	$y = \cot x$	$y = \sec x$	$y = \csc x$
0	0	1.00	0	±∞	1.00	±∞
$\pi/6$	0.50	0.87	0.58	1.73	1.15	2.00
$\pi/4$	0.71	0.71	1.00	1.00	1.41	1.41
$\pi/3$	0.87	0.50	1.73	0.58	2.00	1.15
$\pi/2$	1.00	0	±∞	0	±∞	1.00
$2\pi/3$	0.87	−0.50	−1.73	−0.58	−2.00	1.15
$3\pi/4$	0.71	−0.71	−1.00	−1.00	−1.41	1.41
$5\pi/6$	0.50	−0.87	−0.58	−1.73	−1.15	2.00
π	0	−1.00	0	±∞	−1.00	±∞
$7\pi/6$	−0.50	−0.87	0.58	1.73	−1.15	−2.00
$5\pi/4$	−0.71	−0.71	1.00	1.00	−1.41	−1.41
$4\pi/3$	−0.87	−0.50	1.73	0.58	−2.00	−1.15
$3\pi/2$	−1.00	0	±∞	0	±∞	−1.00
$5\pi/3$	−0.87	0.50	−1.73	−0.58	2.00	−1.15
$7\pi/4$	−0.71	0.71	−1.00	−1.00	1.41	−1.41
$11\pi/6$	−0.50	0.87	−0.58	−1.73	1.15	−2.00
2π	0	1.00	0	±∞	1.00	±∞

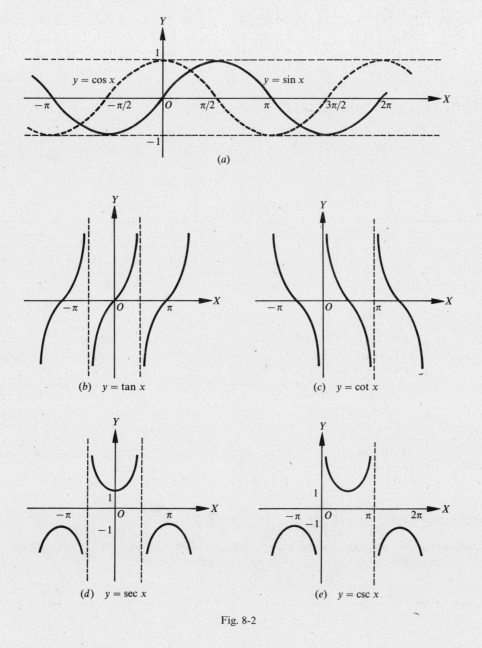

Fig. 8-2

[NOTE 1.   Since $\sin\left(\frac{1}{2}\pi + x\right) = \cos x$, the graph of $y = \cos x$ may be obtained most easily by shifting the graph of $y = \sin x$ a distance $\frac{1}{2}\pi$ to the left.]

[NOTE 2.   Since $\csc\left(\frac{1}{2}\pi + x\right) = \sec x$, the graph of $y = \csc x$ may be obtained by shifting the graph of $y = \sec x$ a distance $\frac{1}{2}\pi$ to the right.]

8.4   HORIZONTAL AND VERTICAL SHIFTS

. The graph of a trigonometric function can be shifted vertically by adding a nonzero constant to the function and horizontally by adding a nonzero constant to the angle of the trigonometric function. Figure 8-3(a) is the graph of $y = \sin x$ and the remaining parts of Fig. 8-3 are the results of shifting this graph.

If c is a positive number, then adding it to a trigonometric function results in the graph being shifted up c units [see Fig. 8-3(b)] and subtracting it from a trigonometric function results in the graph being shifted down c units [see Fig. 8-3(c)].

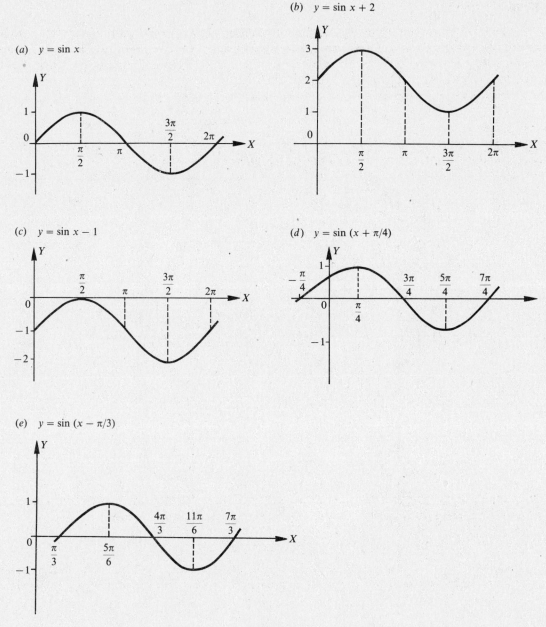

Fig. 8-3

For a positive number d, a trigonometric function is shifted left d units when d is added to the angle [see Fig. 8-3(d)] and shifted right d units when d is subtracted from the angle [see Fig. 8-3(e)].

8.5 PERIODIC FUNCTIONS

Any function of a variable x, $f(x)$, which repeats its values in definite cycles, is called *periodic*. The smallest range of values of x which corresponds to a complete cycle of values of the function is called the *period* of the function. It is evident from the graphs of the trigonometric functions that the sine, cosine, secant, and cosecant are of period 2π while the tangent and cotangent are of period π.

8.6 SINE CURVES

The *amplitude* (maximum ordinate) and period (wavelength) of $y = \sin x$ are respectively 1 and 2π. For a given value of x, the value of $y = a \sin x$, $a > 0$, is a times the value of $y = \sin x$. Thus, the amplitude of $y = a \sin x$ is a and the period is 2π. Since when $bx = 2\pi$, $x = 2\pi/b$, the amplitude of $y = \sin bx$, $b > 0$, is 1 and the period is $2\pi/b$.

The general sine curve (sinusoid) of equation

$$y = a \sin bx \qquad a > 0, b > 0$$

has amplitude a and period $2\pi/b$. Thus the graph of $y = 3 \sin 2x$ has amplitude 3 and period $2\pi/2 = \pi$. Figure 8-4 exhibits the graphs of $y = \sin x$ and $y = 3 \sin 2x$ on the same axes.

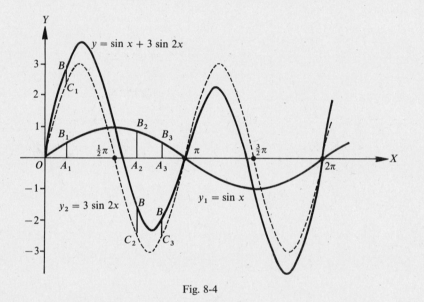

Fig. 8-4

More complicated forms of wave motions are obtained by combining two or more sine curves. The method of adding corresponding ordinates is illustrated in the following example.

EXAMPLE 8.1   Construct the graph of $y = \sin x + 3 \sin 2x$. See Fig. 8-4.

First the graphs of $y_1 = \sin x$ and $y_2 = 3 \sin 2x$ are constructed on the same axes. Then, corresponding to a given value $x = OA_1$, the ordinate A_1B of $y = \sin x + 3 \sin 2x$ is the *algebraic* sum of the ordinates A_1B_1 of $y_1 = \sin x$ and A_1C_1 of $y_2 = 3 \sin 2x$. Also, $A_2B = A_2B_2 + A_2C_2$, $A_3B = A_3B_3 + A_3C_3$, etc.

Solved Problems

8.1   Sketch the graphs of the following for one period.

   (a) $y = 4 \sin x$ (c) $y = 3 \sin \frac{1}{2}x$ (e) $y = 3 \cos \frac{1}{2}x = 3 \sin (\frac{1}{2}x + \frac{1}{2}\pi)$

   (b) $y = \sin 3x$ (d) $y = 2 \cos x = 2 \sin (x + \frac{1}{2}\pi)$

   In each case we use the same curve and then put in the y axis and choose the units on each axis to satisfy the requirements of amplitude and period of each curve.

   (a) $y = 4 \sin x$ has amplitude $= 4$ and period $= 2\pi$.

   (b) $y = \sin 3x$ has amplitude $= 1$ and period $= 2\pi/3$.

(c) $y = 3 \sin \frac{1}{2}x$ has amplitude = 3 and period = $2\pi/\frac{1}{2} = 4\pi$.

(d) $y = 2 \cos x$ has amplitude = 2 and period = 2π. Note the position of the y axis.

(e) $y = 3 \cos \frac{1}{2}x$ has amplitude = 3 and period = 4π.

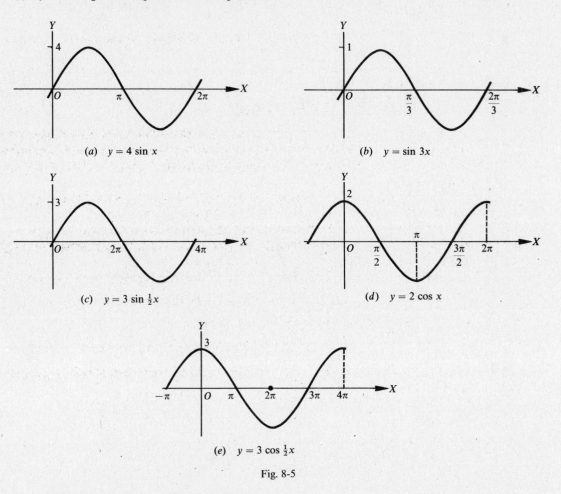

(a) $y = 4 \sin x$

(b) $y = \sin 3x$

(c) $y = 3 \sin \frac{1}{2}x$

(d) $y = 2 \cos x$

(e) $y = 3 \cos \frac{1}{2}x$

Fig. 8-5

8.2   Construct the graph of each of the following.

(a) $y = \frac{1}{2} \tan x$,   (b) $y = 3 \tan x$,   (c) $y = \tan 3x$,   (d) $y = \tan \frac{1}{4}x$

In each case we use the same curve and then put in the y axis and choose the units on the x axis to satisfy the period of the curve.

(a) $y = \frac{1}{2} \tan x$ has period π

(b) $y = 3 \tan x$ has period π

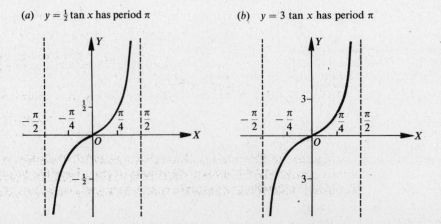

(c) $y = \tan 3x$ has period $\pi/3$ (d) $y = \tan \frac{1}{4}x$ has period $\pi/\frac{1}{4} = 4\pi$

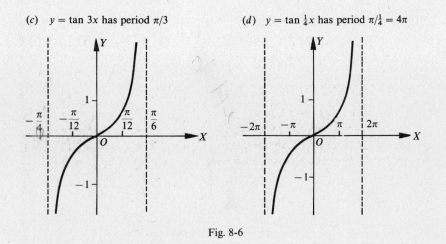

Fig. 8-6

8.3   Construct the graph of each of the following.

(a) $y = \sin x + \cos x$ (c) $y = \sin 2x - \cos 3x$

(b) $y = \sin 2x + \cos 3x$ (d) $y = 3 \sin 2x + 2 \cos 3x$

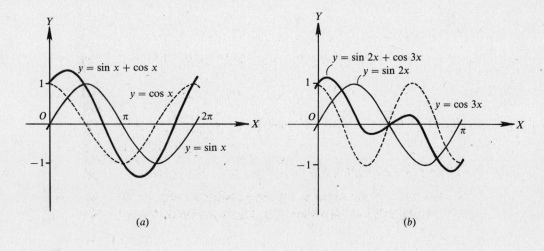

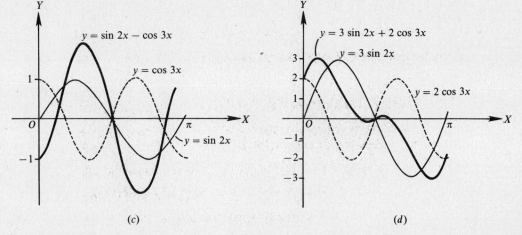

Fig. 8-7

8.4 Construct a graph of each of the following.

  (a) $y = 3 \sin x + 1$ (c) $y = \cos x + 2$

  (b) $y = \sin x - 2$ (d) $y = \frac{1}{2} \cos x - 1$

(a) $y = 3 \sin x$ is shifted up 1 unit

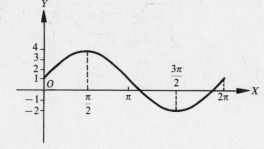

(b) $y = \sin x$ is shifted down 2 units

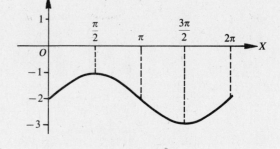

(c) $y = \cos x$ is shifted up 2 units

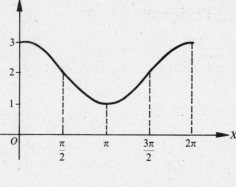

(d) $y = \frac{1}{2} \cos x$ is shifted down 1 unit

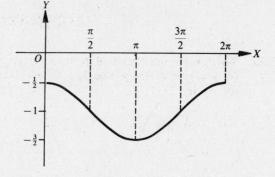

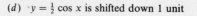

Fig. 8-8

8.5 Construct a graph of each of the following.

  (a) $y = \sin (x - \pi/6)$ (c) $y = \cos (x - \pi/4)$

  (b) $y = \sin (x + \pi/6)$ (d) $y = \cos (x + \pi/3)$

(a) $y = \sin x$ is shifted right $\pi/6$ units

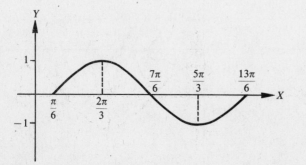

(b) $y = \sin x$ is shifted left $\pi/6$ units

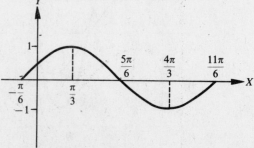

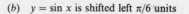

(c) $y = \cos x$ is shifted right $\pi/4$ units

(d) $y = \cos x$ is shifted left $\pi/3$ units

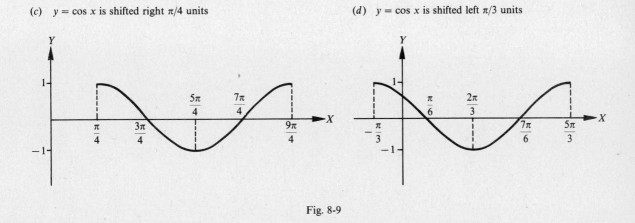

Fig. 8-9

Supplementary Problems

8.6 Sketch the graph of each of the following for one period.
   (a) $y = 3 \sin x$,   (b) $y = \sin 2x$,   (c) $y = 4 \sin x/2$,   (d) $y = 4 \cos x$,   (e) $y = 2 \cos x/3$,   (f) $y = 2 \tan x$,
   (g) $y = \tan 2x$

Ans.   (a) $y = 3 \sin x$

(b) $y = \sin 2x$

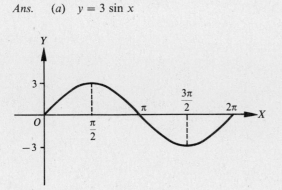

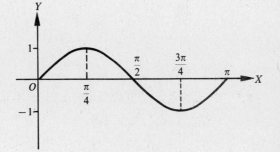

(c) $y = 4 \sin x/2$

(d) $y = 4 \cos x$

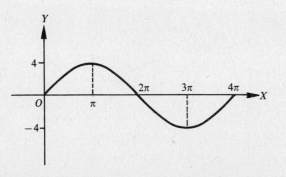

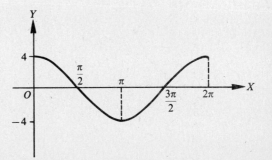

(e) $y = 2 \cos x/3$ (f) $y = 2 \tan x$

(g) $y = \tan 2x$

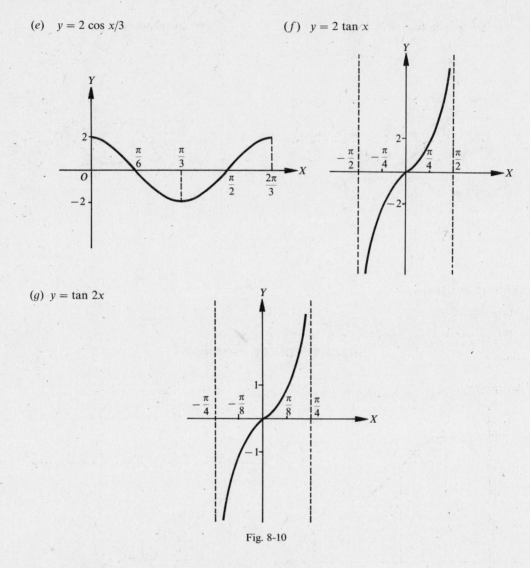

Fig. 8-10

8.7 Construct the graph of each of the following for one period.

(a) $y = \sin x + 2 \cos x$ (c) $y = \sin 2x + \sin 3x$

(b) $y = \sin 3x + \cos 2x$ (d) $y = \sin 3x - \cos 2x$

Ans.   (a) $y = \sin x + 2 \cos x$ (b) $y = \sin 3x + \cos 2x$

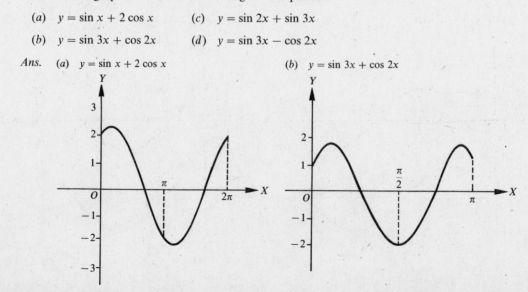

(c) $y = \sin 2x + \sin 3x$ (d) $y = \sin 3x - \cos 2x$

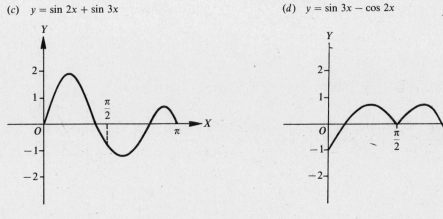

Fig. 8-11

8.8   Construct the graph of each of the following for one period.

(a) $y = \sin x + 3$ (c) $y = \sin (x - \pi/4)$

(b) $y = \cos x - 2$ (d) $y = \cos (x + \pi/6)$

Ans.

(a) $y = \sin x + 3$ (b) $y = \cos x - 2$

(c) $y = \sin (x - \pi/4)$ (d) $y = \cos (x + \pi/6)$

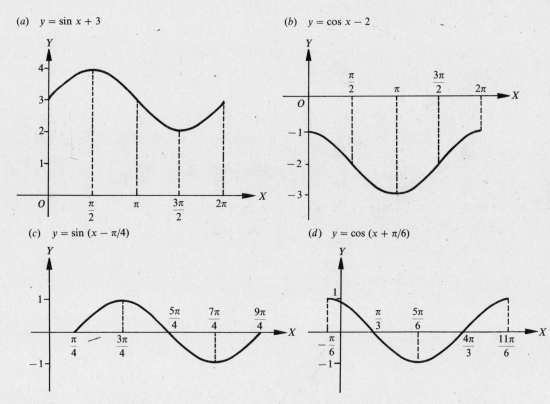

Fig. 8-12

Chapter 9

Basic Relationships and Identities

9.1 BASIC RELATIONSHIPS

Reciprocal Relationships	Quotient Relationships	Pythagorean Relationships
$\csc \theta = \dfrac{1}{\sin \theta}$	$\tan \theta = \dfrac{\sin \theta}{\cos \theta}$	$\sin^2 \theta + \cos^2 \theta = 1$
$\sec \theta = \dfrac{1}{\cos \theta}$	$\cot \theta = \dfrac{\cos \theta}{\sin \theta}$	$1 + \tan^2 \theta = \sec^2 \theta$
$\cot \theta = \dfrac{1}{\tan \theta}$		$1 + \cot^2 \theta = \csc^2 \theta$

The basic relationships hold for every value of θ for which the functions involved are defined.

Thus, $\sin^2 \theta + \cos^2 \theta = 1$ holds for every value of θ while $\tan \theta = \sin \theta / \cos \theta$ holds for all values of θ for which $\tan \theta$ is defined, i.e., for all $\theta \neq n \cdot 90°$ where n is odd. Note that for the excluded values of θ, $\cos \theta = 0$ and $\sin \theta \neq 0$.

For proofs of the quotient and Pythagorean relationships, see Probs. 9.1 and 9.2. The reciprocal relationships were treated in Chap. 2.

(See also Probs. 9.3 to 9.6.)

9.2 SIMPLIFICATION OF TRIGONOMETRIC EXPRESSIONS

It is frequently desirable to transform or reduce a given expression involving trigonometric functions to a simpler form.

EXAMPLE 9.1

(a) Using $\csc \theta = \dfrac{1}{\sin \theta}$, $\cos \theta \csc \theta = \cos \theta \dfrac{1}{\sin \theta} = \dfrac{\cos \theta}{\sin \theta} = \cot \theta$.

(b) Using $\tan \theta = \dfrac{\sin \theta}{\cos \theta}$, $\cos \theta \tan \theta = \cos \theta \dfrac{\sin \theta}{\cos \theta} = \sin \theta$.

EXAMPLE 9.2 Using the relation $\sin^2 \theta + \cos^2 \theta = 1$,

(a) $\sin^3 \theta + \sin \theta \cos^2 \theta = (\sin^2 \theta + \cos^2 \theta) \sin \theta = (1) \sin \theta = \sin \theta$.

(b) $\dfrac{\cos^2 \theta}{1 - \sin \theta} = \dfrac{1 - \sin^2 \theta}{1 - \sin \theta} = \dfrac{(1 - \sin \theta)(1 + \sin \theta)}{1 - \sin \theta} = 1 + \sin \theta$.

(NOTE: The relation $\sin^2 \theta + \cos^2 \theta = 1$ may be written as $\sin^2 \theta = 1 - \cos^2 \theta$ and as $\cos^2 \theta = 1 - \sin^2 \theta$. Each form is equally useful. In Example 9.2 the second of these forms was used.)

(See Probs. 9.7 to 9.9.)

9.3 TRIGONOMETRIC IDENTITIES

An equation involving the trigonometric functions which is valid for all values of the angle for which the functions are defined is called a *trigonometric identity*. The eight basic relationships in Sec. 9.1 are trigonometric identities; so also are

$$\cos\theta\,\csc\theta = \cot\theta \quad \text{and} \quad \cos\theta\,\tan\theta = \sin\theta$$

of Example 9.1.

A trigonometric identity is verified by transforming one member (your choice) into the other. In general, one begins with the more complicated side. In some cases each side is transformed into the same new form.

General Guidelines for Verifying Identities

1. Know the eight basic relationships and recognize alternative forms of each.

2. Know the procedures of adding and subtracting fractions, reducing fractions, and transforming fractions into equivalent fractions.

3. Know factoring and special product techniques.

4. Use only substitution and simplification procedures that allow you to work on exactly one side of an equation.

5. Select the side of the equation that appears more complicated and attempt to transform it into the form of the other side of the equation. (See Example 9.3.)

6. Transform each side of the equation, independently, into the same form. (See Example 9.4.)

7. Avoid substitutions that introduce radicals.

8. Use substitutions to change all trigonometric functions into expressions involving only sine and cosine and then simplify. (See Example 9.5.)

9. Multiply the numerator and denominator of a fraction by the conjugate of either. (See Example 9.6.)

10. Simplify a square root of a fraction by using conjugates to transform it into the quotient of perfect squares. (See Example 9.7.)

EXAMPLE 9.3 Verify the identity $\tan\theta + 2\cot\theta = \dfrac{\sin^2\theta + 2\cos^2\theta}{\sin\theta\cos\theta}$.

We leave the left side unchanged and rewrite the right side as the sum of two fractions, reduce the fractions, and substitute basic relationships to transform the expression.

$$\tan\theta + 2\cot\theta = \frac{\sin^2\theta + 2\cos^2\theta}{\sin\theta\cos\theta}$$

$$= \frac{\sin^2\theta}{\sin\theta\cos\theta} + \frac{2\cos^2\theta}{\sin\theta\cos\theta}$$

$$= \frac{\sin\theta}{\cos\theta} + \frac{2\cos\theta}{\sin\theta}$$

$$\tan\theta + 2\cot\theta = \tan\theta + 2\cot\theta$$

EXAMPLE 9.4 Verify the identity $\tan x + \cot x = \dfrac{\csc x}{\cos x}$.

We transform the right side of the equation until it appears completely simplified. Since the left side is still different from the right side, we now transform the left side into this new form.

$$\tan x + \cot x = \frac{\csc x}{\cos x}$$

$$= \csc x \cdot \frac{1}{\cos x}$$

$$= \frac{1}{\sin x} \cdot \frac{1}{\cos x}$$

$$= \frac{1}{\sin x \cos x}$$

$$\frac{\sin x}{\cos x} + \frac{\cos x}{\sin x} =$$

$$\frac{\sin^2 x}{\sin x \cos x} + \frac{\cos^2 x}{\sin x \cos x} =$$

$$\frac{\sin^2 x + \cos^2 x}{\sin x \cos x} =$$

$$\frac{1}{\sin x \cos x} = \frac{1}{\sin x \cos x}$$

EXAMPLE 9.5   Verify the identity $\dfrac{\sec x}{\cot x + \tan x} = \sin x$.

We change all the functions on the left side into expressions in terms of sines and cosines and then simplify.

$$\frac{\sec x}{\cot x + \tan x} = \sin x$$

$$\frac{\dfrac{1}{\cos x}}{\dfrac{\cos x}{\sin x} + \dfrac{\sin x}{\cos x}} =$$

$$\frac{\cos x \sin x}{\cos x \sin x} \cdot \frac{\dfrac{1}{\cos x}}{\dfrac{\cos x}{\sin x} + \dfrac{\sin x}{\cos x}} =$$

$$\frac{\sin x}{\cos^2 x + \sin^2 x} =$$

$$\frac{\sin x}{1} =$$

$$\sin x = \sin x$$

EXAMPLE 9.6   Verify the identity $\dfrac{\sin x}{1 + \cos x} = \dfrac{1 - \cos x}{\sin x}$.

We multiply the numerator and denominator on the left side by $1 - \cos x$, which is the conjugate of the denominator. (The conjugate of a two-term expression is the expression determined when the sign between the two terms is replaced by its opposite.) The only time we use this procedure is when the product of the expression and its conjugate gives us a form of a Pythagorean relationship.

$$\frac{\sin x}{1 + \cos x} = \frac{1 - \cos x}{\sin x}$$

$$\frac{1 - \cos x}{1 - \cos x} \cdot \frac{\sin x}{1 + \cos x} =$$

$$\frac{(1 - \cos x)\sin x}{1 - \cos^2 x} =$$

$$\frac{(1 - \cos x)\sin x}{\sin^2 x} =$$

$$\frac{1 - \cos x}{\sin x} = \frac{1 - \cos x}{\sin x}$$

EXAMPLE 9.7 Verify the identity $\sqrt{\dfrac{\sec x - \tan x}{\sec x + \tan x}} = \dfrac{1}{\sec x + \tan x}$.

Since the left side has the radical, we want to multiply the numerator and denominator of the fraction under the radical by the conjugate of either. We will use the conjugate of the numerator since this will make the denominator the square of the value we want in the denominator.

$$\sqrt{\frac{\sec x - \tan x}{\sec x + \tan x}} = \frac{1}{\sec x + \tan x}$$

$$\sqrt{\frac{\sec x - \tan x}{\sec x + \tan x} \cdot \frac{\sec x + \tan x}{\sec x + \tan x}} =$$

$$\sqrt{\frac{\sec^2 x - \tan^2 x}{(\sec x + \tan x)^2}} =$$

$$\sqrt{\frac{1}{(\sec x + \tan x)^2}} =$$

$$\frac{1}{\sec x + \tan x} = \frac{1}{\sec x + \tan x}$$

Practice makes deciding which substitutions to make and which procedures to use much easier. The procedures used in Examples 9.3, 9.4, and 9.5 are the most frequently used ones.

(See Probs. 9.10 to 9.18.)

Solved Problems

9.1 Prove the quotient relationships $\tan \theta = \dfrac{\sin \theta}{\cos \theta}$ and $\cot \theta = \dfrac{\cos \theta}{\sin \theta}$.

For any angle θ, $\sin \theta = y/r$, $\cos \theta = x/r$, $\tan \theta = y/x$, and $\cot \theta = x/y$, where $P(x, y)$ is any point on the terminal side of θ at a distance r from the origin.

Then $\tan \theta = \dfrac{y}{x} = \dfrac{y/r}{x/r} = \dfrac{\sin \theta}{\cos \theta}$ and $\cot \theta = \dfrac{x}{y} = \dfrac{x/r}{y/r} = \dfrac{\cos \theta}{\sin \theta}$. $\left(\text{Also, } \cot \theta = \dfrac{1}{\tan \theta} = \dfrac{\cos \theta}{\sin \theta}. \right)$

9.2 Prove the pythagorean relationships (a) $\sin^2 \theta + \cos^2 \theta = 1$, (b) $1 + \tan^2 \theta = \sec^2 \theta$, and (c) $1 + \cot^2 \theta = \csc^2 \theta$.

For $P(x, y)$ defined as in Prob. 9.1, we have $A = (x^2 + y^2 = r^2)$.

(a) Dividing A by r^2, $(x/r)^2 + (y/r)^2 = 1$ and $\sin^2 \theta + \cos^2 \theta = 1$.

(b) Dividing A by x^2, $1 + (y/x)^2 + (r/x)^2$ and $1 + \tan^2 \theta = \sec^2 \theta$.

Also, dividing $\sin^2 \theta + \cos^2 \theta = 1$ by $\cos^2 \theta$, $\left(\dfrac{\sin \theta}{\cos \theta}\right)^2 + 1 = \left(\dfrac{1}{\cos \theta}\right)^2$ or $\tan^2 \theta + 1 = \sec^2 \theta$.

(c) Dividing A by y^2, $(x/y)^2 + 1 = (r/y)^2$ and $\cot^2 \theta + 1 = \csc^2 \theta$.

Also, dividing $\sin^2 \theta + \cos^2 \theta = 1$ by $\sin^2 \theta$, $1 + \left(\dfrac{\cos \theta}{\sin \theta}\right)^2 = \left(\dfrac{1}{\sin \theta}\right)^2$ or $1 + \cot^2 \theta = \csc^2 \theta$.

9.3 Express each of the other functions of θ in terms of $\sin \theta$.

$$\cos^2 \theta = 1 - \sin^2 \theta \qquad \text{and} \qquad \cos \theta = \pm \sqrt{1 - \sin^2 \theta}$$

$$\tan \theta = \frac{\sin \theta}{\cos \theta} = \frac{\sin \theta}{\pm \sqrt{1 - \sin^2 \theta}} \qquad \cot \theta = \frac{1}{\tan \theta} = \frac{\pm \sqrt{1 - \sin^2 \theta}}{\sin \theta}$$

$$\sec \theta = \frac{1}{\cos \theta} = \frac{1}{\pm \sqrt{1 - \sin^2 \theta}} \qquad \csc \theta = \frac{1}{\sin \theta}$$

Note that $\cos \theta = \pm \sqrt{1 - \sin^2 \theta}$. Writing $\cos \theta = \sqrt{1 - \sin^2 \theta}$ limits angle θ to those quadrants (first and fourth) in which the cosine is positive.

9.4 Express each of the other functions of θ in terms of $\tan \theta$.

$$\sec^2 \theta = 1 + \tan^2 \theta \qquad \text{and} \qquad \sec \theta = \pm \sqrt{1 + \tan^2 \theta} \qquad \cos \theta = \frac{1}{\sec \theta} = \frac{1}{\pm \sqrt{1 + \tan^2 \theta}}$$

$$\frac{\sin \theta}{\cos \theta} = \tan \theta \qquad \text{and} \qquad \sin \theta = \tan \theta \cos \theta = \tan \theta \frac{1}{\pm \sqrt{1 + \tan^2 \theta}} = \frac{\tan \theta}{\pm \sqrt{1 + \tan^2 \theta}}$$

$$\csc \theta = \frac{1}{\sin \theta} = \frac{\pm \sqrt{1 + \tan^2 \theta}}{\tan \theta} \qquad \cot \theta = \frac{1}{\tan \theta}$$

9.5 Using the basic relationships, find the values of the functions of θ, given $\sin \theta = 3/5$.

From $\cos^2 \theta = 1 - \sin^2 \theta$, $\cos \theta = \pm \sqrt{1 - \sin^2 \theta} = \pm \sqrt{1 - (3/5)^2} = \pm \sqrt{16/25} = \pm 4/5$.

Now $\sin \theta$ and $\cos \theta$ are both positive when θ is a first-quadrant angle while $\sin \theta = +$ and $\cos \theta = -$ when θ is a second-quadrant angle. Thus,

First-Quadrant		Second-Quadrant	
$\sin \theta = 3/5$	$\cot \theta = 4/3$	$\sin \theta = 3/5$	$\cot \theta = -4/3$
$\cos \theta = 4/5$	$\sec \theta = 5/4$	$\cos \theta = -4/5$	$\sec \theta = -5/4$
$\tan \theta = \dfrac{3/5}{4/5} = 3/4$	$\csc \theta = 5/3$	$\tan \theta = -3/4$	$\csc \theta = 5/3$

9.6 Using the basic relationships, find the values of the functions of θ, given $\tan \theta = -5/12$.

Since $\tan \theta = -$, θ is either a second- or fourth-quadrant angle.

Second-Quadrant	Fourth-Quadrant
$\tan \theta = -5/12$	$\tan \theta = -5/12$
$\cot \theta = 1/\tan \theta = -12/5$	$\cot \theta = -12/5$
$\sec \theta = -\sqrt{1 + \tan^2 \theta} = -13/12$	$\sec \theta = 13/12$
$\cos \theta = 1/\sec \theta = -12/13$	$\cos \theta = 12/13$
$\csc \theta = \sqrt{1 + \cot^2 \theta} = 13/5$	$\csc \theta = -13/5$
$\sin \theta = 1/\csc \theta = 5/13$	$\sin \theta = -5/13$

9.7 Perform the indicated operations.

(a) $(\sin \theta - \cos \theta)(\sin \theta + \cos \theta) = \sin^2 \theta - \cos^2 \theta$

(b) $(\sin A + \cos A)^2 = \sin^2 A + 2 \sin A \cos A + \cos^2 A$

(c) $(\sin x + \cos y)(\sin y - \cos x) = \sin x \sin y - \sin x \cos x + \sin y \cos y - \cos x \cos y$

(d) $(\tan^2 A - \cot A)^2 = \tan^4 A - 2 \tan^2 A \cot A + \cot^2 A$

(e) $1 + \dfrac{\cos \theta}{\sin \theta} = \dfrac{\sin \theta + \cos \theta}{\sin \theta}$

(f) $1 - \dfrac{\sin \theta}{\cos \theta} + \dfrac{2}{\cos^2 \theta} = \dfrac{\cos^2 \theta - \sin \theta \cos \theta + 2}{\cos^2 \theta}$

9.8 Factor.

(a) $\sin^2 \theta - \sin \theta \cos \theta = \sin \theta (\sin \theta - \cos \theta)$

(b) $\sin^2 \theta + \sin^2 \theta \cos^2 \theta = \sin^2 \theta (1 + \cos^2 \theta)$

(c) $\sin^2 \theta + \sin \theta \sec \theta - 6 \sec^2 \theta = (\sin \theta + 3 \sec \theta)(\sin \theta - 2 \sec \theta)$

(d) $\sin^3 \theta \cos^2 \theta - \sin^2 \theta \cos^3 \theta + \sin \theta \cos^2 \theta = \sin \theta \cos^2 \theta (\sin^2 \theta - \sin \theta \cos \theta + 1)$

(e) $\sin^4 \theta - \cos^4 \theta = (\sin^2 \theta + \cos^2 \theta)(\sin^2 \theta - \cos^2 \theta) = (\sin^2 \theta + \cos^2 \theta)(\sin \theta - \cos \theta)(\sin \theta + \cos \theta)$

9.9 Simplify each of the following.

(a) $\sec \theta - \sec \theta \sin^2 \theta = \sec \theta(1 - \sin^2 \theta) = \sec \theta \cos^2 \theta = \dfrac{1}{\cos \theta} \cos^2 \theta = \cos \theta$

(b) $\sin \theta \sec \theta \cot \theta = \sin \theta \dfrac{1}{\cos \theta} \dfrac{\cos \theta}{\sin \theta} = \dfrac{\sin \theta \cos \theta}{\cos \theta \sin \theta} = 1$

(c) $\sin^2 \theta(1 + \cot^2 \theta) = \sin^2 \theta \csc^2 \theta = \sin^2 \theta \dfrac{1}{\sin^2 \theta} = 1$

(d) $\sin^2 \theta \sec^2 \theta - \sec^2 \theta = (\sin^2 \theta - 1) \sec^2 \theta = - \cos^2 \theta \sec^2 \theta = - \cos^2 \theta \dfrac{1}{\cos^2 \theta} = -1$

(e) $(\sin \theta + \cos \theta)^2 + (\sin \theta - \cos \theta)^2 = \sin^2 \theta + 2 \sin \theta \cos \theta + \cos^2 \theta + \sin^2 \theta$
$$- 2 \sin \theta \cos \theta + \cos^2 \theta = 2(\sin^2 \theta + \cos^2 \theta) = 2$$

(f) $\tan^2 \theta \cos^2 \theta + \cot^2 \theta \sin^2 \theta = \dfrac{\sin^2 \theta}{\cos^2 \theta} \cos^2 \theta + \dfrac{\cos^2 \theta}{\sin^2 \theta} \sin^2 \theta = \sin^2 \theta + \cos^2 \theta = 1$

(g) $\tan \theta + \dfrac{\cos \theta}{1 + \sin \theta} = \dfrac{\sin \theta}{\cos \theta} + \dfrac{\cos \theta}{1 + \sin \theta} = \dfrac{\sin \theta \, (1 + \sin \theta) + \cos^2 \theta}{\cos \theta \, (1 + \sin \theta)}$

$\qquad\qquad = \dfrac{\sin \theta + \sin^2 \theta + \cos^2 \theta}{\cos \theta \, (1 + \sin \theta)} = \dfrac{\sin \theta + 1}{\cos \theta \, (1 + \sin \theta)} = \dfrac{1}{\cos \theta} = \sec \theta$

Verify the following identities.

9.10 $\sec^2 \theta \csc^2 \theta = \sec^2 \theta + \csc^2 \theta$

$\sec^2 \theta + \csc^2 \theta = \dfrac{1}{\cos^2 \theta} + \dfrac{1}{\sin^2 \theta} = \dfrac{\sin^2 \theta + \cos^2 \theta}{\sin^2 \theta \cos^2 \theta} = \dfrac{1}{\sin^2 \theta \cos^2 \theta} = \dfrac{1}{\sin^2 \theta} \dfrac{1}{\cos^2 \theta} = \csc^2 \theta \sec^2 \theta$

9.11 $\sec^4 \theta - \sec^2 \theta = \tan^4 \theta + \tan^2 \theta$

$\tan^4 \theta + \tan^2 \theta = \tan^2 \theta \, (\tan^2 \theta + 1) = \tan^2 \theta \sec^2 \theta = (\sec^2 \theta - 1) \sec^2 \theta = \sec^4 \theta - \sec^2 \theta$

or

$\sec^4 \theta - \sec^2 \theta = \sec^2 \theta \, (\sec^2 \theta - 1) = \sec^2 \theta \tan^2 \theta = (1 + \tan^2 \theta) \tan^2 \theta = \tan^2 \theta + \tan^4 \theta$

9.12 $2 \csc x = \dfrac{\sin x}{1 + \cos x} + \dfrac{1 + \cos x}{\sin x}$

$\dfrac{\sin x}{1 + \cos x} + \dfrac{1 + \cos x}{\sin x} = \dfrac{\sin^2 x + (1 + \cos x)^2}{\sin x \, (1 + \cos x)} = \dfrac{\sin^2 x + 1 + 2 \cos x + \cos^2 x}{\sin x \, (1 + \cos x)}$

$\qquad\qquad = \dfrac{2 + 2 \cos x}{\sin x \, (1 + \cos x)} = \dfrac{2(1 + \cos x)}{\sin x \, (1 + \cos x)} = \dfrac{2}{\sin x} = 2 \csc x$

9.13 $\dfrac{1 - \sin x}{\cos x} = \dfrac{\cos x}{1 + \sin x}$

$\dfrac{\cos x}{1 + \sin x} = \dfrac{\cos^2 x}{\cos x \, (1 + \sin x)} = \dfrac{1 - \sin^2 x}{\cos x \, (1 + \sin x)} = \dfrac{(1 - \sin x)(1 + \sin x)}{\cos x \, (1 + \sin x)} = \dfrac{1 - \sin x}{\cos x}$

9.14 $\dfrac{\sec A - \csc A}{\sec A + \csc A} = \dfrac{\tan A - 1}{\tan A + 1}$

$\dfrac{\sec A - \csc A}{\sec A + \csc A} = \dfrac{\dfrac{1}{\cos A} - \dfrac{1}{\sin A}}{\dfrac{1}{\cos A} + \dfrac{1}{\sin A}} = \dfrac{\dfrac{\sin A}{\cos A} - 1}{\dfrac{\sin A}{\cos A} + 1} = \dfrac{\tan A - 1}{\tan A + 1}$

9.15 $\dfrac{\tan x - \sin x}{\sin^3 x} = \dfrac{\sec x}{1 + \cos x}$

$\dfrac{\tan x - \sin x}{\sin^3 x} = \dfrac{\dfrac{\sin x}{\cos x} - \sin x}{\sin^3 x} = \dfrac{\sin x - \sin x \cos x}{\cos x \sin^3 x} = \dfrac{\sin x \, (1 - \cos x)}{\cos x \sin^3 x}$

$\qquad\qquad = \dfrac{1 - \cos x}{\cos x \sin^2 x} = \dfrac{1 - \cos x}{\cos x \, (1 - \cos^2 x)} = \dfrac{1}{\cos x \, (1 + \cos x)} = \dfrac{\sec x}{1 + \cos x}$

9.16 $\dfrac{\cos A \cot A - \sin A \tan A}{\csc A - \sec A} = 1 + \sin A \cos A$

$$\dfrac{\cos A \cot A - \sin A \tan A}{\csc A - \sec A} = \dfrac{\cos A \dfrac{\cos A}{\sin A} - \sin A \dfrac{\sin A}{\cos A}}{\dfrac{1}{\sin A} - \dfrac{1}{\cos A}} = \dfrac{\cos^3 A - \sin^3 A}{\cos A - \sin A}$$

$$= \dfrac{(\cos A - \sin A)(\cos^2 A + \cos A \sin A + \sin^2 A)}{\cos A - \sin A} = \cos^2 A + \cos A \sin A + \sin^2 A = 1 + \cos A \sin A$$

9.17 $\dfrac{\sin \theta - \cos \theta + 1}{\sin \theta + \cos \theta - 1} = \dfrac{\sin \theta + 1}{\cos \theta}$

$$\dfrac{\sin \theta + 1}{\cos \theta} = \dfrac{(\sin \theta + 1)(\sin \theta + \cos \theta - 1)}{\cos \theta(\sin \theta + \cos \theta - 1)} = \dfrac{\sin^2 \theta + \sin \theta \cos \theta + \cos \theta - 1}{\cos \theta (\sin \theta + \cos \theta - 1)}$$

$$= \dfrac{-\cos^2 \theta + \sin \theta \cos \theta + \cos \theta}{\cos \theta (\sin \theta + \cos \theta - 1)} = \dfrac{\cos \theta (\sin \theta - \cos \theta + 1)}{\cos \theta (\sin \theta + \cos \theta - 1)} = \dfrac{\sin \theta - \cos \theta + 1}{\sin \theta + \cos \theta - 1}$$

9.18 $\dfrac{\tan \theta + \sec \theta - 1}{\tan \theta - \sec \theta + 1} = \tan \theta + \sec \theta$

$$\dfrac{\tan \theta + \sec \theta - 1}{\tan \theta - \sec \theta + 1} = \dfrac{\tan \theta + \sec \theta + \tan^2 \theta - \sec^2 \theta}{\tan \theta - \sec \theta + 1} = \dfrac{(\tan \theta + \sec \theta)(1 + \tan \theta - \sec \theta)}{\tan \theta - \sec \theta + 1}$$

$$= \tan \theta + \sec \theta$$

or

$$\tan \theta + \sec \theta = (\tan \theta + \sec \theta) \dfrac{\tan \theta - \sec \theta \mp 1}{\tan \theta - \sec \theta + 1} = \dfrac{\tan^2 \theta - \sec^2 \theta + \tan \theta + \sec \theta}{\tan \theta - \sec \theta + 1}$$

$$= \dfrac{-1 + \tan \theta + \sec \theta}{\tan \theta - \sec \theta + 1}$$

(NOTE: When expressed in terms of $\sin \theta$ and $\cos \theta$, this identity becomes that of Prob. 9.17.)

Supplementary Problems

9.19 Find the values of the trigonometric functions of θ, given $\sin \theta = 2/3$.

Ans. Quad I : $2/3, \sqrt{5}/3, 2/\sqrt{5} = 2\sqrt{5}/5, \sqrt{5}/2, 3/\sqrt{5} = 3\sqrt{5}/5, 3/2$

Quad II: $2/3, -\sqrt{5}/3, -2/\sqrt{5} = -2\sqrt{5}/5, -\sqrt{5}/2, -3/\sqrt{5} = -3\sqrt{5}/5, 3/2$

9.20 Find the values of the trigonometric functions of θ, given $\cos \theta = -5/6$.

Ans. Quad II : $\sqrt{11}/6, -5/6, -\sqrt{11}/5, -5/\sqrt{11} = -5\sqrt{11}/11, -6/5, 6/\sqrt{11} = 6\sqrt{11}/11$

Quad III: $-\sqrt{11}/6, -5/6, \sqrt{11}/5, 5/\sqrt{11} = 5\sqrt{11}/11, -6/5, -6/\sqrt{11} = -6\sqrt{11}/11$

9.21 Find the values of the trigonometric functions of θ, given $\tan \theta = 5/4$.

Ans. Quad I: $5/\sqrt{41} = 5\sqrt{41}/41, 4/\sqrt{41} = 4\sqrt{41}/41, 5/4, 4/5, \sqrt{41}/4, \sqrt{41}/5$

Quad III: $-5/\sqrt{41} = -5\sqrt{41}/41, -4/\sqrt{41} = -4\sqrt{41}/41, 5/4, 4/5, -\sqrt{41}/4, -\sqrt{41}/5$

9.22 Find the values of the trigonometric functions of θ, given $\cot \theta = -\sqrt{3}$.

    Ans.   Quad II: $1/2, -\sqrt{3}/2, -1/\sqrt{3} = -\sqrt{3}/3, -\sqrt{3}, -2/\sqrt{3} = -2\sqrt{3}/3, 2$

                Quad IV: $-1/2, \sqrt{3}/2, -1/\sqrt{3} = -\sqrt{3}/3, -\sqrt{3}, 2/\sqrt{3} = 2\sqrt{3}/3, -2$

9.23 Find the value of $\dfrac{\sin \theta + \cos \theta - \tan \theta}{\sec \theta + \csc \theta - \cot \theta}$ when $\tan \theta = -4/3$.

    Ans.   Quad II: 23/5; Quad IV: 34/35

Verify the following identities.

9.24 $\sin \theta \sec \theta = \tan \theta$

9.25 $(1 - \sin^2 A)(1 + \tan^2 A) = 1$

9.26 $(1 - \cos \theta)(1 + \sec \theta) \cot \theta = \sin \theta$

9.27 $\csc^2 x \,(1 - \cos^2 x) = 1$

9.28 $\dfrac{\sin \theta}{\csc \theta} + \dfrac{\cos \theta}{\sec \theta} = 1$

9.29 $\dfrac{1 - 2\cos^2 A}{\sin A \cos A} = \tan A - \cot A$

9.30 $\tan^2 x \csc^2 x \cot^2 x \sin^2 x = 1$

9.31 $\sin A \cos A \,(\tan A + \cot A) = 1$

9.32 $1 - \dfrac{\cos^2 \theta}{1 + \sin \theta} = \sin \theta$

9.33 $\dfrac{1}{\sec \theta + \tan \theta} = \sec \theta - \tan \theta$

9.34 $\dfrac{1}{1 - \sin A} + \dfrac{1}{1 + \sin A} = 2 \sec^2 A$

9.35 $\dfrac{1 - \cos x}{1 + \cos x} = \dfrac{\sec x - 1}{\sec x + 1} = (\cot x - \csc x)^2$

9.36 $\tan \theta \sin \theta + \cos \theta = \sec \theta$

9.37 $\tan \theta - \csc \theta \sec \theta \,(1 - 2\cos^2 \theta) = \cot \theta$

9.38 $\dfrac{\sin \theta}{\sin \theta + \cos \theta} = \dfrac{\sec \theta}{\sec \theta + \csc \theta}$

9.39 $\dfrac{\sin x + \tan x}{\cot x + \csc x} = \sin x \tan x$

9.40 $\dfrac{\sec x + \csc x}{\tan x + \cot x} = \sin x + \cos x$

9.41 $\dfrac{\sin^3 \theta + \cos^3 \theta}{\sin \theta + \cos \theta} = 1 - \sin \theta \cos \theta$

9.42 $\cot \theta + \dfrac{\sin \theta}{1 + \cos \theta} = \csc \theta$

9.43 $\dfrac{\sin \theta \cos \theta}{\cos^2 \theta - \sin^2 \theta} = \dfrac{\tan \theta}{1 - \tan^2 \theta}$

9.44 $(\tan x + \tan y)(1 - \cot x \cot y) + (\cot x + \cot y)(1 - \tan x \tan y) = 0$

9.45 $(x \sin \theta - y \cos \theta)^2 + (x \cos \theta + y \sin \theta)^2 = x^2 + y^2$

9.46 $(2r \sin \theta \cos \theta)^2 + r^2(\cos^2 \theta - \sin^2 \theta)^2 = r^2$

9.47 $(r \sin \theta \cos \phi)^2 + (r \sin \theta \sin \phi)^2 + (r \cos \theta)^2 = r^2$

Chapter 10

Trigonometric Functions of Two Angles

10.1 ADDITION FORMULAS

$$\sin(\alpha + \beta) = \sin \alpha \cos \beta + \cos \alpha \sin \beta$$

$$\cos(\alpha + \beta) = \cos \alpha \cos \beta - \sin \alpha \sin \beta$$

$$\tan(\alpha + \beta) = \frac{\tan \alpha + \tan \beta}{1 - \tan \alpha \tan \beta}$$

For a proof of these formulas, see Probs. 10.1 to 10.3.

10.2 SUBTRACTION FORMULAS

$$\sin(\alpha - \beta) = \sin \alpha \cos \beta - \cos \alpha \sin \beta$$

$$\cos(\alpha - \beta) = \cos \alpha \cos \beta + \sin \alpha \sin \beta$$

$$\tan(\alpha - \beta) = \frac{\tan \alpha - \tan \beta}{1 + \tan \alpha \tan \beta}$$

For a proof of these formulas, see Prob. 10.4.

10.3 DOUBLE-ANGLE FORMULAS

$$\sin 2\alpha = 2 \sin \alpha \cos \alpha$$

$$\cos 2\alpha = \cos^2 \alpha - \sin^2 \alpha = 1 - 2 \sin^2 \alpha = 2 \cos^2 \alpha - 1$$

$$\tan 2\alpha = \frac{2 \tan \alpha}{1 - \tan^2 \alpha}$$

For a proof of these formulas, see Prob. 10.14.

10.4 HALF-ANGLE FORMULAS

$$\sin \tfrac{1}{2}\theta = \pm \sqrt{\frac{1 - \cos \theta}{2}}$$

$$\cos \tfrac{1}{2}\theta = \pm \sqrt{\frac{1 + \cos \theta}{2}}$$

$$\tan \tfrac{1}{2}\theta = \pm \sqrt{\frac{1 - \cos \theta}{1 + \cos \theta}} = \frac{\sin \theta}{1 + \cos \theta} = \frac{1 - \cos \theta}{\sin \theta}$$

For a proof of these formulas, see Prob. 10.15.

Solved Problems

10.1 Prove   (1) $\sin(\alpha + \beta) = \sin\alpha\cos\beta + \cos\alpha\sin\beta$

and   (2) $\cos(\alpha + \beta) = \cos\alpha\cos\beta - \sin\alpha\sin\beta$ when α and β are positive acute angles.

Let α and β be positive acute angles such that $\alpha + \beta < 90°$ [Fig. 10-1(a)] and $\alpha + \beta > 90°$ [Fig. 10-1(b)].

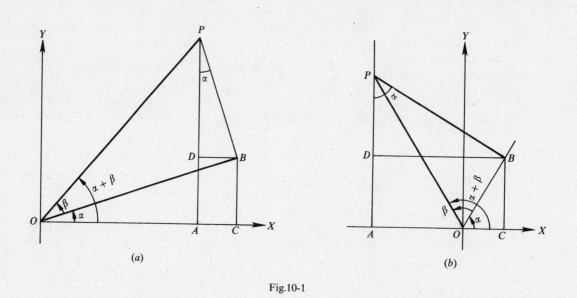

Fig.10-1

To construct these figures, place angle α in standard position and then place angle β with its vertex at O and its initial side along the terminal side of angle α. Let P be any point on the terminal side of angle $(\alpha + \beta)$. Draw PA perpendicular to OX, PB perpendicular to the terminal side of angle α, BC perpendicular to OX, and BD perpendicular to AP.

Now $\angle APB = \alpha$ since corresponding sides (OA and AP, and OB and BP) are perpendicular. Then

$$\sin(\alpha + \beta) = \frac{AP}{OP} = \frac{AD + DP}{OP} = \frac{CB + DP}{OP} = \frac{CB}{OP} + \frac{DP}{OP} = \frac{CB}{OB}\cdot\frac{OB}{OP} + \frac{DP}{BP}\cdot\frac{BP}{OP}$$

$$= \sin\alpha\cos\beta + \cos\alpha\sin\beta$$

and

$$\cos(\alpha + \beta) = \frac{OA}{OP} = \frac{OC - AC}{OP} = \frac{OC - DB}{OP} = \frac{OC}{OP} - \frac{DB}{OP} = \frac{OC}{OB}\cdot\frac{OB}{OP} - \frac{DB}{BP}\cdot\frac{BP}{OP}$$

$$= \cos\alpha\cos\beta - \sin\alpha\sin\beta$$

10.2 Show that (1) and (2) of Prob. 10.1 are valid when α and β are any angles.

First check the formulas for the case $\alpha = 0°$ and $\beta = 0°$. Since

$$\sin(0° + 0°) = \sin 0°\cos 0° + \cos 0°\sin 0° = 0\cdot 1 + 1\cdot 0 = 0 = \sin 0°$$

and

$$\cos(0° + 0°) = \cos 0°\cos 0° - \sin 0°\sin 0° = 1\cdot 1 - 0\cdot 0 = 1 = \cos 0°$$

the formulas are valid for this case.

Next, it will be shown that if (1) and (2) are valid for any two given angles α and β, the formulas are also valid when, say, α is increased by 90°. Let α and β be two angles for which (1) and (2) hold and consider

(a) $\sin(\alpha + \beta + 90°) = \sin(\alpha + 90°)\cos\beta + \cos(\alpha + 90°)\sin\beta$

and

(b) $\cos(\alpha + \beta + 90°) = \cos(\alpha + 90°)\cos\beta - \sin(\alpha + 90°)\sin\beta$

From the graphs in Sec. 8.3 we see that $\sin(\theta + 90°) = \cos\theta$ and $\cos(\theta + 90°) = -\sin\theta$. It follows that $\sin(\alpha + \beta + 90°) = \cos(\alpha + \beta)$ and $\cos(\alpha + \beta + 90°) = -\sin(\alpha + \beta)$. Then ($a$) and ($b$) reduce to

$$(a') \quad \cos(\alpha + \beta) = \cos\alpha\cos\beta + (-\sin\alpha)\sin\beta = \cos\alpha\cos\beta - \sin\alpha\sin\beta$$

and

$$(b') -\sin(\alpha + \beta) = -\sin\alpha\cos\beta - \cos\alpha\sin\beta$$

or

$$\sin(\alpha + \beta) = \sin\alpha\cos\beta + \cos\alpha\sin\beta$$

which, by assumption, are valid relations. Thus, (a) and (b) are valid relations.

The same argument may be made to show that if (1) and (2) are valid for two angles α and β, they are also valid when β is increased by 90°. Thus, the formulas are valid when both α and β are increased by 90°. Now any positive angle can be expressed as a multiple of 90° plus θ, where θ is either 0° or an acute angle. Thus, by a finite number of repetitions of the argument, we show that the formulas are valid for any two given positive angles.

It will be left for the reader to carry through the argument when, instead of an increase, there is a decrease of 90° and thus to show that (1) and (2) are valid when one angle is positive and the other negative, and when both are negative.

10.3 Prove $\tan(\alpha + \beta) = \dfrac{\tan\alpha + \tan\beta}{1 - \tan\alpha\tan\beta}$.

$$\tan(\alpha + \beta) = \frac{\sin(\alpha + \beta)}{\cos(\alpha + \beta)} = \frac{\sin\alpha\cos\beta + \cos\alpha\sin\beta}{\cos\alpha\cos\beta - \sin\alpha\sin\beta}$$

$$= \frac{\dfrac{\sin\alpha\cos\beta}{\cos\alpha\cos\beta} + \dfrac{\cos\alpha\sin\beta}{\cos\alpha\cos\beta}}{\dfrac{\cos\alpha\cos\beta}{\cos\alpha\cos\beta} - \dfrac{\sin\alpha\sin\beta}{\cos\alpha\cos\beta}} = \frac{\tan\alpha + \tan\beta}{1 - \tan\alpha\tan\beta}$$

10.4 Prove the subtraction formulas.

$$\sin(\alpha - \beta) = \sin[\alpha + (-\beta)] = \sin\alpha\cos(-\beta) + \cos\alpha\sin(-\beta)$$
$$= \sin\alpha(\cos\beta) + \cos\alpha(-\sin\beta) = \sin\alpha\cos\beta - \cos\alpha\sin\beta$$

$$\cos(\alpha - \beta) = \cos[\alpha + (-\beta)] = \cos\alpha\cos(-\beta) - \sin\alpha\sin(-\beta)$$
$$= \cos\alpha(\cos\beta) - \sin\alpha(-\sin\beta) = \cos\alpha\cos\beta + \sin\alpha\sin\beta$$

$$\tan(\alpha - \beta) = \tan[\alpha + (-\beta)] = \frac{\tan\alpha + \tan(-\beta)}{1 - \tan\alpha\tan(-\beta)}$$

$$= \frac{\tan\alpha + (-\tan\beta)}{1 - \tan\alpha(-\tan\beta)} = \frac{\tan\alpha - \tan\beta}{1 + \tan\alpha\tan\beta}$$

10.5 Find the values of the sine, cosine, and tangent of 15°, using (a) $15° = 45° - 30°$ and (b) $15° = 60° - 45°$.

(a) $\sin 15° = \sin(45° - 30°) = \sin 45°\cos 30° - \cos 45°\sin 30°$

$$= \frac{1}{\sqrt{2}} \cdot \frac{\sqrt{3}}{2} - \frac{1}{\sqrt{2}} \cdot \frac{1}{2} = \frac{\sqrt{3} - 1}{2\sqrt{2}} = \frac{\sqrt{2}}{4}(\sqrt{3} - 1) = \frac{\sqrt{6} - \sqrt{2}}{4}$$

$\cos 15° = \cos(45° - 30°) = \cos 45°\cos 30° + \sin 45°\sin 30°$

$$= \frac{1}{\sqrt{2}} \cdot \frac{\sqrt{3}}{2} + \frac{1}{\sqrt{2}} \cdot \frac{1}{2} = \frac{\sqrt{2}}{4}(\sqrt{3} + 1) = \frac{\sqrt{6} + \sqrt{2}}{4}$$

$$\tan 15° = \tan(45° - 30°) = \frac{\tan 45° - \tan 30°}{1 + \tan 45°\tan 30°} = \frac{1 - 1/\sqrt{3}}{1 + 1(1/\sqrt{3})} = \frac{\sqrt{3} - 1}{\sqrt{3} + 1} = 2 - \sqrt{3}$$

(b) $\sin 15° = \sin(60° - 45°) = \sin 60° \cos 45° - \cos 60° \sin 45° = \dfrac{\sqrt{3}}{2} \cdot \dfrac{1}{\sqrt{2}} - \dfrac{1}{2} \cdot \dfrac{1}{\sqrt{2}} = \dfrac{\sqrt{2}}{4}(\sqrt{3} - 1)$

$\qquad\qquad = \dfrac{\sqrt{6} - \sqrt{2}}{4}$

$\cos 15° = \cos(60° - 45°) = \cos 60° \cos 45° + \sin 60° \sin 45° = \dfrac{1}{2} \cdot \dfrac{1}{\sqrt{2}} + \dfrac{\sqrt{3}}{2} \cdot \dfrac{1}{\sqrt{2}} = \dfrac{\sqrt{2}}{4}(\sqrt{3} + 1)$

$\qquad\qquad = \dfrac{\sqrt{6} + \sqrt{2}}{4}$

$\tan 15° = \tan(60° - 45°) = \dfrac{\tan 60° - \tan 45°}{1 + \tan 60° \tan 45°} = \dfrac{\sqrt{3} - 1}{\sqrt{3} + 1} = 2 - \sqrt{3}$

10.6 Find the values of the sine, cosine, and tangent of $\pi/12$ radian.

Since $\pi/3$ and $\pi/4$ are special angles and $\pi/3 - \pi/4 = \pi/12$, they can be used to find the values needed.

$\sin \dfrac{\pi}{12} = \sin\left(\dfrac{\pi}{3} - \dfrac{\pi}{4}\right) = \sin \dfrac{\pi}{3} \cos \dfrac{\pi}{4} - \cos \dfrac{\pi}{3} \sin \dfrac{\pi}{4} = \dfrac{\sqrt{3}}{2} \cdot \dfrac{\sqrt{2}}{2} - \dfrac{1}{2} \cdot \dfrac{\sqrt{2}}{2} = \dfrac{\sqrt{6}}{4} - \dfrac{\sqrt{2}}{4} = \dfrac{\sqrt{6} - \sqrt{2}}{4}$

$\cos \dfrac{\pi}{12} = \cos\left(\dfrac{\pi}{3} - \dfrac{\pi}{4}\right) = \cos \dfrac{\pi}{3} \cos \dfrac{\pi}{4} + \sin \dfrac{\pi}{3} \sin \dfrac{\pi}{4} = \dfrac{1}{2} \cdot \dfrac{\sqrt{2}}{2} + \dfrac{\sqrt{3}}{2} \cdot \dfrac{\sqrt{2}}{2} = \dfrac{\sqrt{2}}{4} + \dfrac{\sqrt{6}}{4} = \dfrac{\sqrt{2} + \sqrt{6}}{4}$

$\tan \dfrac{\pi}{12} = \tan\left(\dfrac{\pi}{3} - \dfrac{\pi}{4}\right) = \dfrac{\tan \dfrac{\pi}{3} - \tan \dfrac{\pi}{4}}{1 + \tan \dfrac{\pi}{3} \tan \dfrac{\pi}{4}} = \dfrac{\sqrt{3} - 1}{1 + \sqrt{3}(1)} = \dfrac{\sqrt{3} - 1}{1 + \sqrt{3}}$

$\qquad = \dfrac{\sqrt{3} - 1}{\sqrt{3} + 1} = \dfrac{\sqrt{3} - 1}{\sqrt{3} + 1} \cdot \dfrac{\sqrt{3} - 1}{\sqrt{3} - 1} = \dfrac{3 - 2\sqrt{3} + 1}{3 - 1} = \dfrac{4 - 2\sqrt{3}}{2} = 2 - \sqrt{3}$

10.7 Find the values of the sine, cosine, and tangent of $5\pi/12$ radians.

Since $\pi/6$ and $\pi/4$ are special angles and $\pi/6 + \pi/4 = 5\pi/12$, they can be used to find the values needed.

$\sin \dfrac{5\pi}{12} = \sin\left(\dfrac{\pi}{6} + \dfrac{\pi}{4}\right) = \sin \dfrac{\pi}{6} \cos \dfrac{\pi}{4} + \cos \dfrac{\pi}{6} \sin \dfrac{\pi}{4} = \dfrac{1}{2} \cdot \dfrac{\sqrt{2}}{2} + \dfrac{\sqrt{3}}{2} \cdot \dfrac{\sqrt{2}}{2} = \dfrac{\sqrt{2}}{4} + \dfrac{\sqrt{6}}{4} = \dfrac{\sqrt{2} + \sqrt{6}}{4}$

$\cos \dfrac{5\pi}{12} = \cos\left(\dfrac{\pi}{6} + \dfrac{\pi}{4}\right) = \cos \dfrac{\pi}{6} \cos \dfrac{\pi}{4} - \sin \dfrac{\pi}{6} \sin \dfrac{\pi}{4} = \dfrac{\sqrt{3}}{2} \cdot \dfrac{\sqrt{2}}{2} - \dfrac{1}{2} \cdot \dfrac{\sqrt{2}}{2} = \dfrac{\sqrt{6}}{4} - \dfrac{\sqrt{2}}{4} = \dfrac{\sqrt{6} - \sqrt{2}}{4}$

$\tan \dfrac{5\pi}{12} = \tan\left(\dfrac{\pi}{6} + \dfrac{\pi}{4}\right) = \dfrac{\tan \dfrac{\pi}{6} + \tan \dfrac{\pi}{4}}{1 - \tan \dfrac{\pi}{6} \tan \dfrac{\pi}{4}} = \dfrac{\dfrac{\sqrt{3}}{3} + 1}{1 - \dfrac{\sqrt{3}}{3} \cdot 1} = \dfrac{\sqrt{3} + 3}{3 - \sqrt{3}}$

$\qquad = \dfrac{3 + \sqrt{3}}{3 - \sqrt{3}} \cdot \dfrac{3 + \sqrt{3}}{3 + \sqrt{3}} = \dfrac{9 + 6\sqrt{3} + 3}{9 - 3} = \dfrac{12 + 6\sqrt{3}}{6} = 2 + \sqrt{3}$

10.8 Rewrite each expression as a single function of an angle.

  (a) $\sin 75° \cos 28° - \cos 75° \sin 28°$

  (b) $\cos 31° \cos 48° - \sin 31° \sin 48°$

  (c) $2 \sin 75° \cos 75°$

  (d) $1 - 2 \sin^2 37°$

ans. (a) $\sin 75° \cos 28° - \cos 75° \sin 28° = \sin (75° - 28°) = \sin 47°$

  (b) $\cos 31° \cos 48° - \sin 31° \sin 48° = \cos (31° + 48°) = \cos 79°$

  (c) $2 \sin 75° \cos 75° = \sin 2(75°) = \sin 150°$

  (d) $1 - 2 \sin^2 37° = \cos 2(37°) = \cos 74°$

10.9 Rewrite each expression as a single function of an angle.

  (a) $\dfrac{\tan 37° + \tan 68°}{1 - \tan 37° \tan 68°}$ $\qquad$ (d) $\sqrt{\dfrac{1 + \cos 160°}{2}}$

  (b) $\dfrac{2 \tan 31°}{1 - \tan^2 31°}$ $\qquad$ (e) $\dfrac{\sin 142°}{1 + \cos 142°}$

  (c) $\sqrt{\dfrac{1 - \cos 84°}{2}}$ $\qquad$ (f) $\dfrac{1 - \cos 184°}{\sin 184°}$

ans. (a) $\dfrac{\tan 37° + \tan 68°}{1 - \tan 37° \tan 68°} = \tan (37° + 68°) = \tan 105°$

  (b) $\dfrac{2 \tan 31°}{1 - \tan^2 31°} = \tan 2(31°) = \tan 62°$

  (c) $\sqrt{\dfrac{1 - \cos 84°}{2}} = \sin \tfrac{1}{2}(84°) = \sin 42°$

  (d) $\sqrt{\dfrac{1 + \cos 160°}{2}} = \cos \tfrac{1}{2}(160°) = \cos 80°$

  (e) $\dfrac{\sin 142°}{1 + \cos 142°} = \tan \tfrac{1}{2}(142°) = \tan 71°$

  (f) $\dfrac{1 - \cos 184°}{\sin 184°} = \tan \tfrac{1}{2}(184°) = \tan 92°$

10.10 Prove (a) $\sin (45° + \theta) - \sin (45° - \theta) = \sqrt{2} \sin \theta$ and (b) $\sin (30° + \theta) + \cos (60° + \theta) = \cos \theta$.

  (a) $\sin (45° + \theta) - \sin (45° - \theta) = (\sin 45° \cos \theta + \cos 45° \sin \theta) - (\sin 45° \cos \theta - \cos 45° \sin \theta)$

$$= 2 \cos 45° \sin \theta = 2 \frac{1}{\sqrt{2}} \sin \theta = \sqrt{2} \sin \theta$$

  (b) $\sin (30° + \theta) + \cos (60° + \theta) = (\sin 30° \cos \theta + \cos 30° \sin \theta) + (\cos 60° \cos \theta - \sin 60° \sin \theta)$

$$= \left(\frac{1}{2} \cos \theta + \frac{\sqrt{3}}{2} \sin \theta \right) + \left(\frac{1}{2} \cos \theta - \frac{\sqrt{3}}{2} \sin \theta \right) = \cos \theta$$

10.11 Simplify: (a) $\sin(\alpha + \beta) + \sin(\alpha - \beta)$,   (b) $\cos(\alpha + \beta) - \cos(\alpha - \beta)$,

  (c) $\dfrac{\tan(\alpha + \beta) - \tan\alpha}{1 + \tan(\alpha + \beta)\tan\alpha}$,

  (d) $(\sin\alpha\cos\beta - \cos\alpha\sin\beta)^2 + (\cos\alpha\cos\beta + \sin\alpha\sin\beta)^2$

(a) $\sin(\alpha + \beta) + \sin(\alpha - \beta) = (\sin\alpha\cos\beta + \cos\alpha\sin\beta) + (\sin\alpha\cos\beta - \cos\alpha\sin\beta)$
$$= 2\sin\alpha\cos\beta$$

(b) $\cos(\alpha + \beta) - \cos(\alpha - \beta) = (\cos\alpha\cos\beta - \sin\alpha\sin\beta) - (\cos\alpha\cos\beta + \sin\alpha\sin\beta)$
$$= -2\sin\alpha\sin\beta$$

(c) $\dfrac{\tan(\alpha + \beta) - \tan\alpha}{1 + \tan(\alpha + \beta)\tan\alpha} = \tan[(\alpha + \beta) - \alpha] = \tan\beta$

(d) $(\sin\alpha\cos\beta - \cos\alpha\sin\beta)^2 + (\cos\alpha\cos\beta + \sin\alpha\sin\beta)^2 = \sin^2(\alpha - \beta) + \cos^2(\alpha - \beta) = 1$

10.12 Find $\sin(\alpha + \beta)$, $\cos(\alpha + \beta)$, $\sin(\alpha - \beta)$, and $\cos(\alpha - \beta)$ and determine the quadrants in which $(\alpha + \beta)$ and $(\alpha - \beta)$ terminate, given

(a) $\sin\alpha = 4/5$, $\cos\beta = 5/13$; α and β in quadrant I

(b) $\sin\alpha = 2/3$, $\cos\beta = 3/4$; α in quadrant II, β in quadrant IV

(a) $\cos\alpha = 3/5$, see Fig. 10-2(a), and $\sin\beta = 12/13$, see Fig. 10-2(b).

$\sin(\alpha + \beta) = \sin\alpha\cos\beta + \cos\alpha\sin\beta = \dfrac{4}{5}\cdot\dfrac{5}{13} + \dfrac{3}{5}\cdot\dfrac{12}{13} = \dfrac{56}{65}$ $\left.\begin{array}{c}\\ \\ \\ \\\end{array}\right\}$ $(\alpha + \beta)$ in quadrant II

$\cos(\alpha + \beta) = \cos\alpha\cos\beta - \sin\alpha\sin\beta = \dfrac{3}{5}\cdot\dfrac{5}{13} - \dfrac{4}{5}\cdot\dfrac{12}{13} = -\dfrac{33}{65}$

$\sin(\alpha - \beta) = \sin\alpha\cos\beta - \cos\alpha\sin\beta = \dfrac{4}{5}\cdot\dfrac{5}{13} - \dfrac{3}{5}\cdot\dfrac{12}{13} = -\dfrac{16}{65}$ $\left.\begin{array}{c}\\ \\ \\ \\\end{array}\right\}$ $(\alpha - \beta)$ in quadrant IV

$\cos(\alpha - \beta) = \cos\alpha\cos\beta + \sin\alpha\sin\beta = \dfrac{3}{5}\cdot\dfrac{5}{13} + \dfrac{4}{5}\cdot\dfrac{12}{13} = \dfrac{63}{65}$

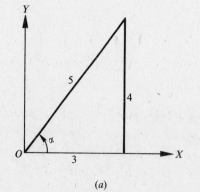

(a)

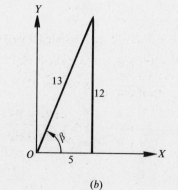

(b)

Fig.10-2

(b) $\cos \alpha = -\sqrt{5}/3$, see Fig. 10-3(a), and $\sin \beta = -\sqrt{7}/4$, see Fig. 10-3(b).

$$\sin (\alpha + \beta) = \sin \alpha \cos \beta + \cos \alpha \sin \beta = \frac{2}{3} \cdot \frac{3}{4} + \left(-\frac{\sqrt{5}}{3}\right)\left(-\frac{\sqrt{7}}{4}\right) = \frac{6 + \sqrt{35}}{12}$$

$$\cos (\alpha + \beta) = \cos \alpha \cos \beta - \sin \alpha \sin \beta = \left(-\frac{\sqrt{5}}{3}\right)\frac{3}{4} - \frac{2}{3}\left(-\frac{\sqrt{7}}{4}\right) = \frac{-3\sqrt{5} + 2\sqrt{7}}{12}$$
$\left.\right\}$ $(\alpha + \beta)$ in quadrant II

$$\sin (\alpha - \beta) = \sin \alpha \cos \beta - \cos \alpha \sin \beta = \frac{2}{3} \cdot \frac{3}{4} - \left(-\frac{\sqrt{5}}{3}\right)\left(-\frac{\sqrt{7}}{4}\right) = \frac{6 - \sqrt{35}}{12}$$

$$\cos (\alpha - \beta) = \cos \alpha \cos \beta + \sin \alpha \sin \beta = \left(-\frac{\sqrt{5}}{3}\right)\frac{3}{4} + \frac{2}{3}\left(-\frac{\sqrt{7}}{4}\right) = \frac{-3\sqrt{5} - 2\sqrt{7}}{12}$$
$\left.\right\}$ $(\alpha - \beta)$ in quadrant II

10.13   Prove (a) $\cot (\alpha + \beta) = \dfrac{\cot \alpha \cot \beta - 1}{\cot \beta + \cot \alpha}$ and (b) $\cot (\alpha - \beta) = \dfrac{\cot \alpha \cot \beta + 1}{\cot \beta - \cot \alpha}$.

(a) $\cot (\alpha + \beta) = \dfrac{1}{\tan (\alpha + \beta)} = \dfrac{1 - \tan \alpha \tan \beta}{\tan \alpha + \tan \beta} = \dfrac{1 - \dfrac{1}{\cot \alpha \cot \beta}}{\dfrac{1}{\cot \alpha} + \dfrac{1}{\cot \beta}} = \dfrac{\cot \alpha \cot \beta - 1}{\cot \beta + \cot \alpha}$

(b) $\cot (\alpha - \beta) = \cot [\alpha + (-\beta)] = \dfrac{\cot \alpha \cot (-\beta) - 1}{\cot (-\beta) + \cot \alpha} = \dfrac{-\cot \alpha \cot \beta - 1}{-\cot \beta + \cot \alpha} = \dfrac{\cot \alpha \cot \beta + 1}{\cot \beta - \cot \alpha}$

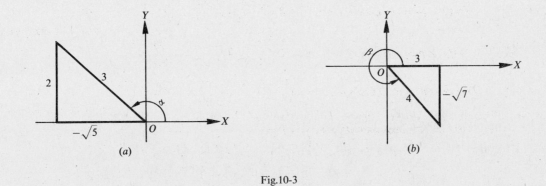

(a)   (b)

Fig.10-3

10.14   Prove the double-angle formulas.

In $\sin (\alpha + \beta) = \sin \alpha \cos \beta + \cos \alpha \sin \beta$, $\cos (\alpha + \beta) = \cos \alpha \cos \beta - \sin \alpha \sin \beta$,   and $\tan (\alpha + \beta) = \dfrac{\tan \alpha + \tan \beta}{1 - \tan \alpha \tan \beta}$, put $\beta = \alpha$. Then

$$\sin 2\alpha = \sin \alpha \cos \alpha + \cos \alpha \sin \alpha = 2 \sin \alpha \cos \alpha$$

$$\cos 2\alpha = \cos \alpha \cos \alpha - \sin \alpha \sin \alpha$$
$$= \cos^2 \alpha - \sin^2 \alpha = (1 - \sin^2 \alpha) - \sin^2 \alpha = 1 - 2 \sin^2 \alpha$$
$$= \cos^2 \alpha - (1 - \cos^2 \alpha) = 2 \cos^2 \alpha - 1$$

$$\tan 2\alpha = \frac{\tan \alpha + \tan \alpha}{1 - \tan \alpha \tan \alpha} = \frac{2 \tan \alpha}{1 - \tan^2 \alpha}$$

10.15 Prove the half-angle formulas.

In $\cos 2\alpha = 1 - 2 \sin^2 \alpha$, put $\alpha = \frac{1}{2}\theta$. Then

$$\cos \theta = 1 - 2 \sin^2 \tfrac{1}{2}\theta \qquad \sin^2 \tfrac{1}{2}\theta = \frac{1 - \cos \theta}{2} \quad \text{and} \quad \sin \tfrac{1}{2}\theta = \pm \sqrt{\frac{1 - \cos \theta}{2}}$$

In $\cos 2\alpha = 2 \cos^2 \alpha - 1$, put $\alpha = \frac{1}{2}\theta$. Then

$$\cos \theta = 2 \cos^2 \tfrac{1}{2}\theta - 1 \qquad \cos^2 \tfrac{1}{2}\theta = \frac{1 + \cos \theta}{2} \quad \text{and} \quad \cos \tfrac{1}{2}\theta = \pm \sqrt{\frac{1 + \cos \theta}{2}}$$

Finally,
$$\tan \tfrac{1}{2}\theta = \frac{\sin \tfrac{1}{2}\theta}{\cos \tfrac{1}{2}\theta} = \pm \sqrt{\frac{1 - \cos \theta}{1 + \cos \theta}}$$

$$= \pm \sqrt{\frac{(1 - \cos \theta)(1 + \cos \theta)}{(1 + \cos \theta)(1 + \cos \theta)}} = \pm \sqrt{\frac{1 - \cos^2 \theta}{(1 + \cos \theta)^2}} = \frac{\sin \theta}{1 + \cos \theta}$$

$$= \pm \sqrt{\frac{(1 - \cos \theta)(1 - \cos \theta)}{(1 + \cos \theta)(1 - \cos \theta)}} = \pm \sqrt{\frac{(1 - \cos \theta)^2}{1 - \cos^2 \theta}} = \frac{1 - \cos \theta}{\sin \theta}$$

The signs $\pm$ are not needed here since $\tan \frac{1}{2}\theta$ and $\sin \theta$ always have the same sign (Prob. 7.8, Chap. 7) and $1 - \cos \theta$ is always positive.

10.16 Using the half-angle formulas, find the exact values of (a) $\sin 15°$, (b) $\sin 292\frac{1}{2}°$, and (c) $\sin \pi/8$.

(a) $\sin 15° = \sqrt{\dfrac{1 - \cos 30°}{2}} = \sqrt{\dfrac{1 - \sqrt{3}/2}{2}} = \tfrac{1}{2}\sqrt{2 - \sqrt{3}}$

(b) $\sin 292\tfrac{1}{2}° = -\sqrt{\dfrac{1 - \cos 585°}{2}} = -\sqrt{\dfrac{1 - \cos 225°}{2}} = -\sqrt{\dfrac{1 + 1/\sqrt{2}}{2}} = -\tfrac{1}{2}\sqrt{2 + \sqrt{2}}$

(c) $\sin \dfrac{\pi}{8} = \sqrt{\dfrac{1 - \cos \pi/4}{2}} = \sqrt{\dfrac{1 - \sqrt{2}/2}{2}} = \sqrt{\dfrac{2 - \sqrt{2}}{4}} = \tfrac{1}{2}\sqrt{2 - \sqrt{2}}$

10.17 Find the values of the sine, cosine, and tangent of $\frac{1}{2}\theta$, given (a) $\sin \theta = 5/13$, θ in quadrant II and (b) $\cos \theta = 3/7$, θ in quadrant IV.

(a) $\sin \theta = 5/13$, $\cos \theta = -12/13$, and $\frac{1}{2}\theta$ in quadrant I, see Fig. 10-4(a).

$$\sin \tfrac{1}{2}\theta = \sqrt{\frac{1 - \cos \theta}{2}} = \sqrt{\frac{1 + 12/13}{2}} = \sqrt{\frac{25}{26}} = \frac{5\sqrt{26}}{26}$$

$$\cos \tfrac{1}{2}\theta = \sqrt{\frac{1 + \cos \theta}{2}} = \sqrt{\frac{1 - 12/13}{2}} = \sqrt{\frac{1}{26}} = \frac{\sqrt{26}}{26}$$

$$\tan \tfrac{1}{2}\theta = \frac{1 - \cos \theta}{\sin \theta} = \frac{1 + 12/13}{5/13} = 5$$

(b) $\sin \theta = -2\sqrt{10}/7$, $\cos \theta = 3/7$, and $\frac{1}{2}\theta$ in quadrant II, see Fig. 10-4(b).

$$\sin \tfrac{1}{2}\theta = \sqrt{\frac{1 - \cos \theta}{2}} = \sqrt{\frac{1 - 3/7}{2}} = \frac{\sqrt{14}}{7}$$

$$\cos \tfrac{1}{2}\theta = -\sqrt{\frac{1 + \cos \theta}{2}} = -\sqrt{\frac{1 + 3/7}{2}} = -\frac{\sqrt{35}}{7},$$

$$\tan \tfrac{1}{2}\theta = \frac{1 - \cos \theta}{\sin \theta} = \frac{1 - 3/7}{-2\sqrt{10}/7} = -\frac{\sqrt{10}}{5}$$

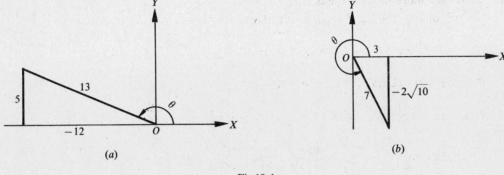

Fig.10-4

10.18 Show that   (a) $\sin\theta = 2\sin\frac{1}{2}\theta\cos\frac{1}{2}\theta$ (d) $\cos 6\theta = 1 - 2\sin^2 3\theta$

(b) $\sin A = \pm\sqrt{\dfrac{1 - \cos 2A}{2}}$ (e) $\sin^2\frac{1}{2}\theta = \frac{1}{2}(1 - \cos\theta),\ \cos^2\frac{1}{2}\theta = \frac{1}{2}(1 + \cos\theta).$

(c) $\tan 4x = \dfrac{\sin 8x}{1 + \cos 8x}$

(a)   This is obtained from $\sin 2\alpha = 2\sin\alpha\cos\alpha$ by putting $\alpha = \frac{1}{2}\theta$.

(b)   This is obtained from $\sin\frac{1}{2}\theta = \pm\sqrt{\dfrac{1 - \cos\theta}{2}}$ by putting $\theta = 2A$.

(c)   This is obtained from $\tan\frac{1}{2}\theta = \dfrac{\sin\theta}{1 + \cos\theta}$ by putting $\theta = 8x$.

(d)   This is obtained from $\cos 2\alpha = 1 - 2\sin^2\alpha$ by putting $\alpha = 3\theta$.

(e)   These formulas are obtained by squaring $\sin\frac{1}{2}\theta = \pm\sqrt{\dfrac{1 - \cos\theta}{2}}$ and $\cos\frac{1}{2}\theta = \pm\sqrt{\dfrac{1 + \cos\theta}{2}}$.

10.19 Express (a) $\sin 3\alpha$ in terms of $\sin\alpha$ and (b) $\cos 4\alpha$ in terms of $\cos\alpha$.

(a) $\sin 3\alpha = \sin(2\alpha + \alpha) = \sin 2\alpha\cos\alpha + \cos 2\alpha\sin\alpha$

$= (2\sin\alpha\cos\alpha)\cos\alpha + (1 - 2\sin^2\alpha)\sin\alpha = 2\sin\alpha\cos^2\alpha + (1 - 2\sin^2\alpha)\sin\alpha$

$= 2\sin\alpha(1 - \sin^2\alpha) + (1 - 2\sin^2\alpha)\sin\alpha = 3\sin\alpha - 4\sin^3\alpha$

(b) $\cos 4\alpha = \cos 2(2\alpha) = 2\cos^2 2\alpha - 1 = 2(2\cos^2\alpha - 1)^2 - 1 = 8\cos^4\alpha - 8\cos^2\alpha + 1$

10.20 Prove $\cos 2x = \cos^4 x - \sin^4 x$.

$\cos^4 x - \sin^4 x = (\cos^2 x + \sin^2 x)(\cos^2 x - \sin^2 x) = \cos^2 x - \sin^2 x = \cos 2x$

10.21 Prove $1 - \frac{1}{2}\sin 2x = \dfrac{\sin^3 x + \cos^3 x}{\sin x + \cos x}$.

$$\frac{\sin^3 x + \cos^3 x}{\sin x + \cos x} = \frac{(\sin x + \cos x)(\sin^2 x - \sin x\cos x + \cos^2 x)}{\sin x + \cos x}$$

$$= 1 - \sin x\cos x = 1 - \frac{1}{2}(2\sin x\cos x) = 1 - \frac{1}{2}\sin 2x$$

10.22 Prove $\cos \theta = \sin (\theta + 30°) + \cos (\theta + 60°)$.

$$\sin (\theta + 30°) + \cos (\theta + 60°) = (\sin \theta \cos 30° + \cos \theta \sin 30°) + (\cos \theta \cos 60° - \sin \theta \sin 60°)$$

$$= \frac{\sqrt{3}}{2} \sin \theta + \frac{1}{2} \cos \theta + \frac{1}{2} \cos \theta - \frac{\sqrt{3}}{2} \sin \theta = \cos \theta$$

10.23 Prove $\cos x = \dfrac{1 - \tan^2 \frac{1}{2}x}{1 + \tan^2 \frac{1}{2}x}$.

$$\frac{1 - \tan^2 \frac{1}{2}x}{1 + \tan^2 \frac{1}{2}x} = \frac{1 - \dfrac{\sin^2 \frac{1}{2}x}{\cos^2 \frac{1}{2}x}}{\sec^2 \frac{1}{2}x} = \frac{\left(1 - \dfrac{\sin^2 \frac{1}{2}x}{\cos^2 \frac{1}{2}x}\right) \cos^2 \frac{1}{2}x}{\sec^2 \frac{1}{2}x \cos^2 \frac{1}{2}x} = \cos^2 \tfrac{1}{2}x - \sin^2 \tfrac{1}{2}x = \cos x$$

10.24 Prove $2 \tan 2x = \dfrac{\cos x + \sin x}{\cos x - \sin x} - \dfrac{\cos x - \sin x}{\cos x + \sin x}$.

$$\frac{\cos x + \sin x}{\cos x - \sin x} - \frac{\cos x - \sin x}{\cos x + \sin x} = \frac{(\cos x + \sin x)^2 - (\cos x - \sin x)^2}{(\cos x - \sin x)(\cos x + \sin x)}$$

$$= \frac{(\cos^2 x + 2 \sin x \cos x + \sin^2 x) - (\cos^2 x - 2 \sin x \cos x + \sin^2 x)}{\cos^2 x - \sin^2 x}$$

$$= \frac{4 \sin x \cos x}{\cos^2 x - \sin^2 x} = \frac{2 \sin 2x}{\cos 2x} = 2 \tan 2x$$

10.25 Prove $\sin^4 A = \dfrac{3}{8} - \dfrac{1}{2} \cos 2A + \dfrac{1}{8} \cos 4A$.

$$\sin^4 A = (\sin^2 A)^2 = \left(\frac{1 - \cos 2A}{2}\right)^2 = \frac{1 - 2 \cos 2A + \cos^2 2A}{4}$$

$$= \tfrac{1}{4}\left(1 - 2 \cos 2A + \frac{1 + \cos 4A}{2}\right) = \tfrac{3}{8} - \tfrac{1}{2} \cos 2A + \tfrac{1}{8} \cos 4A$$

10.26 Prove $\tan^6 x = \tan^4 x \sec^2 x - \tan^2 x \sec^2 x + \sec^2 x - 1$.

$$\tan^6 x = \tan^4 x \tan^2 x = \tan^4 x (\sec^2 x - 1) = \tan^4 x \sec^2 x - \tan^2 x \tan^2 x$$
$$= \tan^4 x \sec^2 x - \tan^2 x (\sec^2 x - 1) = \tan^4 x \sec^2 x - \tan^2 x \sec^2 x + \tan^2 x$$
$$= \tan^4 x \sec^2 x - \tan^2 x \sec^2 x + \sec^2 x - 1$$

10.27 When $A + B + C = 180°$, show that $\sin 2A + \sin 2B + \sin 2C = 4 \sin A \sin B \sin C$.

Since $C = 180° - (A + B)$,

$$\sin 2A + \sin 2B + \sin 2C = \sin 2A + \sin 2B + \sin [360° - 2(A + B)]$$
$$= \sin 2A + \sin 2B - \sin 2(A + B)$$
$$= \sin 2A + \sin 2B - \sin 2A \cos 2B - \cos 2A \sin 2B$$
$$= (\sin 2A)(1 - \cos 2B) + (\sin 2B)(1 - \cos 2A)$$
$$= 2 \sin 2A \sin^2 B + 2 \sin 2B \sin^2 A$$
$$= 4 \sin A \cos A \sin^2 B + 4 \sin B \cos B \sin^2 A$$
$$= 4 \sin A \sin B (\sin A \cos B + \cos A \sin B)$$
$$= 4 \sin A \sin B \sin (A + B)$$
$$= 4 \sin A \sin B \sin [180° - (A + B)] = 4 \sin A \sin B \sin C$$

10.28 When $A + B + C = 180°$, show that $\tan A + \tan B + \tan C = \tan A \tan B \tan C$.

Since $C = 180° - (A + B)$,

$\tan A + \tan B + \tan C$

$$= \tan A + \tan B + \tan [180° - (A + B)] = \tan A + \tan B - \tan (A + B)$$

$$= \tan A + \tan B - \frac{\tan A + \tan B}{1 - \tan A \tan B} = (\tan A + \tan B)\left(1 - \frac{1}{1 - \tan A \tan B}\right)$$

$$= (\tan A + \tan B)\left(-\frac{\tan A \tan B}{1 - \tan A \tan B}\right) = -\tan A \tan B \frac{\tan A + \tan B}{1 - \tan A \tan B}$$

$$= -\tan A \tan B \tan (A + B) = \tan A \tan B \tan [180° - (A + B)] = \tan A \tan B \tan C$$

Supplementary Problems

10.29 Find the values of the sine, cosine, and tangent of (a) $75°$ and (b) $255°$.

Ans.   (a) $\dfrac{\sqrt{6} + \sqrt{2}}{4}, \dfrac{\sqrt{6} - \sqrt{2}}{4}, 2 + \sqrt{3}$ (b) $-\dfrac{\sqrt{6} + \sqrt{2}}{4}, -\dfrac{\sqrt{6} - \sqrt{2}}{4}, 2 + \sqrt{3}$

10.30 Find the values of the sine, cosine, and tangent of (a) $7\pi/12$ and (b) $11\pi/12$.

Ans.   (a) $\dfrac{\sqrt{6} + \sqrt{2}}{4}, \dfrac{-\sqrt{6} + \sqrt{2}}{4}, -2 - \sqrt{3}$

   (b) $\dfrac{\sqrt{6} - \sqrt{2}}{4}, -\dfrac{\sqrt{6} + \sqrt{2}}{4}, -2 + \sqrt{3}$

10.31 Rewrite each expression as a single function of an angle.

(a) $\sin 173° \cos 82° + \cos 173° \sin 82°$ *Ans.* $\sin 255°$

(b) $\cos 86° \cos 73° + \sin 86° \sin 73°$ *Ans.* $\cos 13°$

(c) $\dfrac{\tan 87° - \tan 21°}{1 + \tan 87° \tan 21°}$ *Ans.* $\tan 66°$

(d) $\sin 87° \cos 87°$ *Ans.* $\frac{1}{2} \sin 174°$

(e) $2 \cos^2 151° - 1$ *Ans.* $\cos 302°$

(f) $1 - 2 \sin^2 100°$ *Ans.* $\cos 200°$

(g) $\dfrac{\tan 42°}{1 - \tan^2 42°}$ *Ans.* $\frac{1}{2} \tan 84°$

(h) $\cos^2 81° - \sin^2 81°$ *Ans.* $\cos 162°$

(i) $\dfrac{\sin 56°}{1 + \cos 56°}$ *Ans.* $\tan 28°$

(j) $\sqrt{\dfrac{1 + \cos 76°}{2}}$ *Ans.* $\cos 38°$

10.32   Find the values of $\sin(\alpha + \beta)$, $\cos(\alpha + \beta)$, and $\tan(\alpha + \beta)$, given:

     (a) $\sin \alpha = 3/5$, $\cos \beta = 5/13$, α and β in Quadrant I   *Ans.* $63/65$, $-16/65$, $-63/16$

     (b) $\sin \alpha = 8/17$, $\tan \beta = 5/12$, α and β in Quadrant I   *Ans.* $171/221$, $140/221$, $171/140$

     (c) $\cos \alpha = -12/13$, $\cot \beta = 24/7$, α in Quadrant II, β in Quadrant III

                                                  Ans. $-36/325$, $323/325$, $-36/323$

     (d) $\sin \alpha = 1/3$, $\sin \beta = 2/5$, α in Quadrant I, β in Quadrant II

$$Ans. \quad \frac{4\sqrt{2} - \sqrt{21}}{15}, \quad -\frac{2 + 2\sqrt{42}}{15}, \quad -\frac{4\sqrt{2} - \sqrt{21}}{2 + 2\sqrt{42}} = \frac{-25\sqrt{2} + 9\sqrt{2}}{82}$$

10.33   Find the values of $\sin(\alpha - \beta)$, $\cos(\alpha - \beta)$, and $\tan(\alpha - \beta)$, given:

     (a) $\sin \alpha = 3/5$, $\sin \beta = 5/13$, α and β in Quadrant I   *Ans.* $16/65$, $63/65$, $16/63$

     (b) $\sin \alpha = 8/17$, $\tan \beta = 5/12$, α and β in Quadrant I   *Ans.* $21/221$, $220/221$, $21/220$

     (c) $\cos \alpha = -12/13$, $\cot \beta = 24/7$, α in Quadrant II, β is Quadrant I

                                                  Ans. $204/325$, $-253/325$, $-204/253$

     (d) $\sin \alpha = 1/3$, $\sin \beta = 2/5$, α in Quadrant II, β in Quadrant I

$$Ans. \quad \frac{4\sqrt{2} + \sqrt{21}}{15}, \quad -\frac{2\sqrt{42} - 2}{15}, \quad -\frac{4\sqrt{2} + \sqrt{21}}{2\sqrt{42} - 2} = -\frac{25\sqrt{2} + 9\sqrt{2}}{82}$$

10.34   Prove:

     (a) $\sin(\alpha + \beta) - \sin(\alpha - \beta) = 2\cos\alpha\sin\beta$

     (b) $\cos(\alpha + \beta) + \cos(\alpha - \beta) = 2\cos\alpha\cos\beta$

     (c) $\tan(45° - \theta) = \dfrac{1 - \tan\theta}{1 + \tan\theta}$

     (d) $\dfrac{\tan(\alpha + \beta)}{\cot(\alpha - \beta)} = \dfrac{\tan^2\alpha - \tan^2\beta}{1 - \tan^2\alpha\tan^2\beta}$

     (f) $\dfrac{\sin(x + y)}{\cos(x - y)} = \dfrac{\tan x + \tan y}{1 + \tan x \tan y}$

     (g) $\tan(45° + \theta) = \dfrac{\cos\theta + \sin\theta}{\cos\theta - \sin\theta}$

     (h) $\sin(\alpha + \beta)\sin(\alpha - \beta) = \sin^2\alpha - \sin^2\beta$

     (e) $\tan(\alpha + \beta + \gamma) = \tan[(\alpha + \beta) + \gamma] = \dfrac{\tan\alpha + \tan\beta + \tan\gamma - \tan\alpha\tan\beta\tan\gamma}{1 - \tan\alpha\tan\beta - \tan\beta\tan\gamma - \tan\gamma\tan\alpha}$

10.35   If A and B are acute angles, find $A + B$ given:

     (a) $\tan A = 1/4$, $\tan B = 3/5$ Hint: $\tan(A + B) = 1$.   *Ans.* $45°$

     (b) $\tan A = 5/3$, $\tan B = 4$ *Ans.* $135°$

10.36   If $\tan(x + y) = 33$ and $\tan x = 3$, show that $\tan y = 0.3$.

10.37   Find the values of $\sin 2\theta$, $\cos 2\theta$, and $\tan 2\theta$, given:

     (a) $\sin \theta = 3/5$, θ in Quadrant I   *Ans.* $24/25$, $7/25$, $24/7$

     (b) $\sin \theta = 3/5$, θ in Quadrant II   *Ans.* $-24/25$, $7/25$, $-24/7$

     (c) $\sin \theta = -1/2$, θ in Quadrant IV   *Ans.* $-\sqrt{3}/2$, $1/2$, $-\sqrt{3}$

     (d) $\tan \theta = -1/5$, θ in Quadrant II   *Ans.* $-5/13$, $12/13$, $-5/12$

     (e) $\tan \theta = u$, θ in Quadrant I   *Ans.* $\dfrac{2u}{1 + u^2}$, $\dfrac{1 - u^2}{1 + u^2}$, $\dfrac{2u}{1 - u^2}$

10.38 Prove:

(a) $\tan \theta \sin 2\theta = 2 \sin^2 \theta$

(b) $\cot \theta \sin 2\theta = 1 + \cos 2\theta$

(e) $\cos 2\theta = \dfrac{1 - \tan^2 \theta}{1 + \tan^2 \theta}$

(c) $\dfrac{\sin^3 x - \cos^3 x}{\sin x - \cos x} = 1 + \frac{1}{2} \sin 2x$

(f) $\dfrac{1 + \cos 2\theta}{\sin 2\theta} = \cot \theta$

(d) $\dfrac{1 - \sin 2A}{\cos 2A} = \dfrac{1 - \tan A}{1 + \tan A}$

(g) $\cos 3\theta = 4 \cos^3 \theta - 3 \cos \theta$

(h) $\cos^4 x = \frac{3}{8} + \frac{1}{2} \cos 2x + \frac{1}{8} \cos 4x$

10.39 Find the values of the sine, cosine, and tangent of

(a) $30°$, given $\cos 60° = 1/2$ *Ans.* $1/2, \sqrt{3}/2, 1/\sqrt{3} = \sqrt{3}/3$

(b) $105°$, given $\cos 210° = -\sqrt{3}/2$ *Ans.* $\frac{1}{2}\sqrt{2 + \sqrt{3}}, -\frac{1}{2}\sqrt{2 - \sqrt{3}}, -(2 + \sqrt{3})$

(c) $\frac{1}{2}\theta$, given $\sin \theta = 3/5$, θ in Quadrant I   *Ans.* $1/\sqrt{10} = \sqrt{10}/10, 3/\sqrt{10} = 3\sqrt{10}/10, 1/3$

(d) θ, given $\cot 2\theta = 7/24$, 2θ in Quadrant I   *Ans.* $3/5, 4/5, 3/4$

(e) θ, given $\cot 2\theta = -5/12$, 2θ in Quadrant II   *Ans.* $3/\sqrt{13} = 3\sqrt{13}/13, 2/\sqrt{13} = 2\sqrt{13}/13, 3/2$

10.40 Find the values of the sine, cosine, and tangent of

(a) $7\pi/8$, given $\cos 7\pi/4 = \sqrt{2}/2$ *Ans.* $\frac{1}{2}\sqrt{2 - \sqrt{2}}, -\frac{1}{2}\sqrt{2 + \sqrt{2}}, -\sqrt{3 - 2\sqrt{2}}$

(b) $5\pi/8$, given $\sin 5\pi/4 = -\sqrt{2}/2$ *Ans.* $\frac{1}{2}\sqrt{2 + \sqrt{2}}, -\frac{1}{2}\sqrt{2 - \sqrt{2}}, -\sqrt{3 + 2\sqrt{2}}$

10.41 Prove:

(a) $\cos x = 2 \cos^2 \frac{1}{2}x - 1 = 1 - 2 \sin^2 \frac{1}{2}x$

(b) $\sin x = 2 \sin \frac{1}{2}x \cos \frac{1}{2}x$

(c) $(\sin \frac{1}{2}\theta - \cos \frac{1}{2}\theta)^2 = 1 - \sin \theta$

(d) $\tan \frac{1}{2}\theta = \csc \theta - \cot \theta$

(e) $\dfrac{1 - \tan \frac{1}{2}\theta}{1 + \tan \frac{1}{2}\theta} = \dfrac{1 - \sin \theta}{\cos \theta} = \dfrac{\cos \theta}{1 + \sin \theta}$

(f) $\dfrac{2 \tan \frac{1}{2}x}{1 + \tan^2 \frac{1}{2}x} = \sin x$

10.42 In the right triangle ABC, in which C is the right angle, prove:

$$\sin 2A = \frac{2ab}{c^2} \qquad \cos 2A = \frac{b^2 - a^2}{c^2} \qquad \sin \frac{1}{2}A = \sqrt{\frac{c - b}{2c}} \qquad \cos \frac{1}{2}A = \sqrt{\frac{c + b}{2c}}$$

10.43 Prove (a) $\dfrac{\sin 3x}{\sin x} - \dfrac{\cos 3x}{\cos x} = 2$ and (b) $\tan 50° - \tan 40° = 2 \tan 10°$.

10.44 If $A + B + C = 180°$, prove:

(a) $\sin A + \sin B + \sin C = 4 \cos \frac{1}{2}A \cos \frac{1}{2}B \cos \frac{1}{2}C$

(b) $\cos A + \cos B + \cos C = 1 + 4 \sin \frac{1}{2}A \sin \frac{1}{2}B \sin \frac{1}{2}C$

(c) $\sin^2 A + \sin^2 B - \sin^2 C = 2 \sin A \sin B \cos C$

(d) $\tan \frac{1}{2}A \tan \frac{1}{2}B + \tan \frac{1}{2}B \tan \frac{1}{2}C + \tan \frac{1}{2}C \tan \frac{1}{2}A = 1$

Chapter 11

Sum, Difference, and Product Formulas

11.1 PRODUCTS OF SINES AND COSINES

$$\sin \alpha \cos \beta = \tfrac{1}{2}[\sin (\alpha + \beta) + \sin (\alpha - \beta)]$$

$$\cos \alpha \sin \beta = \tfrac{1}{2}[\sin (\alpha + \beta) - \sin (\alpha - \beta)]$$

$$\cos \alpha \cos \beta = \tfrac{1}{2}[\cos (\alpha + \beta) + \cos (\alpha - \beta)]$$

$$\sin \alpha \sin \beta = -\tfrac{1}{2}[\cos (\alpha + \beta) - \cos (\alpha - \beta)]$$

For proofs of these formulas, see Prob. 11.1.

11.2 SUM AND DIFFERENCE OF SINES AND COSINES

$$\sin A + \sin B = 2 \sin \tfrac{1}{2}(A + B) \cos \tfrac{1}{2}(A - B)$$

$$\sin A - \sin B = 2 \cos \tfrac{1}{2}(A + B) \sin \tfrac{1}{2}(A - B)$$

$$\cos A + \cos B = 2 \cos \tfrac{1}{2}(A + B) \cos \tfrac{1}{2}(A - B)$$

$$\cos A - \cos B = -2 \sin \tfrac{1}{2}(A + B) \sin \tfrac{1}{2}(A - B)$$

For proofs of these formulas, see Prob. 11.2.

Solved Problems

11.1 Derive the product formulas.

Since
$$\sin (\alpha + \beta) + \sin (\alpha - \beta) = (\sin \alpha \cos \beta + \cos \alpha \cos \beta) + (\sin \alpha \cos \beta - \cos \alpha \sin \beta)$$

$$= 2 \sin \alpha \cos \beta$$

$$\sin \alpha \cos \beta = \tfrac{1}{2}[\sin (\alpha + \beta) + \sin (\alpha - \beta)]$$

Since
$$\sin (\alpha + \beta) - \sin (\alpha - \beta) = 2 \cos \alpha \sin \beta,$$

$$\cos \alpha \sin \beta = \tfrac{1}{2}[\sin (\alpha + \beta) - \sin (\alpha - \beta)]$$

Since
$$\cos (\alpha + \beta) + \cos (\alpha - \beta) = (\cos \alpha \cos \beta - \sin \alpha \sin \beta) + (\cos \alpha \cos \beta + \sin \alpha \sin \beta)$$

$$= 2 \cos \alpha \cos \beta$$

$$\cos \alpha \cos \beta = \tfrac{1}{2}[\cos (\alpha + \beta) + \cos (\alpha - \beta)]$$

Since
$$\cos (\alpha + \beta) - \cos (\alpha - \beta) = -2 \sin \alpha \sin \beta$$

$$\sin \alpha \sin \beta = -\tfrac{1}{2}[\cos (\alpha + \beta) - \cos (\alpha - \beta)]$$

11.2 Derive the sum and difference formulas.

Let $\alpha + \beta = A$ and $\alpha - \beta = B$ so that $\alpha = \tfrac{1}{2}(A + B)$ and $\beta = \tfrac{1}{2}(A - B)$. Then (see Prob. 11.1)

$$\sin (\alpha + \beta) + \sin (\alpha + \beta) = 2 \sin \alpha \cos \beta \qquad \text{becomes} \qquad \sin A + \sin B = 2 \sin \tfrac{1}{2}(A + B) \cos \tfrac{1}{2}(A - B)$$

$$\sin (\alpha + \beta) - \sin (\alpha - \beta) = 2 \cos \alpha \sin \beta \qquad \text{becomes} \qquad \sin A - \sin B = 2 \cos \tfrac{1}{2}(A + B) \sin \tfrac{1}{2}(A - B)$$

$$\cos(\alpha + \beta) + \cos(\alpha - \beta) = 2\cos\alpha\cos\beta \quad \text{becomes} \quad \cos A + \cos B = 2\cos\tfrac{1}{2}(A + B)\cos\tfrac{1}{2}(A - B)$$

$$\cos(\alpha + \beta) - \cos(\alpha - \beta) = -2\sin\alpha\cos\beta \quad \text{becomes} \quad \cos A - \cos B = -2\sin\tfrac{1}{2}(A + B)\sin\tfrac{1}{2}(A - B)$$

11.3 Express each of the following as a sum or difference:

(a) $\sin 40° \cos 30°$, (b) $\cos 110° \sin 55°$, (c) $\cos 50° \cos 35°$, (d) $\sin 55° \sin 40°$

(a) $\sin 40° \cos 30° = \tfrac{1}{2}[\sin(40° + 30°) + \sin(40° - 30°)] = \tfrac{1}{2}(\sin 70° + \sin 10°)$

(b) $\cos 110° \sin 55° = \tfrac{1}{2}[\sin(110° + 55°) - \sin(110° - 55°)] = \tfrac{1}{2}(\sin 165° - \sin 55°)$

(c) $\cos 50° \cos 35° = \tfrac{1}{2}[\cos(50° + 35°) + \cos(50° - 35°)] = \tfrac{1}{2}(\cos 85° + \cos 15°)$

(d) $\sin 55° \sin 40° = -\tfrac{1}{2}[\cos(55° + 40°) - \cos(55° - 40°)] = -\tfrac{1}{2}(\cos 95° - \cos 15°)$

11.4 Express each of the following as a product:

(a) $\sin 50° + \sin 40°$, (b) $\sin 70° - \sin 20°$, (c) $\cos 55° + \cos 25°$, (d) $\cos 35° - \cos 75°$

(a) $\sin 50° + \sin 40° = 2\sin\tfrac{1}{2}(50° + 40°)\cos\tfrac{1}{2}(50° - 40°) = 2\sin 45° \cos 5°$

(b) $\sin 70° - \sin 20° = 2\cos\tfrac{1}{2}(70° + 20°)\sin\tfrac{1}{2}(70° - 20°) = 2\cos 45° \sin 25°$

(c) $\cos 55° + \cos 25° = 2\cos\tfrac{1}{2}(55° + 25°)\cos\tfrac{1}{2}(55° - 25°) = 2\cos 40° \cos 15°$

(d) $\cos 35° - \cos 75° = -2\sin\tfrac{1}{2}(35° + 75°)\sin\tfrac{1}{2}(35° - 75°) = -2\sin 55° \sin(-20°)$

$$= 2\sin 55° \sin 20°$$

11.5 Prove $\dfrac{\sin 4A + \sin 2A}{\cos 4A + \cos 2A} = \tan 3A.$

$$\frac{\sin 4A + \sin 2A}{\cos 4A + \cos 2A} = \frac{2\sin\tfrac{1}{2}(4A + 2A)\cos\tfrac{1}{2}(4A - 2A)}{2\cos\tfrac{1}{2}(4A + 2A)\cos\tfrac{1}{2}(4A - 2A)} = \frac{\sin 3A}{\cos 3A} = \tan 3A$$

11.6 Prove $\dfrac{\sin A - \sin B}{\sin A + \sin B} = \dfrac{\tan\tfrac{1}{2}(A - B)}{\tan\tfrac{1}{2}(A + B)}.$

$$\frac{\sin A - \sin B}{\sin A + \sin B} = \frac{2\cos\tfrac{1}{2}(A + B)\sin\tfrac{1}{2}(A - B)}{2\sin\tfrac{1}{2}(A + B)\cos\tfrac{1}{2}(A - B)} = \cot\tfrac{1}{2}(A + B)\tan\tfrac{1}{2}(A - B) = \frac{\tan\tfrac{1}{2}(A - B)}{\tan\tfrac{1}{2}(A + B)}$$

11.7 Prove $\cos^3 x \sin^2 x = \tfrac{1}{16}(2\cos x - \cos 3x - \cos 5x).$

$$\cos^3 x \sin^2 x = (\sin x \cos x)^2 \cos x = \frac{1}{4}\sin^2 2x \cos x = \frac{1}{4}(\sin 2x)(\sin 2x \cos x)$$

$$= \frac{1}{4}(\sin 2x)[\tfrac{1}{2}(\sin 3x + \sin x)] = \frac{1}{8}(\sin 3x \sin 2x + \sin 2x \sin x)$$

$$= \frac{1}{8}\{-\tfrac{1}{2}(\cos 5x - \cos x) + [-\tfrac{1}{2}(\cos 3x - \cos x)]\}$$

$$= \frac{1}{16}(2\cos x - \cos 3x - \cos 5x)$$

11.8 Prove $1 + \cos 2x + \cos 4x + \cos 6x = 4 \cos x \cos 2x \cos 3x$.

$$1 + (\cos 2x + \cos 4x) + \cos 6x = 1 + 2 \cos 3x \cos x + \cos 6x = (1 + \cos 6x) + 2 \cos 3x \cos x$$
$$= 2 \cos^2 3x + 2 \cos 3x \cos x = 2 \cos 3x (\cos 3x + \cos x)$$
$$= 2 \cos 3x (2 \cos 2x \cos x) = 4 \cos x \cos 2x \cos 3x$$

11.9 Transform $4 \cos x + 3 \sin x$ into the form $c \cos (x - \alpha)$.

Since $c \cos (x - \alpha) = c(\cos x \cos \alpha + \sin x \sin \alpha)$, set $c \cos \alpha = 4$ and $c \sin \alpha = 3$.
Then $\cos \alpha = 4/c$ and $\sin \alpha = 3/c$. Since $\sin^2 \alpha + \cos^2 \alpha = 1$, $c = 5$ and -5.
Using $c = 5$, $\cos \alpha = 4/5$, $\sin \alpha = 3/5$, and $\alpha = 0.6435$ rad. Thus,

$$4 \cos x + 3 \sin x = 5 \cos (x - 0.6435).$$

Using $c = -5$, $\alpha = 3.7851$ rad and

$$4 \cos x + 3 \sin x = -5 \cos (x - 3.7851)$$

11.10 Find the maximum and minimum values of $4 \cos x + 3 \sin x$ on the interval $0 \leq x \leq 2\pi$.

From Prob. 11.9, $4 \cos x + 3 \sin x = 5 \cos (x - 0.6435)$.
Now on the prescribed interval, $\cos \theta$ attains its maximum value 1 when $\theta = 0$ and its minimum value -1 when $\theta = \pi$. Thus, the maximum value of $4 \cos x + 3 \sin x$ is 5 which occurs when $x - 0.6435 = 0$ or when $x = 0.6435$ while the minimum value is -5 which occurs when $x - 0.6435 = \pi$ or when $x = 3.7851$.

Supplementary Problems

11.11 Express each of the following products as a sum or difference of sines or of cosines.

(a) $\sin 35° \cos 25° = \frac{1}{2}(\sin 60° + \sin 10°)$

(b) $\sin 25° \cos 75° = \frac{1}{2}(\sin 100° - \sin 50°)$

(c) $\cos 50° \cos 70° = \frac{1}{2}(\cos 120° + \cos 20°)$

(d) $\sin 130° \sin 55° = -\frac{1}{2}(\cos 185° - \cos 75°)$

(e) $\sin 4x \cos 2x = \frac{1}{2}(\sin 6x + \sin 2x)$

(f) $\sin x/2 \cos 3x/2 = \frac{1}{2}(\sin 2x - \sin x)$

(g) $\cos 7x \cos 4x = \frac{1}{2}(\cos 11x + \cos 3x)$

(h) $\sin 5x \sin 4x = -\frac{1}{2}(\cos 9x - \cos x)$

11.12 Show that

(a) $2 \sin 45° \cos 15° = \dfrac{\sqrt{3} + 1}{2}$ and $\cos 15° = \dfrac{\sqrt{6} + \sqrt{2}}{4}$

(b) $2 \sin 82\frac{1}{2}° \cos 37\frac{1}{2}° = \dfrac{\sqrt{3} + \sqrt{2}}{2}$ (c) $2 \sin 127\frac{1}{2}° \sin 97\frac{1}{2}° = \dfrac{\sqrt{3} + \sqrt{2}}{2}$

11.13 Express each of the following as a product.

(a) $\sin 50° + \sin 20° = 2 \sin 35° \cos 15°$ (e) $\sin 4x + \sin 2x = 2 \sin 3x \cos x$

(b) $\sin 75° - \sin 35° = 2 \cos 55° \sin 20°$ (f) $\sin 7\theta - \sin 3\theta = 2 \cos 5\theta \sin 2\theta$

(c) $\cos 65° + \cos 15° = 2 \cos 40° \cos 25°$ (g) $\cos 6\theta + \cos 2\theta = 2 \cos 4\theta \cos 2\theta$

(d) $\cos 80° - \cos 70° = -2 \sin 75° \sin 5°$ (h) $\cos 3x/2 - \cos 9x/2 = 2 \sin 3x \sin 3x/2$

11.14   Show that

   (a) $\sin 40° + \sin 20° = \cos 10°$ (c) $\cos 465° + \cos 165° = -\sqrt{6}/2$

   (b) $\sin 105° + \sin 15° = \sqrt{6}/2$ (d) $\dfrac{\sin 75° - \sin 15°}{\cos 75° + \cos 15°} = \sqrt{3}/3$

11.15   Prove:

   (a) $\dfrac{\sin A + \sin 3A}{\cos A + \cos 3A} = \tan 2A$ (c) $\dfrac{\sin A + \sin B}{\sin A - \sin B} = \dfrac{\tan\frac{1}{2}(A + B)}{\tan\frac{1}{2}(A - B)}$

   (b) $\dfrac{\sin 2A + \sin 4A}{\cos 2A + \cos 4A} = \tan 3A$ (d) $\dfrac{\cos A + \cos B}{\cos A - \cos B} = -\cot\frac{1}{2}(A - B)\cot\frac{1}{2}(A + B)$

   (e) $\sin \theta + \sin 2\theta + \sin 3\theta = \sin 2\theta + (\sin \theta + \sin 3\theta) = \sin 2\theta\,(1 + 2 \cos \theta)$

   (f) $\cos \theta + \cos 2\theta + \cos 3\theta = \cos 2\theta\,(1 + 2 \cos \theta)$

   (g) $\sin 2\theta + \sin 4\theta + \sin 6\theta = (\sin 2\theta + \sin 4\theta) + 2 \sin 3\theta \cos 3\theta$
   $$= 4 \cos \theta \cos 2\theta \sin 3\theta$$

   (h) $\dfrac{\sin 3x + \sin 5x + \sin 7x + \sin 9x}{\cos 3x + \cos 5x + \cos 7x + \cos 9x} = \dfrac{(\sin 3x + \sin 9x) + (\sin 5x + \sin 7x)}{(\cos 3x + \cos 9x) + (\cos 5x + \cos 7x)} = \tan 6x$

11.16   Prove:

   (a) $\cos 130° + \cos 110° + \cos 10° = 0$ (b) $\cos 220° + \cos 100° + \cos 20° = 0$

11.17   Prove:

   (a) $\cos^2 \theta \sin^3 \theta = \frac{1}{16}(2 \sin \theta + \sin 3\theta - \sin 5\theta)$

   (b) $\cos^2 \theta \sin^4 \theta = \frac{1}{32}(2 - \cos 2\theta - 2 \cos 4\theta + \cos 6\theta)$

   (c) $\cos^5 \theta = \frac{1}{16}(10 \cos \theta + 5 \cos 3\theta + \cos 5\theta)$

   (d) $\sin^5 \theta = \frac{1}{16}(10 \sin \theta - 5 \sin 3\theta + \sin 5\theta)$

11.18   Transform (using radians):

   (a) $4 \cos x + 3 \sin x$ into the form $c \sin (x + \alpha)$ *Ans.* $5 \sin (x + 0.9273)$

   (b) $4 \cos x + 3 \sin x$ into the form $c \sin (x - \alpha)$ *Ans.* $5 \sin (x - 5.3559)$

   (c) $\sin x - \cos x$ into the form $c \sin (x - \alpha)$ *Ans.* $\sqrt{2} \sin (x - \pi/4)$

   (d) $5 \cos 3t + 12 \sin 3t$ into the form $c \cos (3t - \alpha)$ *Ans.* $13 \cos (3t - 1.1760)$

11.19   Find the maximum and minimum values of each sum of Prob. 11.18 and a value of x or t between 0 and 2π at which each occurs.

   Ans.   (a)   Maximum = 5, when $x = 0.6435$ (i.e., when $x + 0.9273 = \pi/2$); minimum = -5, when $x = 3.7851$.

   (b)   Same as (a).

   (c)   Maximum = $\sqrt{2}$, when $x = 3\pi/4$; minimum = $-\sqrt{2}$, when $x = 7\pi/4$.

   (d)   Maximum = 13, when $t = 0.3920$; minimum = -13, when $t = 1.4392$.

Chapter 12

Oblique Triangles

12.1 OBLIQUE TRIANGLES

An *oblique triangle* is one which does not contain a right angle. Such a triangle contains either three acute angles or two acute angles and one obtuse angle.

The convention of denoting the angles by A, B, and C and the lengths of the corresponding opposite sides by a, b, and c will be used here. See Fig. 12-1.

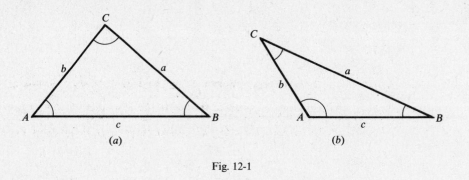

Fig. 12-1

12.2 LAW OF SINES

In any triangle ABC, the ratio of a side and the sine of the opposite angle is a constant; i.e.,

$$\frac{a}{\sin A} = \frac{b}{\sin B} = \frac{c}{\sin C} \qquad \text{or} \qquad \frac{\sin A}{a} = \frac{\sin B}{b} = \frac{\sin C}{c}$$

For a proof of the law of sines see Prob. 12.1.

12.3 LAW OF COSINES

In any triangle ABC, the square of any side is equal to the sum of the squares of the other two sides diminished by twice the product of these sides and the cosine of the included angle; i.e.,

$$a^2 = b^2 + c^2 - 2bc \cos A$$
$$b^2 = a^2 + c^2 - 2ac \cos B$$
$$c^2 = a^2 + b^2 - 2ab \cos C$$

For a proof of the law of cosines see Prob. 12.3.

12.4 SOLUTION OF OBLIQUE TRIANGLES

When three parts of a triangle, not all angles, are known, the triangle is uniquely determined, except in one case noted below. The five cases of oblique triangles are

▲ Case I: Given two angles and the side opposite one of them

▲ Case II: Given two angles and the included side

▲ Case III: Given two sides and the angle opposite one of them

120

▲ Case IV: Given two sides and the included angle

▲ Case V: Given the three sides

Case	Use Law of	First Part to Find
I	Sines	Side opposite second given angle
II	Sines	Third angle, then either of remaining sides
III	Sines	Angle opposite second given side
IV	Cosines	Third side
V	Cosines	Any angle can be found

In Case III there is not always a unique solution. It is possible to have no solution for the angle, one solution for the angle, or two solutions—an angle and its supplement. See Example 12.3 and Prob. 12.2 for a complete discussion of this case.

12.5 CHECKING OBLIQUE-TRIANGLE SOLUTIONS

There are two procedures available to check your solution to an oblique triangle. Both Mollweide's formulas and the projection formulas serve as ways, independent of the solution procedure used, to check your results.

In any triangle ABC, Mollweide's formulas are

$$\frac{a+b}{c} = \frac{\cos\frac{1}{2}(A-B)}{\sin\frac{1}{2}C} \qquad \frac{a-b}{c} = \frac{\sin\frac{1}{2}(A-B)}{\cos\frac{1}{2}C}$$

together with those obtained by cyclic changes of the letters, i.e.,

$$\frac{b+c}{a} = \frac{\cos\frac{1}{2}(B-C)}{\sin\frac{1}{2}A} \qquad \frac{b-c}{a} = \frac{\sin\frac{1}{2}(B-C)}{\cos\frac{1}{2}A}$$

$$\frac{c+a}{b} = \frac{\cos\frac{1}{2}(C-A)}{\sin\frac{1}{2}B} \qquad \frac{c-a}{b} = \frac{\sin\frac{1}{2}(C-A)}{\cos\frac{1}{2}B}$$

and those obtained by interchanging the two letters (small and capital) in the numerators of each relation.

For derivations of these formulas, see Prob. 12.4.

In any triangle ABC, the projection formulas are

$$a = b\cos C + c\cos B$$

$$b = c\cos A + a\cos C$$

$$c = a\cos B + b\cos A$$

For the derivation of these formulas, see Prob. 12.5.

CASE I. Given two angles and the side opposite one of them

EXAMPLE 12.1 Suppose b, B, and C are given.

To find c, use $\dfrac{c}{\sin C} = \dfrac{b}{\sin B}$; then $c = \dfrac{b\sin C}{\sin B}$.

To find A, use $A = 180° - (B + C)$.

To find a, use $\dfrac{a}{\sin A} = \dfrac{b}{\sin B}$; then $a = \dfrac{b\sin A}{\sin B}$.

To check, use one of the Mollweide formulas or one of the projection formulas.

(See Prob. 12.6.)

CASE II.   Given two angles and the included side

EXAMPLE 12.2   Suppose a, B, and C are given.
  To find A, use $A = 180° - (B + C)$.
  To find b, use $\dfrac{b}{\sin B} = \dfrac{a}{\sin A}$; then $b = \dfrac{a \sin B}{\sin A}$.
  To find c, use $\dfrac{c}{\sin C} = \dfrac{a}{\sin A}$; then $c = \dfrac{a \sin C}{\sin A}$.
  To check, use one of the Mollweide formulas or one of the projection formulas.

(See Prob. 12.7.)

CASE III.   Given two sides and the angle opposite one of them

EXAMPLE 12.3   Suppose b, c, and B are given.

  From $\dfrac{\sin C}{c} = \dfrac{\sin B}{b}$, $\sin C = \dfrac{c \sin B}{b}$.

  If $\sin C > 1$, no angle C is determined.
  If $\sin C = 1$, $C = 90°$ and a right triangle is determined.
  If $\sin C < 1$, two angles are determined: an acute angle C and an obtuse angle $C' = 180° - C$. Thus, there may be one or two triangles determined.

  This case is discussed geometrically in Prob. 12.2. The results obtained may be summarized as follows:

  When the given angle is *acute*, there will be

  (a)   One solution if the side opposite the given angle is equal to or greater than the other given side

  (b)   No solution, *one* solution (right triangle), or *two* solutions if the side opposite the given angle is less than the other given side

  When the given angle is *obtuse*, there will be

  (c)   *No* solution when the side opposite the given angle is less than or equal to the other given side

  (d)   *One* solution if the side opposite the given angle is greater than the other given side

EXAMPLE 12.4
  (1)   When $b = 30$, $c = 20$, and $B = 40°$, there is one solution since B is acute and $b > c$.

  (2)   When $b = 20$, $c = 30$, and $B = 40°$, there is either no solution, one solution, or two solutions. The particular subcase is determined after computing $\sin C = \dfrac{c \sin B}{b}$.

  (3)   When $b = 30$, $c = 20$, and $B = 140°$, there is one solution.

  (4)   When $b = 20$, $c = 30$, and $B = 140°$, there is no solution.

  This, the so-called ambiguous case, is solved by the law of sines and may be checked by either the Mollweide formulas or the projection formulas.

(See Probs. 12.11 to 12.13.)

CASE IV.   Given two sides and the included angle

EXAMPLE 12.5   Suppose a, b, and C are given.
  To find c, use $c^2 = a^2 + b^2 - 2ab \cos C$.
  To find A, use $\sin A = \dfrac{a \sin C}{c}$. To find B, use $\sin B = \dfrac{b \sin C}{c}$.
  To check, use $A + B + C = 180°$.

(See Probs. 12.15 and 12.16.)

CASE V   Given the three sides

EXAMPLE 12.6   With a, b, and c given, solve the law of cosines for each of the angles.

To find the angles, use $\cos A = \dfrac{b^2 + c^2 - a^2}{2bc}$, $\cos B = \dfrac{c^2 + a^2 - b^2}{2ca}$, and $\cos C = \dfrac{a^2 + b^2 - c^2}{2ab}$.

To check, use $A + B + C = 180°$.

(See Probs. 12.19 and 12.20.)

12.6   LAW OF TANGENTS

The law of cosines of Sec. 12.3 is not well-adapted for logarithmic computation. In solving Case IV, the law of tangents

$$\frac{a - b}{a + b} = \frac{\tan \frac{1}{2}(A - B)}{\tan \frac{1}{2}(A + B)} \qquad \frac{b - c}{b + c} = \frac{\tan \frac{1}{2}(B - C)}{\tan \frac{1}{2}(B + C)} \qquad \frac{c - a}{c + a} = \frac{\tan \frac{1}{2}(C - A)}{\tan \frac{1}{2}(C + A)}$$

will be used. For a proof of the law, see Prob. 12.26.

[NOTE: If, for example, $b > a$ it will be more convenient to write the first formula with the letters a and b (also A and B) interchanged.]

(NOTE: See Appendix 3 to review logarithm rules and problem-solving procedures.)

CASE VI.   Given two sides and the included angle

The triangle is solved by using the law of tangents to find the unknown angles and the law of sines to find the unknown side. The solution is checked by using one of the Mollweide formulas.

EXAMPLE 12.7   Let $c > b$ and A be given. Then, in Fig. 12-2

$\frac{1}{2}(C + B) = 90° - \frac{1}{2}A,$

$\tan \frac{1}{2}(C - B) = \dfrac{c - b}{c + b} \tan \frac{1}{2}(C + B), \quad a = \dfrac{c \sin A}{\sin C}.$

Check: $(c + b) \sin \frac{1}{2}A = a \cos \frac{1}{2}(C - B).$

(See Probs. 12.27 and 12.28.)

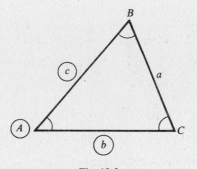

Fig. 12-2

12.7   HALF-ANGLE FORMULAS

In any triangle ABC, see Fig. 12-2,

$$\tan \frac{1}{2}A = \frac{r}{s - a} \qquad \tan \frac{1}{2}B = \frac{t}{s - b} \qquad \tan \frac{1}{2}C = \frac{r}{s - c}$$

where $s = \frac{1}{2}(a + b + c)$ is the semiperimeter of the triangle and $r = \sqrt{\dfrac{(s - a)(s - b)(s - c)}{s}}$.

For a proof of the formulas, see Prob. 12.30.

CASE V.   Given the three sides

The triangle is solved by using the half-angle formulas and is checked by using the angle relation.

(See Prob. 12.31.)

Solved Problems

12.1   Derive the law of sines.

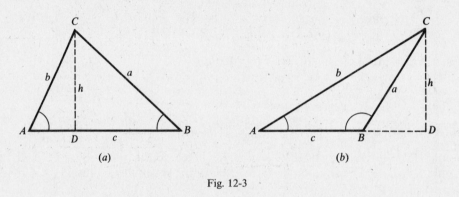

Fig. 12-3

Let ABC be any oblique triangle. In Fig. 12-3(a), angles A and B are acute while in Fig. 12-3(b), angle B is obtuse. Draw CD perpendicular to AB or AB extended and denote its length by h.

In the right triangle ACD of either figure, $h = b \sin A$ while in the right triangle BCD, $h = a \sin B$ since in Fig. 12-3(b), $h = a \sin \angle DBC = a \sin (180° - B) = a \sin B$. Thus,

$$a \sin B = b \sin A \qquad \text{or} \qquad \frac{a}{\sin A} = \frac{b}{\sin B}$$

In a similar manner (by drawing a perpendicular from B to AC or a perpendicular from A to BC), we obtain

$$\frac{a}{\sin A} = \frac{c}{\sin C} \qquad \text{or} \qquad \frac{b}{\sin B} = \frac{c}{\sin C}$$

Thus, finally,

$$\frac{a}{\sin A} = \frac{b}{\sin B} = \frac{c}{\sin C}$$

12.2   Discuss the several special cases when two sides and the angle opposite one of them are given.

Let b, c, and B be the given parts. Construct the given angle B and lay off the side $BA = c$. With A as center and radius equal to b (the side opposite the given angle), describe an arc. Figure 12-4(a) to (e) illustrates the special cases which may occur when the given angle B is acute while Fig. 12-4(f) and (g) illustrates the cases when B is obtuse.

The given angle B is acute.

Fig. 12-4(a). When $b < AD = c \sin B$, the arc does not meet BX and no triangle is determined.

Fig. 12-4(b). When $b = AD$, the arc is tangent to BX and one triangle—a right triangle with the right angle at C—is determined.

Fig. 12-4(c). When $b > AD$ and $b < c$, the arc meets BX in two points C and C' on the same side of B. Two triangles ABC, in which C is acute, and ABC', in which $C' = 180° - C$ is obtuse, are determined.

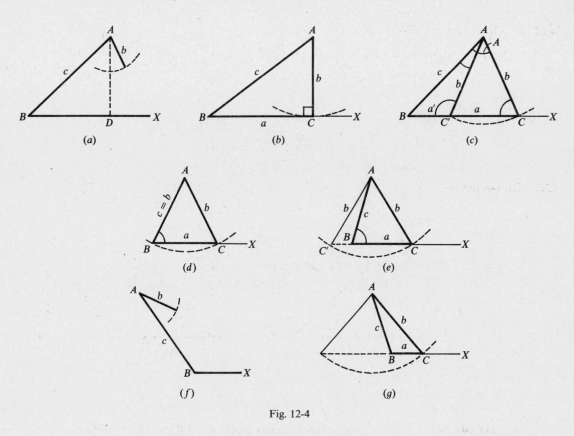

Fig. 12-4

Fig. 12-4(d). When $b > AD$ and $b = c$, the arc meets BX in C and B. One triangle (isosceles) is determined.

Fig. 12-4(e). When $b > c$, the arc meets BX in C and BX extended in C'. Since the triangle ABC' does not contain the given angle B, only one triangle ABC is determined.

The given angle is obtuse.

Fig. 12-4(f). When $b < c$ or $b = c$, no triangle is formed.

Fig. 12-4(g). When $b > c$, only one triangle is formed as in Fig. 12-4(e).

12.3 Derive the law of cosines.

In either right triangle ACD of Fig. 12-5, $b^2 = h^2 + (AD)^2$.

In the right triangle BCD of Fig. 12-5(a), $h = a \sin B$ and $DB = a \cos B$.

Then
$$AD = AB - DB = c - a \cos B$$

and
$$b^2 = h^2 + (AD)^2 = a^2 \sin^2 B + c^2 - 2ca \cos B + a^2 \cos^2 B$$
$$= a^2(\sin^2 B + \cos^2 B) + c^2 - 2ca \cos B = c^2 + a^2 - 2ca \cos B$$

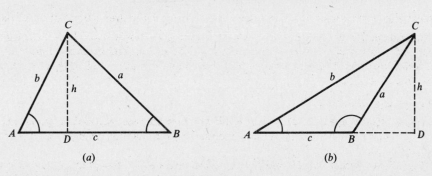

Fig. 12-5

In the right triangle BCD of Fig. 12-5(b),

$$h = a \sin \angle CBD = a \sin (180° - B) = a \sin B$$

and $$BD = a \cos \angle CBD = a \cos (180° - B) = -a \cos B$$

Then $$AD = AB + BD = c - a \cos B \qquad \text{and} \qquad b^2 = c^2 + a^2 - 2ca \cos B$$

The remaining equations may be obtained by cyclic changes of the letters.

12.4   Derive a pair of Mollweide's formulas. See Fig. 12-3.

By the law of sines, $\dfrac{a}{c} = \dfrac{\sin A}{\sin C}$ and $\dfrac{b}{c} = \dfrac{\sin B}{\sin C}$.

Then $\dfrac{a+b}{c} = \dfrac{\sin A + \sin B}{\sin C} = \dfrac{2 \sin \frac{1}{2}(A+B) \cos \frac{1}{2}(A-B)}{2 \sin \frac{1}{2}C \cos \frac{1}{2}C} = \dfrac{\cos \frac{1}{2}(A-B)}{\sin \frac{1}{2}C}$,

since $\sin \frac{1}{2}(A+B) = \sin \frac{1}{2}(180° - C) = \sin (90° - \frac{1}{2}C) = \cos \frac{1}{2}C$.

Similarly, $\dfrac{a-b}{c} = \dfrac{\sin A - \sin B}{\sin C} = \dfrac{2 \cos \frac{1}{2}(A+B) \sin \frac{1}{2}(A-B)}{2 \sin \frac{1}{2}C \cos \frac{1}{2}C} = \dfrac{\sin \frac{1}{2}(A-B)}{\cos \frac{1}{2}C}$,

since $\cos \frac{1}{2}(A+B) = \cos (90° - \frac{1}{2}C) = \sin \frac{1}{2}C$.

12.5   Derive one of the projection formulas.

Refer to Fig. 12-3. In the right triangle ACD of either figure, $AD = b \cos A$.
In the right triangle BCD of Fig. 12-3(a), $DB = a \cos B$. Thus, in Fig. 12-3(a),

$$c = AB = AD + DB = b \cos A + a \cos B = a \cos B + b \cos A$$

In the right triangle BCD of Fig. 12-3(b), $BD = a \cos \angle DBC = a \cos (180° - B) = -a \cos B$. Thus, in Fig. 12-3($b$),

$$c = AB = AD - BD = b \cos A - (-a \cos B) = a \cos B + b \cos A$$

(NOTE: See the chart in Sec. 4.7 for the rules guiding the number of significant figures in the sides and the accuracy to which angles are to be found.)

CASE I

12.6   Solve the triangle ABC, given $a = 62.5$, $A = 112°20'$, and $C = 42°10'$. See Fig. 12-6.

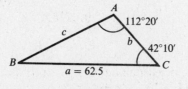

Fig. 12-6

For B: $B = 180° - (C + A) = 180° - 154°30' = 25°30'$.

For b: $b = \dfrac{a \sin B}{\sin A} = \dfrac{62.5 \sin 25°30'}{\sin 112°20'} = \dfrac{62.5(0.4305)}{0.9250} = 29.1$

($\sin 112°20' = \sin (180° - 112°20') = \sin 67°40'$)

For c: $c = \dfrac{a \sin C}{\sin A} = \dfrac{62.5 \sin 42°10'}{\sin 112°20'} = \dfrac{62.5(0.6713)}{0.9250} = 45.4$.

Check: $(c + b) \sin \frac{1}{2}A = a \cos \frac{1}{2}(C - B)$

$$(c + b) \sin \frac{1}{2}A = 74.5 \sin 56°10' = 74.5(0.8307) = 61.89$$

$$a \cos \frac{1}{2}(C - B) = 62.5 \cos 8°20' = 62.5(0.9894) = 61.84$$

or $a = b \cos C + c \cos B = 29.1(0.7412) + 45.4(0.9026) = 62.55$

The required parts are $b = 29.1$, $c = 45.4$, and $B = 25°30'$.

CASE II

12.7 Solve the triangle ABC, given $c = 25$, $A = 35°$, and $B = 68°$. See Fig. 12-7.

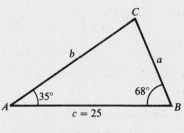

Fig. 12-7

For C: $C = 180° - (A + B) = 180° - 103° = 77°$.

For a: $a = \dfrac{c \sin A}{\sin C} = \dfrac{25 \sin 35°}{\sin 77°} = \dfrac{25(0.5736)}{0.9744} = 15$.

For b: $b = \dfrac{c \sin B}{\sin C} = \dfrac{25 \sin 68°}{\sin 77°} = \dfrac{25(0.9272)}{0.9744} = 24$.

Check by Mollweide's formula:

$$\frac{b + a}{c} = \frac{\cos \frac{1}{2}(B - A)}{\sin \frac{1}{2}C} \quad \text{or} \quad (b + a) \sin \frac{1}{2}C = c \cos \frac{1}{2}(B - A)$$

$$(b + a) \sin \frac{1}{2}C = 39 \sin 38°30' = 39(0.6225) = 24.3$$

$$c \cos \frac{1}{2}(B - A) = 25 \cos 16°30' = 25(0.9588) = 24.0$$

Check by projection formula: $c = a \cos B + b \cos A = 15 \cos 68° + 24 \cos 35°$
$$= 15(0.3746) + 24(0.8192) = 25.3$$

The required parts are $a = 15$, $b = 24$, and $C = 77°$.

12.8 A and B are two points on opposite banks of a river. From A a line $AC = 275$ m is laid off and the angles $CAB = 125°40'$ and $ACB = 48°50'$ are measured. Find the length of AB.

In the triangle ABC of Fig. 12-8(a), $B = 180° - (C + A) = 5°30'$ and

$$AB = c = \frac{b \sin C}{\sin B} = \frac{275 \sin 48°50'}{\sin 5°30'} = \frac{275(0.7528)}{0.0958} = 2160 \text{ m}$$

12.9 A ship is sailing due east when a light is observed bearing N62°10′E. After the ship has traveled 2250 m, the light bears N48°25′E. If the course is continued, how close will the ship approach the light? (See Prob. 5.5, Chap. 5.)

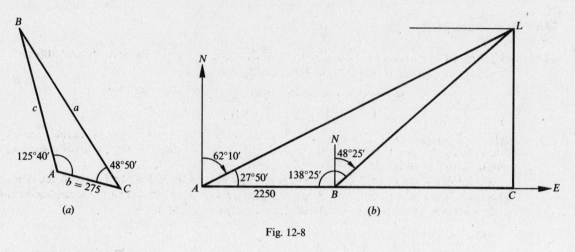

Fig. 12-8

Refer to Fig. 12-8(b).

In the oblique triangle ABL: $AB = 2250$, $\angle BAL = 27°50'$, and $\angle ABL = 138°25'$.

$$\angle ALB = 180° - (\angle BAL + \angle ABL) = 13°45'.$$

$$BL = \frac{AB \sin \angle BAL}{\sin \angle ALB} = \frac{2250 \sin 27°50'}{\sin 13°45'} = \frac{2250(0.4669)}{0.2377} = 4420.$$

In the right triangle BLC: $BL = 4420$ and $\angle CBL = 90° - 48°25' = 41°35'.$

$$CL = BL \sin \angle CBL = 4420 \sin 41°35' = 4420(0.6637) = 2934 \text{ m}.$$

For an alternative solution, find AL in the oblique triangle ABL and then CL in the right triangle ALC.

12.10 A tower 125 ft high is on a cliff on the bank of a river. From the top of the tower the angle of depression of a point on the opposite shore is $28°40'$ and from the base of the tower the angle of depression of the same point is $18°20'$. Find the width of the river and the height of the cliff.

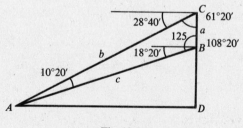

Fig. 12-9

In Fig. 12-9 BC represents the tower, DB represents the cliff, and A is the point on the opposite shore.

In triangle ABC: $\angle ACB = 90° - 28°40' = 61°20'.$
$\angle CBA = 90° + 18°20' = 108°20'.$
$\angle BAC = 180° - (\angle CBA + \angle ACB) = 10°20'.$

$$c = \frac{a \sin \angle ACB}{\sin \angle BAC} = \frac{125 \sin 61°20'}{\sin 10°20'} = \frac{125(0.8774)}{0.1794} = 611$$

In right triangle ABD:

$$DB = c \sin 18°20' = 611(0.3145) = 192$$
$$AD = c \cos 18°20' = 611(0.9492) = 580$$

The river is 580 ft wide and the cliff is 192 ft high.

CASE III

12.11 Solve the triangle ABC, given $c = 628$, $b = 480$, and $C = 55°10'$. Refer to Fig. 12-10(a).

Since C is acute and $c > b$, there is only one solution.

For B: $\sin B = \dfrac{b \sin C}{c} = \dfrac{480 \sin 55°10'}{628} = \dfrac{480(0.8208)}{628} = 0.6274$ and $B = 38°50'$.

(NOTE: If $\sin B = 0.6274$, then $B = 38°50'$ or $B' = 180° - 38°50' = 141°10'$ and each could be an angle in a triangle. Since $A + B + C = 180°$, it follows that $C + B < 180°$; thus $B' = 141°10'$ is not a solution in this problem because $C + B = 55°10' + 141°10' = 196°20' > 180°$.

Whenever $0 < \sin x < 1$, it is possible to find angles x in quadrants I and II that satisfy the value of $\sin x$ and could be angles in a triangle. The first-quadrant angle is always a solution but the second-quadrant angle is only a solution when its sum with the given angle is less than $180°$.)

For A: $A = 180° - (B + C) = 86°0'$.

For a: $a = \dfrac{b \sin A}{\sin B} = \dfrac{480 \sin 86°0'}{\sin 38°50'} = \dfrac{480(0.9976)}{0.6271} = 764$.

Check: $(a + b) \sin \tfrac{1}{2}C = c \cos \tfrac{1}{2}(A - B)$

$(a + b) \sin \tfrac{1}{2}C = 1244 \sin 27°35' = 1244(0.4630) = 576.0$

$c \cos \tfrac{1}{2}(A - B) = 628 \cos 23°35' = 628(0.9165) = 575.6$

If preferred, a projection formula may be used for the check.

The required parts are $B = 38°50'$, $A = 86°0'$, and $a = 764$.

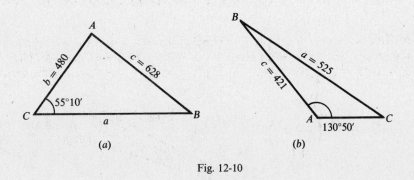

Fig. 12-10

12.12 Solve the triangle ABC, given $a = 525$, $c = 421$, and $A = 130°50'$. Refer to Fig. 12-10(b).

Since A is obtuse and $a > c$, there is one solution.

For C: $\sin C = \dfrac{c \sin A}{a} = \dfrac{421 \sin 130°50'}{525} = \dfrac{421(0.7566)}{525} = 0.6067$ and $C = 37°20'$.

For B: $B = 180° - (C + A) = 11°50'$.

For b: $b = \dfrac{a \sin B}{\sin A} = \dfrac{525 \sin 11°50'}{\sin 130°50'} = \dfrac{525(0.2051)}{0.7566} = 142$.

Check: $(c + b) \sin \tfrac{1}{2}A = a \cos \tfrac{1}{2}(C - B)$

$(c + b) \sin \tfrac{1}{2}A = 563 \sin 65°25' = 563(0.9094) = 512.0$

$a \cos \tfrac{1}{2}(C - B) = 525 \cos 12°45' = 525(0.9754) = 512.1$

The required parts are $C = 37°20'$, $B = 11°50'$, and $b = 142$.

12.13 Solve the triangle ABC, given $a = 31.5$, $b = 51.8$, and $A = 33°40'$. Refer to Fig. 12-11(a).

Since A is acute and $a < b$, there is the possibility of two solutions.

For B: $\sin B = \dfrac{b \sin A}{a} = \dfrac{51.8 \sin 33°40'}{31.5} = \dfrac{51.8(0.5544)}{31.5} = 0.9117.$

There are two solutions, $B = 65°40'$ and $B' = 180° - 65°40' = 114°20'$.

For C: $C = 180° - (A + B) = 80°40'.$ For C': $C' = 180° - (A + B') = 32°0'.$

For c: $c = \dfrac{a \sin C}{\sin A} = \dfrac{31.5 \sin 80°40'}{\sin 33°40'}$ For c': $c' = \dfrac{a \sin C'}{\sin A} = \dfrac{31.5 \sin 32°0'}{\sin 33°40'}$

          $= \dfrac{31.5(0.9868)}{0.5544} = 56.1.$ $= \dfrac{31.5(0.5299)}{0.5544} = 30.1.$

Check: $(c + b) \sin \tfrac{1}{2}A = a \cos \tfrac{1}{2}(C - B)$ Check: $(b + c') \sin \tfrac{1}{2}A = a \cos \tfrac{1}{2}(B' - C')$

        $(c + b) \sin \tfrac{1}{2}A = 107.9 \sin 16°50'$ $(b + c') \sin \tfrac{1}{2}A = 81.9 \sin 16°50'$

                $= 107.9(0.2896)$ $= 81.9(0.2896)$

                $= 31.25$ $= 23.72$

      $a \cos \tfrac{1}{2}(C - B) = 31.5 \cos 7°30'$ $a \cos \tfrac{1}{2}(B' - C') = 31.5 \cos 41°10'$

                $= 31.5(0.9914)$ $= 31.5(0.7528)$

                $= 31.23$ $= 23.71$

The required parts are

        for triangle ABC: $B = 65°40'$, $C = 80°40'$, and $c = 56.1$,

and   for triangle ABC': $B' = 114°20'$, $C' = 32°0'$, and $c' = 30.1$.

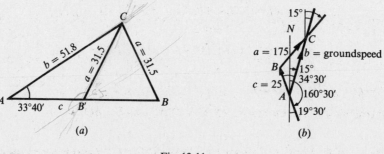

Fig. 12-11

12.14 A pilot wishes a course $15°0'$ against a wind of 25 mi/h from $160°30'$. Find his required heading and the groundspeed when the airspeed is 175 mi/h. Refer to Fig. 12-11(b).

Since $\angle BAC$ is acute and $a > c$, there is one solution.

$$\sin C = \frac{c \sin \angle BAC}{a} = \frac{25 \sin 34°30'}{175} = \frac{25(0.5664)}{175} = 0.0809 \text{ and } \angle ACB = 4°40'.$$

$$B = 180° - (\angle BAC + \angle ACB) = 140°50'. \quad b = \frac{a \sin B}{\sin \angle BAC} = \frac{175 \sin 140°50'}{\sin 34°30'} = \frac{175(0.6316)}{0.5664} = 195.$$

The groundspeed is 195 mi/h and the required heading is $19°40'$.

CASE IV

12.15 Solve the triangle ABC, given $a = 132$, $b = 224$, and $C = 28°40'$. Refer to Fig. 12-12(a).

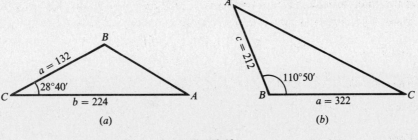

Fig. 12-12

For c: $c^2 = a^2 + b^2 - 2ab \cos C$
$$= (132)^2 + (224)^2 - 2(132)(224) \cos 28°40'$$
$$= (132)^2 + (224)^2 - 2(132)(224)(0.8774) = 15{,}714 \quad \text{and} \quad c = 125.$$

For A: $\sin A = \dfrac{a \sin C}{c} = \dfrac{132 \sin 28°40'}{125} = \dfrac{132(0.4797)}{125} = 0.5066$ and $A = 30°30'$.

For B: $\sin B = \dfrac{b \sin C}{c} = \dfrac{224 \sin 28°40'}{125} = \dfrac{224(0.4797)}{125} = 0.8596$ and $B = 120°40'$.

(Since $b > a$, A is acute; since $A + C < 90°$, $B > 90°$.)

Check: $A + B + C = 179°50'$. The required parts are $A = 30°30'$, $B = 120°40'$, and $c = 125$.

12.16 Solve the triangle ABC, given $a = 322$, $c = 212$, and $B = 110°50'$. Refer to Fig. 12-12(b).

For b: $b^2 = c^2 + a^2 - 2ca \cos B$ [$\cos 110°50' = -\cos(180° - 110°50') = -\cos 69°10'$.]
$$= (212)^2 + (322)^2 - 2(212)(322)(-0.3557) = 197{,}191 \quad \text{and} \quad b = 444.$$

For A: $\sin A = \dfrac{a \sin B}{b} = \dfrac{322 \sin 110°50'}{444} = \dfrac{322(0.9346)}{444} = 0.6778$ and $A = 42°40'$.

For C: $\sin C = \dfrac{c \sin B}{b} = \dfrac{212 \sin 110°50'}{444} = \dfrac{212(0.9346)}{444} = 0.4463$ and $C = 26°30'$.

Check: $A + B + C = 180°$.

The required parts are $A = 42°40'$, $C = 26°30'$, and $b = 444$.

12.17 Two forces of 17.5 and 22.5 lb act on a body. If their directions make an angle of 50°10' with each other, find the magnitude of their resultant and the angle that it makes with the larger force.

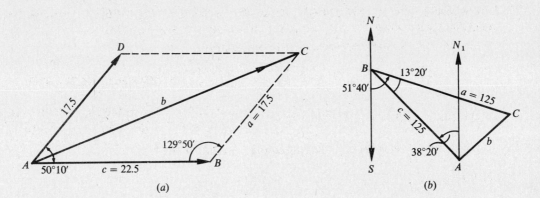

Fig. 12-13

Refer to Fig. 12-13(a).

In the parallelogram $ABCD$, $\angle DAC + \angle B = \angle BCD + \angle D = 180°$ and $B = 180° - 50°10' = 129°50'$.

In the triangle ABC,

$$b^2 = c^2 + a^2 - 2ca \cos B \qquad [\cos 129°50' = -\cos(180° - 129°50') = -\cos 50°10'.]$$

$$= (22.5)^2 + (17.5)^2 - 2(22.5)(17.5)(-0.6406) = 1317 \quad \text{and} \quad b = 36.3.$$

$$\sin \angle BAC = \frac{a \sin B}{b} = \frac{17.5 \sin 129°50'}{36.3} = \frac{17.5(0.7679)}{36.3} = 0.3702 \quad \text{and} \quad \angle BAC = 21°40'.$$

The resultant is a force of 36.3 lb and the required angle is 21°40'.

12.18 From A a pilot flies 125 km in the direction N38°20'W and turns back. Through an error, the pilot then flies 125 km in the direction S51°40'E. How far and in what direction must the pilot now fly to reach the intended destination A?

Refer to Fig. 12-13(b).

Denote the turn back point as B and the final position as C.

In the triangle ABC,

$$b^2 = c^2 + a^2 - 2ca \cos \angle ABC$$

$$= (125)^2 + (125)^2 - 2(125)(125) \cos 13°20'$$

$$= 2(125)^2(1 - 0.9730) = 843.7 \quad \text{and} \quad b = 29.0$$

$$\sin \angle BAC = \frac{a \sin \angle ABC}{b} = \frac{125 \sin 13°20'}{29.0} = \frac{125(0.2306)}{29.0} = 0.9940 \quad \text{and} \quad \angle BAC = 83°40'.$$

Since $\angle CAN_1 = \angle BAC - \angle N_1AB = 45°20'$, the pilot must fly a course S45°20'W for 29.0 km in going from C to A.

CASE V

12.19 Solve the triangle ABC, given $a = 25.2$, $b = 37.8$, and $c = 43.4$. Refer to Fig. 12-14(a).

For A: $\cos A = \dfrac{b^2 + c^2 - a^2}{2bc} = \dfrac{(37.8)^2 + (43.4)^2 - (25.2)^2}{2(37.8)(43.4)} = 0.8160$ and $A = 35°20'$.

For B: $\cos B = \dfrac{c^2 + a^2 - b^2}{2ca} = \dfrac{(43.4)^2 + (25.2)^2 - (37.8)^2}{2(43.4)(25.2)} = 0.4982$ and $B = 60°10'$.

For C: $\cos C = \dfrac{a^2 + b^2 - c^2}{2ab} = \dfrac{(25.2)^2 + (37.8)^2 - (43.4)^2}{2(25.2)(37.8)} = 0.0947$ and $C = 84°30'$.

Check: $A + B + C = 180°$.

12.20 Solve the triangle ABC, given $a = 30.3$, $b = 40.4$, and $c = 62.6$. Refer to Fig. 12-14(b):

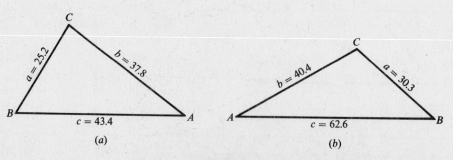

(a)   (b)

Fig. 12-14

For A: $\cos A = \dfrac{b^2 + c^2 - a^2}{2bc} = \dfrac{(40.4)^2 + (62.6)^2 - (30.3)^2}{2(40.4)(62.6)} = 0.9159$ and $A = 23°40'$.

For B: $\cos B = \dfrac{c^2 + a^2 - b^2}{2ca} = \dfrac{(62.6)^2 + (30.3)^2 - (40.4)^2}{2(62.6)(30.3)} = 0.8448$ and $B = 32°20'$.

For C: $\cos C = \dfrac{a^2 + b^2 - c^2}{2ab} = \dfrac{(30.3)^2 + (40.4)^2 - (62.6)^2}{2(30.3)(40.4)} = -0.5590$ and $C = 124°0'$.

Check: $A + B + C = 180°$.

12.21 The distances of a point C from two points A and B, which cannot be measured directly, are required. The line CA is continued through A for a distance 175 m to D, the line CB is continued through B for 225 m to E, and the distances $AB = 300$ m, $DB = 326$ m, and $DE = 488$ m are measured. Find AC and BC. See Fig. 12-15.

Triangle ABC may be solved for the required parts after the angles $\angle BAC$ and $\angle ABC$ have been found. The first angle is the supplement of $\angle BAD$ and the second is the supplement of the sum of $\angle ABD$ and $\angle DBE$.

In the triangle ABD whose sides are known,

$$\cos \angle BAD = \frac{(175)^2 + (300)^2 - (326)^2}{2(175)(300)} = 0.1367$$

and $\angle BAD = 82°10'$

and $\cos \angle ABD = \dfrac{(300)^2 + (326)^2 - (175)^2}{2(300)(326)} = 0.8469$

and $\angle ABD = 32°10'$.

In the triangle BDE whose sides are known,

$$\cos \angle DBE = \frac{(225)^2 + (326)^2 - (488)^2}{2(225)(326)} = -0.5538 \quad \text{and} \quad \angle DBE = 123°40'.$$

In the triangle ABC: $AB = 300$, $\angle BAC = 180° - \angle BAD = 97°50'$,
$\angle ABC = 180° - (\angle ABD + \angle DBE) = 24°10'$,
and $\angle ACB = 180° - (\angle BAC + \angle ABC) = 58°0'$.

Then $AC = \dfrac{AB \sin \angle ABC}{\sin \angle ACB} = \dfrac{300 \sin 24°10'}{\sin 58°10'} = \dfrac{300(0.4094)}{0.8480} = 145$

and $BC = \dfrac{AB \sin \angle BAC}{\sin \angle ACB} = \dfrac{300 \sin 97°50'}{\sin 58°0'} = \dfrac{300(0.9907)}{0.8480} = 350$.

The required distances are $AC = 145$ m and $BC = 350$ m.

(NOTE: Logarithmic solution of oblique triangles is the focus of Probs. 12.22 to 12.31. If you are not using logarithms go directly to the Supplementary Problems.)

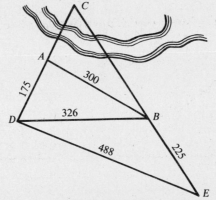

Fig. 12-15

CASE II

12.22 Solve the triangle ABC, given $a = 38.12$, $A = 46°32'$, and $C = 79°17'$. See Fig. 12-16.

$B = 180° - (A + C) = 54°11'$.

$$c = \frac{a \sin C}{\sin A} \qquad\qquad b = \frac{a \sin B}{\sin A}$$

$\log a = 1.5812$	$\log a = 1.5812$
$\log \sin C = 9.9924 - 10$	$\log \sin B = 9.9090 - 10$
$\operatorname{colog} \sin A = 0.1392$	$\operatorname{colog} \sin A = 0.1392$
$\log c = 1.7128$	$\log b = 1.6294$
$c = 51.62$	$b = 42.60$

Check: $(c + b) \sin \tfrac{1}{2}A = a \cos \tfrac{1}{2}(C - B)$

$c + b = 94.22$, $\tfrac{1}{2}A = 23°16'$ $a = 38.12$, $\tfrac{1}{2}(C - B) = 12°33'$

$\log (c + b) = 1.9741$	$\log a = 1.5812$
$\log \sin \tfrac{1}{2}A = 9.5966 - 10$	$\log \cos \tfrac{1}{2}(C - B) = 9.9895 - 10$
1.5707	1.5707

[NOTE: $\operatorname{colog} \sin A = \log (1/\sin A) = -\log \sin A$.]

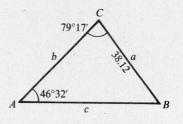

Fig. 12-16

12.23 Solve the triangle ABC, given $b = 282.7$, $A = 111°43'$, and $C = 24°26'$. See Fig. 12-17.

$B = 180° - (C + A)43°51'$.

$$a = \frac{b \sin A}{\sin B} \qquad\qquad c = \frac{b \sin C}{\sin B}$$

$\log b = 2.4513$	$\log b = 2.4513$
$\log \sin A = 9.9680 - 10$	$\log \sin C = 9.6166 - 10$
$\operatorname{colog} \sin B = 0.1594$	$\operatorname{colog} \sin B = 0.1594$
$\log a = 2.5787$	$\log c = 2.2273$
$a = 379.1$	$c = 168.8$

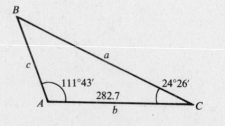

Fig. 12-17

Check: $$(a + c) \sin \tfrac{1}{2}B = b \cos \tfrac{1}{2}(A - C)$$

$$a + c = 547.9, \tfrac{1}{2}B = 21°56' \qquad b = 282.7, \tfrac{1}{2}(A - C) = 43°38'$$

$\log (a + c) = 2.7387$	$\log b = 2.4513$
$\log \sin \tfrac{1}{2}B = 9.5723 - 10$	$\log \cos \tfrac{1}{2}(A - C) = 9.8596 - 10$
$\overline{2.3111}$	$\overline{2.3109}$

CASE III

12.24 Solve the triangle ABC, given $b = 67.25$, $c = 56.92$, and $B = 65°16'$. See Fig. 12-18.

Since B is acute and $b > c$, there is one solution.

$$\sin C = \frac{c \sin B}{b} \qquad\qquad\qquad a = \frac{b \sin A}{\sin B}$$

$\log c = 1.7553$	$\log b = 1.8277$
$\log \sin B = 9.9582 - 10$	$\log \sin A = 9.9554 - 10$
$\text{colog } b = 8.1723 - 10$	$\text{colog } \sin B = 0.0418$
$\log \sin C = 9.8858 - 10$	$\log a = 1.8249$
$C = 50°15'$	$a = 66.82$
$A = 180° - (B + C) = 64°29'$	

Check: $$(b + c) \sin \tfrac{1}{2}A = a \cos \tfrac{1}{2}(B - C)$$

$$b + c = 124.17, \tfrac{1}{2}A = 32°15' \qquad a = 66.82, \tfrac{1}{2}(B - C) = 7°31'$$

$\log (b + c) = 2.0940$	$\log a = 1.8249$
$\log \sin \tfrac{1}{2}A = 9.7272 - 10$	$\log \cos \tfrac{1}{2}(B - C) = 9.9963 - 10$
$\overline{1.8212}$	$\overline{1.8212}$

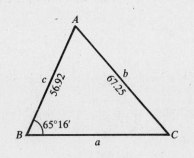

Fig. 12-18

12.25 Solve the triangle ABC, given $a = 123.2$, $b = 155.4$, $A = 16°34'$. See Fig. 12-19.

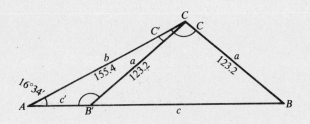

Fig. 12-19

Since A is acute and $a < b$, there may be two solutions.

$$\sin B = \frac{b \sin A}{a}$$

$$\log b = 2.1914$$
$$\log \sin A = 9.4550 - 10$$
$$\text{colog } a = 7.9094 - 10$$
$$\overline{\log \sin B = 9.5558 - 10}$$

$B = 21°4'$	$B' = 180° - B = 158°56'$
$C = 180° - (A + B) = 142°22'$	$C' = 180° - (A + B') = 4°30'$

$$c = \frac{a \sin C}{\sin A} \qquad\qquad c' = \frac{a \sin C'}{\sin A}$$

$\log a = 2.0906$	$\log a = 2.0906$
$\log \sin C = 9.7858 - 10$	$\log \sin C' = 8.8946 - 10$
$\text{colog } \sin A = 0.5450$	$\text{colog } \sin A = 0.5450$
$\overline{\log c = 2.4214}$	$\overline{\log c' = 1.5302}$
$c = 263.9$	$c' = 33.90$

Check: $(b + a)\sin \tfrac{1}{2}C = c \cos \tfrac{1}{2}(B - A)$ Check: $(b + c)\sin \tfrac{1}{2}C' = c' \cos \tfrac{1}{2}(B - A)$

$b + a = 278.6,\ \tfrac{1}{2}C = 71°11'$	$b + a = 278.6,\ \tfrac{1}{2}C' = 2°15'$
$c = 263.9,\ \tfrac{1}{2}(B - A) = 2°15'$	$c' = 33.9°,\ \tfrac{1}{2}(B' - A) = 71°11'$

$\log (b + a) = 2.4449$	$\log (b + a) = 2.4449$
$\log \sin \tfrac{1}{2}C = 9.9761 - 10$	$\log \sin \tfrac{1}{2}C' = 8.5939 - 10$
$\overline{2.4210}$	$\overline{1.0388}$

$\log c = 2.4214$	$\log c' = 1.5302$
$\log \cos \tfrac{1}{2}(B - A) = 9.9997 - 10$	$\log \cos \tfrac{1}{2}(B' - A) = 9.5086 - 10$
$\overline{2.4211}$	$\overline{1.0388}$

12.26   Derive the law of tangents.

In any triangle ABC, we have the Mollweide formulas

$$\frac{a - b}{c} = \frac{\sin \tfrac{1}{2}(A - B)}{\cos \tfrac{1}{2}C} \quad \text{and} \quad \frac{a + b}{c} = \frac{\cos \tfrac{1}{2}(A - B)}{\sin \tfrac{1}{2}C}$$

Dividing the first by the second,

$$\frac{a - b}{c} \cdot \frac{c}{a + b} = \frac{\sin \tfrac{1}{2}(A - B)}{\cos \tfrac{1}{2}C} \cdot \frac{\sin \tfrac{1}{2}C}{\cos \tfrac{1}{2}(A - B)} = \tan \tfrac{1}{2}(A - B) \cdot \tan \tfrac{1}{2}C$$

Since $C = 180° - (A + B)$, $\tfrac{1}{2}C = 90° - \tfrac{1}{2}(A + B)$ and $\tan \tfrac{1}{2}C = \cot \tfrac{1}{2}(A + B) = \dfrac{1}{\tan \tfrac{1}{2}(A + B)}$.

Thus, $$\frac{a - b}{a + b} = \tan \tfrac{1}{2}(A - B) \cdot \tan \tfrac{1}{2}C = \frac{\tan \tfrac{1}{2}(A - B)}{\tan \tfrac{1}{2}(A + B)}$$

The two other forms may be obtained in a similar manner or by cyclic changes of letters on the above form.

CASE IV

12.27   Solve the triangle ABC, given $a = 2526$, $c = 1388$, $B = 54°24'$. See Fig. 12-20.

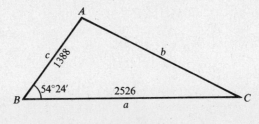

Fig. 12-20

$$A + C = 180° - B = 125°36'$$
$$\tfrac{1}{2}(A + C) = 62°48'$$

$$a = 2526$$
$$c = 1388$$
$$\overline{}$$
$$a - c = 1138$$
$$a + c = 3914$$

$$\tan \tfrac{1}{2}(A - C) = \frac{a - c}{a + c} \tan \tfrac{1}{2}(A + C) \qquad\qquad b = \frac{c \sin B}{\sin C}$$

$\log (a - c) = 3.0561$	$\log c = 3.1424$
$\operatorname{colog} (a + c) = 6.4074 - 10$	$\log \sin B = 9.9101 - 10$
$\log \tan \tfrac{1}{2}(A + C) = 0.2891$	$\operatorname{colog} \sin C = 0.2604$
$\log \tan \tfrac{1}{2}(A - C) = 9.7526 - 10$	$\log b = 3.3129$
$\tfrac{1}{2}(A - C) = 29°30'$	$b = 2055$
$\tfrac{1}{2}(A + C) = 62°48'$	
$A = 92°18'$	
$C = 33°18'$	

Check: It is left to the student to check the solution by using the Mollweide formula

$$(a + c) \sin \tfrac{1}{2}B = b \cos \tfrac{1}{2}(A - C)$$

12.28 Solve the triangle ABC, given $b = 472.1$, $c = 607.4$, $A = 125°14'$. See Fig. 12-21.

$$C + B = 180° - A = 54°46'$$
$$\tfrac{1}{2}(C + B) = 27°23'$$

$$c = 607.4$$
$$b = 472.1$$
$$\overline{}$$
$$c - b = 135.3$$
$$c + b = 1079.5 = 1080$$

$$\tan \tfrac{1}{2}(C - B) = \frac{c - b}{c + b} \tan \tfrac{1}{2}(C + B) \qquad\qquad a = \frac{b \sin A}{\sin B}$$

$\log (c - b) = 2.1313$	$\log b = 2.6740$
$\operatorname{colog} (c + b) = 6.9666 - 10$	$\log \sin A = 9.9121 - 10$
$\log \tan \tfrac{1}{2}(C + B) = 9.7143 - 10$	$\operatorname{colog} \sin B = 0.3964$
$\log \tan \tfrac{1}{2}(C - B) = 8.8122 - 10$	$\log a = 2.9825$
$\tfrac{1}{2}(C - B) = 3°43'$	$s = 960.5$
$\tfrac{1}{2}(C + B) = 27°23'$	
$C = 31°6'$	
$B = 23°40'$	

Check: To check the solution use the Mollweide formula $(c + b) \sin \tfrac{1}{2}A = a \cos \tfrac{1}{2}(C - B)$.

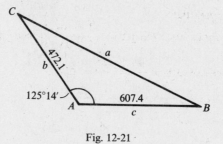

Fig. 12-21

12.29 Two adjacent sides of a parallelogram are 3473 and 4822 ft respectively and the angle between them is $72°14'$. Find the length of the longer diagonal. See Fig. 12-22.

In triangle ABC: $B = 180° - 72°14' = 107°46'$

$$\angle ACB + \angle CAB = 72°14' \text{ and } \tfrac{1}{2}(\angle ACB + \angle CAB) = 36°7'$$

$$
\begin{aligned}
c &= 4822 \\
a &= 3473 \\
\hline
c - a &= 1349 \\
c + a &= 8295
\end{aligned}
$$

$$\tan \tfrac{1}{2}(\angle ACB - \angle CAB) = \frac{c-a}{c+a} \tan \tfrac{1}{2}(\angle ACB + \angle CAB) \qquad\qquad b = \frac{c \sin B}{\sin \angle ACB}$$

$$
\begin{aligned}
\log (c - a) &= 3.1300 \\
\operatorname{colog} (c + a) &= 6.0812 - 10 \\
\log \tan \tfrac{1}{2}(\angle ACB + \angle CAB) &= 9.8631 - 10 \\
\hline
\log \tan \tfrac{1}{2}(\angle ACB - \angle CAB) &= 9.07460 - 10 \\
\tfrac{1}{2}(\angle ACB - \angle CAB) &= \ 6°46' \\
\tfrac{1}{2}(\angle ACB - \angle CAB) &= 36°7' \\
\hline
\angle ACB &= 42°53' \\
\angle CAB &= 29°21'
\end{aligned}
$$

$$
\begin{aligned}
\log c &= 3.6832 \\
\log \sin B &= 9.9788 - 10 \\
\operatorname{colog} \sin \angle ACB &= 0.1671 \\
\hline
\log b &= 3.8291 \\
b &= 6747 \text{ ft}
\end{aligned}
$$

Check: $$(c + a) \sin \tfrac{1}{2}B = b \cos \tfrac{1}{2}(\angle ACB - \angle CAB)$$

$$
\begin{aligned}
\log (c + a) &= 3.9188 \\
\log \sin \tfrac{1}{2}B &= 9.9073 - 10 \\
\hline
& 3.8261
\end{aligned}
\qquad\qquad
\begin{aligned}
\log b &= 3.8291 \\
\log \cos \tfrac{1}{2}(\angle ACB - \angle CAB) &= 9.9970 - 10 \\
\hline
& 3.8261
\end{aligned}
$$

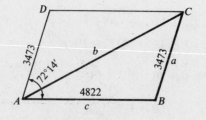

Fig. 12-22

12.30 Derive the half-angle formulas.

Let ABC be any triangle. Then $\tan \frac{1}{2}A = \sqrt{\dfrac{1 - \cos A}{1 + \cos A}}$ since $\frac{1}{2}A$ is always acute.

By the law of cosines, $\cos A = \dfrac{b^2 + c^2 - a^2}{2bc}$ so that

$$1 - \cos A = 1 - \frac{b^2 + c^2 - a^2}{2bc} = \frac{2bc - b^2 - c^2 + a^2}{2bc} = \frac{a^2 - (b - c)^2}{2bc} = \frac{(a - b + c)(a + b - c)}{2bc}$$

and $1 + \cos A = 1 + \dfrac{b^2 + c^2 - a^2}{2bc} = \dfrac{2bc + b^2 + c^2 - a^2}{2bc} = \dfrac{(b + c)^2 - a^2}{2bc} = \dfrac{(b + c + a)(b + c - a)}{2bc}$

Let $a + b + c = 2s$; then $a - b + c = (a + b + c) - 2b = 2s - 2b = 2(s - b)$, $a + b - c = 2(s - c)$, $b + c - a = 2(s - a)$, and

$$\tan \tfrac{1}{2}A = \sqrt{\frac{1 - \cos A}{1 + \cos A}} = \sqrt{\frac{(a - b + c)(a + b - c)}{2bc} \cdot \frac{2bc}{(b + c + a)(b + c - a)}} = \sqrt{\frac{2(s - b) \cdot 2(s - c)}{2s \cdot 2(s - a)}}$$

$$= \sqrt{\frac{(s - b)(s - c)}{s(s - a)}} = \sqrt{\frac{(s - a)(s - b)(s - c)}{s(s - a)^2}} = \frac{1}{s - a}\sqrt{\frac{(s - a)(s - b)(s - c)}{s}}$$

Finally, setting $r = \sqrt{\dfrac{(s - a)(s - b)(s - c)}{s}}$, $\tan \frac{1}{2}A = \dfrac{r}{s - a}$. The remaining formulas may be obtained by cyclic changes of letters.

CASE V

12.31 Solve the triangle ABC, given $a = 643.8$, $b = 778.7$, and $c = 912.3$. See Fig. 12-23.

$$s = \tfrac{1}{2}(a + b + c) \qquad\qquad r = \sqrt{\frac{(s - a)(s - b)(s - c)}{s}}$$

$a = 643.8$	$s - a = 523.6$	$\log (s - a) = 2.7190$
$b = 778.7$	$s - b = 388.7$	$\log (s - b) = 2.5896$
$c = 912.3$	$s - c = 255.1$	$\log (s - c) = 2.4067$
$2s = 2334.8$	$s = 1167.4$	$\text{colog } s = 6.9328 - 10$
$s = 1167.4$		$2 \log r = 4.6485$
		$\log r = 2.32425 = 2.3242$

$$\tan \tfrac{1}{2}A = \frac{r}{s - a} \qquad\qquad \tan \tfrac{1}{2}B = \frac{r}{s - b} \qquad\qquad \tan \tfrac{1}{2}C = \frac{r}{s - c}$$

$\log r = 2.3242$	$\log r = 2.3242$	$\log r = 2.3242$
$\log (s - a) = 2.7190$	$\log (s - b) = 2.5896$	$\log (s - c) = 2.4067$
$\log \tan \tfrac{1}{2}A = 9.6052 - 10$	$\log \tan \tfrac{1}{2}B = 9.7346 - 10$	$\log \tan \tfrac{1}{2}C = 9.9175 - 10$
$\tfrac{1}{2}A = 21°57'$	$\tfrac{1}{2}B = 28°29'$	$\tfrac{1}{2}C = 39°35'$
$A = 43°54'$	$B = 56°58'$	$C = 79°10'$

Check: $A + B + C = 180°2'$

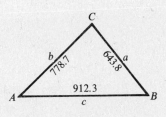

Fig. 12-23

Supplementary Problems

12.32 Consider the given parts of triangle ABC. State whether the law of sines or cosines should be used to solve for the required part and then find its value.

     (a) $a = 17$, $c = 14$, and $B = 30°$; find b.   *Ans.*   cosines, 8.5

     (b) $b = 17$, $a = 12$, and $A = 24°$; find B.   *Ans.*   sines, 35° and 145°

     (c) $c = 189$, $a = 150$, and $C = 85.18''$; find A.   *Ans.*   sines, 52°17'

     (d) $A = 24°18'$, $B = 56°48'$, and $a = 32.3$; find b.   *Ans.*   sines, 65.7

     (e) $c = 0.5$, $b = 0.8$, and $A = 70°$; find a.   *Ans.*   log cosines, 0.79

     (f) $a = 315.2$, $b = 457.8$, and $A = 42.45°$; find B.   *Ans.*   sines, 78.61° and 101.39°

     (g) $a = 25.7$, $b = 38.7$, and $C = 10.8°$; find c.   *Ans.*   cosines, 14.3

     (h) $a = 7.6$, $b = 4.8$, and $c = 7.1$; find B.   *Ans.*   cosines, 38°

Solve each of the following oblique triangles ABC, given:

12.33 $a = 125$, $A = 54°40'$, $B = 65°10'$ *Ans.* $b = 139$, $c = 133$, $C = 60°10'$

12.34 $b = 321$, $A = 75°20'$, $C = 38°30'$ *Ans.* $a = 339$, $c = 218$, $B = 66°10'$

12.35 $b = 215$, $c = 150$, $B = 42°40'$ *Ans.* $a = 300$, $A = 109°10'$, $C = 28°10'$

12.36 $a = 512$, $b = 426$, $A = 48°50'$ *Ans.* $c = 680$, $B = 38°50'$, $C = 92°20'$

12.37 $b = 50.4$, $c = 33.3$, $B = 118°30'$ *Ans.* $a = 25.1$, $A = 26°0'$, $C = 35°30'$

12.38 $b = 40.2$, $a = 31.5$, $B = 112°20'$ *Ans.* $c = 15.7$, $A = 46°30'$, $C = 21°10'$

12.39 $b = 51.5$, $a = 62.5$, $B = 40°40'$ *Ans.* $c = 78.9$, $A = 52°20'$, $C = 87°0'$
                  $c' = 16.0$, $A' = 127°40'$, $C' = 11°40'$

12.40 $a = 320$, $c = 475$, $A = 35°20'$ *Ans.* $b = 552$, $B = 85°30'$, $C = 59°10'$
                  $b' = 224$, $B' = 23°50'$, $C' = 120°50'$

12.41 $b = 120$, $c = 270$, $A = 118°40'$ *Ans.* $a = 344$, $B = 17°50'$, $C = 43°30'$

12.42 $a = 24.5$, $b = 18.6$, $c = 26.4$ *Ans.* $A = 63°10'$, $B = 42°40'$, $C = 74°10'$

12.43 $a = 6.34$, $b = 7.30$, $c = 9.98$ *Ans.* $A = 39°20'$, $B = 46°50'$, $C = 93°50'$

12.44 Two ships have radio equipment with a range of 200 km. One is 155 km N42°40'E and the other is 165 km N45°10'W of a shore station. Can the two ships communicate directly?

    Ans.   No; they are 222 km apart.

12.45 A ship sails 15.0 mi on a course S40°10'W and then 21.0 mi on a course N28°20'W. Find the distance and direction of the last position from the first.

    Ans.   20.9 mi, N70°30'W

12.46 A lighthouse is 10 km northwest of a dock. A ship leaves the dock at 9 A.M. and steams west at 12 km/h. At what time will it be 8 km from the lighthouse?

    Ans.   9:17 A.M. and 9:54 A.M.

12.47 Two forces of 115 and 215 lb acting on an object have a resultant of magnitude 275 lb. Find the angle between the directions in which the given forces act.

Ans. 70°50′

12.48 A tower 150 m high is situated at the top of a hill. At a point 650 m down the hill the angle between the surface of the hill and the line of sight to the top of the tower is 12°30′. Find the inclination of the hill to a horizontal plane.

Ans. 7°50′

12.49 Three circles of radii 115, 150, and 225 m respectively are tangent to each other externally. Find the angles of the triangle formed by joining the centers of the circles.

Ans. 43°10′, 61°20′, 75°30′

Use logarithms to solve each of the oblique triangles *ABC*, given:

12.50 $c = 78.75, A = 33°10′, C = 81°25′$ *Ans.* $a = 43.57, b = 72.43, B = 65°25′$

12.51 $b = 730.8, B = 42°13′, C = 109°33′$ *Ans.* $a = 514.5, c = 1025, A = 28°14′$

12.52 $a = 31.26, A = 58°, C = 23°37′$ *Ans.* $b = 36.47, c = 14.77, B = 98°23′$

12.53 $b = 13.22, c = 10.00, B = 25°57′$ *Ans.* $a = 21.47, A = 134°43′, C = 19°20′$

12.54 $b = 10.88, c = 35.73, C = 115°34′$ *Ans.* $a = 29.66, A = 48°29′, B = 15°57′$

12.55 $b = 86.43, c = 73.46, C = 49°19′$ *Ans.* $a = 89.52, B = 63°10′, A = 67°31′$
$a' = 23.19, B' = 116°50′, A' = 13°51′$

12.56 $a = 12.70, c = 15.87, A = 24°7′$ *Ans.* $b = 25.40, B = 125°11′, C = 30°42′$
$b' = 3.56, B' = 6°35′, C' = 149°18′$

12.57 $a = 482.3, c = 395.7, B = 137°32′$ *Ans.* $b = 819.2, A = 23°26′, C = 19°2′$

12.58 $b = 561.2, c = 387.2, A = 56°44′$ *Ans.* $a = 475.9, B = 80°24′, C = 42°52′$

12.59 $a = 123.8, b = 264.2, c = 256.0$ *Ans.* $A = 27°28′, B = 79°56′, C = 72°34′$

12.60 $a = 1894, b = 2246, c = 3549$ *Ans.* $A = 28°10′, B = 34°2′, C = 117°48′$

Chapter 13

Area of a Triangle

13.1 AREA OF A TRIANGLE

The area K of any triangle equals one-half the product of its base and altitude. In general if enough information about a triangle is known so that it can be solved, then its area can be found.

13.2 AREA FORMULAS

CASES I and II. Given two angles and a side of triangle ABC

The third angle is found using the fact that $A + B + C = 180°$. The area of the triangle equals a side squared times the product of the sines of the angles including the side divided by twice the sine of the angle opposite the side; i.e.,

$$K = \frac{a^2 \sin B \sin C}{2 \sin A} = \frac{b^2 \sin A \sin C}{2 \sin B} = \frac{c^2 \sin A \sin B}{2 \sin C}$$

For a derivation of these formulas see Prob. 13.2. (See also Probs. 13.4 and 13.5.)

CASE III. Given two sides and the angle opposite one of them in triangle ABC

A second angle is found by using the law of sines and the appropriate formula from Case I. Since there are sometimes two solutions for the second angle, there will be times when the area of two triangles must be found.

(See Probs. 13.6 and 13.7.)

CASE IV. Given two sides and the included angle of triangle ABC

The area of the triangle is equal to one-half the product of the two sides times the sine of the included angle; i.e.,

$$K = \tfrac{1}{2}ab \sin C = \tfrac{1}{2}ac \sin B = \tfrac{1}{2}bc \sin A$$

For a derivation of these formulas see Prob. 13.1.

(See also Probs. 13.8 and 13.9.)

CASE V. Given the three sides of triangle ABC

The area of a triangle is equal to the square root of the product of the semiperimeter and the semiperimeter minus one side times the semiperimeter minus a second side times the semiperimeter minus a third side; i.e.,

$$K = \sqrt{s(s-a)(s-b)(s-c)} \qquad \text{where } s = \tfrac{1}{2}(a+b+c)$$

[NOTE: The formula is known as Heron's (or Hero's) formula. For a derivation of the formula see Prob. 13.3.]

(See Probs. 13.10 and 13.11.)

Solved Problems

13.1 Derive the formula $K = \tfrac{1}{2}bc \sin A$. See Fig. 13-1.

Denoting the altitude drawn to side b of the triangle ABC by h, we have from either figure $h = c \sin A$. Thus, $K = \tfrac{1}{2}bh = \tfrac{1}{2}bc \sin A$.

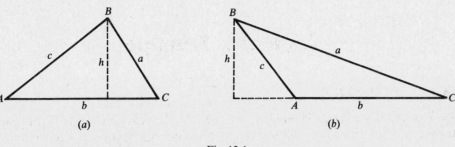

Fig. 13-1

13.2 Derive the formula $K = \dfrac{c^2 \sin A \sin B}{2 \sin C}$.

From Prob. 13.1, $K = \frac{1}{2}bc \sin A$; and by the law of sines $b = \dfrac{c \sin B}{\sin C}$.

Then $K = \frac{1}{2}bc \sin A = \dfrac{1}{2}\dfrac{c \sin B}{\sin C} c \sin A = \dfrac{c^2 \sin A \sin B}{2 \sin C}$.

13.3 Derive the formula $K = \sqrt{s(s-a)(s-b)(s-c)}$.

From the derivations in Prob. 12.50, Chap. 12,

$$\sin^2 \tfrac{1}{2}A = \tfrac{1}{2}(1 - \cos A) = \frac{(a-b+c)(a+b-c)}{4bc} = \frac{2(s-b)\cdot 2(s-c)}{4bc} = \frac{(s-b)(s-c)}{bc}$$

and

$$\cos^2 \tfrac{1}{2}A = \tfrac{1}{2}(1 + \cos A) = \frac{(b+c+a)(b+c-a)}{4bc} = \frac{2s\cdot 2(s-a)}{4bc} = \frac{s(s-a)}{bc}$$

Since $\frac{1}{2}A < 90°$, $\sin \frac{1}{2}A = \sqrt{\dfrac{(s-b)(s-c)}{bc}}$ and $\cos \frac{1}{2}A = \sqrt{\dfrac{s(s-a)}{bc}}$. Then

$$K = \tfrac{1}{2}bc \sin A = bc \sin \tfrac{1}{2}A \cos \tfrac{1}{2}A = bc \sqrt{\frac{(s-b)(s-c)}{bc}} \sqrt{\frac{s(s-a)}{bc}} = \sqrt{s(s-a)(s-b)(s-c)}$$

13.4 Case I. Find the area of triangle ABC, given $c = 23$ cm, $A = 20°$, and $C = 15°$.

$$B = 180° - (A + C) = 145°$$

$$K = \frac{c^2 \sin A \sin B}{2 \sin C}$$

$$= \frac{23^2 \sin 20° \sin 145°}{2 \sin 15°}$$

$$= 200 \text{ cm}^2$$

13.5 Case II. Find the area of triangle ABC, given $c = 23$ cm, $A = 20°$, and $B = 15°$.

$$C = 180° - (A + B) = 145°$$

$$K = \frac{c^2 \sin A \sin B}{2 \sin C}$$

$$= \frac{23^2 \sin 20° \sin 15°}{2 \sin 145°}$$

$$= 41 \text{ cm}^2$$

13.6 Case III. Find the area of triangle ABC, given $a = 112$ m, $b = 219$ m, and $A = 20°$.

$$\sin B = \frac{b \sin A}{a} = \frac{219 \sin 20°}{112} = 0.6688; \; B = 42° \text{ and } B' = 138°.$$

$$C = 180° - (A + B) = 118° \qquad\qquad C' = 180° - (A + B') = 22°$$

$$K = \frac{a^2 \sin B \sin C}{2 \sin A} \qquad\qquad K' = \frac{a^2 \sin B' \sin C'}{2 \sin A}$$

$$= \frac{112^2 \sin 42° \sin 118°}{2 \sin 20°} \qquad\qquad = \frac{112^2 \sin 138° \sin 22°}{2 \sin 20°}$$

$$= 10{,}800 \text{ m}^2 \qquad\qquad\qquad = 4600 \text{ m}^2$$

13.7 Case III. Find the area of triangle ABC, given $A = 41°50'$, $a = 123$ ft, and $b = 96.2$ ft.

$$\sin B = \frac{b \sin A}{a} = \frac{96.2 \sin 41°50'}{123} = 0.5216; \; B = 31°30'.$$

$$C = 180° - (A + B) = 106°40'$$

$$K = \frac{b^2 \sin A \sin C}{2 \sin B}$$

$$= \frac{96.2^2 \sin 41°50' \sin 106°40'}{2 \sin 31°30'}$$

$$= 5660 \text{ ft}^2$$

13.8 Case IV. Find the area of triangle ABC, given $b = 27$ yd, $c = 14$ yd, and $A = 43°$.

$$K = \tfrac{1}{2}bc \sin A$$

$$= \tfrac{1}{2}(27)(14) \sin 43°$$

$$= 130 \text{ yd}^2$$

13.9 Case IV. Find the area of triangle ABC, given $a = 14.27$ cm, $c = 17.23$ cm, and $B = 86°14'$.

$$K = \tfrac{1}{2}ac \sin B$$

$$= \tfrac{1}{2}(14.27)(17.23) \sin 86°14'$$

$$= 122.7 \text{ cm}^2$$

13.10 Case V. Find the area of triangle ABC, given $a = 5.00$ m, $b = 7.00$ m, and $c = 10.0$ m.

$$s = \tfrac{1}{2}(a + b + c) = \tfrac{1}{2}(5 + 7 + 10) = 11 \text{ m.}$$

$$K = \sqrt{s(s - a)(s - b)(s - c)}$$

$$= \sqrt{11(11 - 5)(11 - 7)(11 - 10)}$$

$$= \sqrt{264}$$

$$= 16.2 \text{ m}^2$$

13.11 Case V. Find the area of triangle ABC, given $a = 1.017$ cm, $b = 2.032$ cm, and $c = 2.055$ cm.

$$s = \tfrac{1}{2}(a + b + c) = \tfrac{1}{2}(1.017 + 2.032 + 2.055) = 2.552 \text{ cm}$$

$$\begin{aligned}
K &= \sqrt{s(s - a)(s - b)(s - c)} \\
&= \sqrt{2.552(2.552 - 1.017)(2.552 - 2.032)(2.552 - 2.055)} \\
&= \sqrt{1.012392} \\
&= 1.006 \text{ cm}^2
\end{aligned}$$

13.12 Find the area of an isosceles triangle with a base of 19.2 in and base angles of $23°10'$ each.

In Fig. 13-2, $b = 19.2$ in, $A = 23°10'$, and $C = 23°10'$. Then

$$B = 180° - 2(23°10') = 133°40'$$

$$\begin{aligned}
K &= \frac{b^2 \sin A \sin C}{2 \sin B} \\
&= \frac{19.2^2 \sin 23°10' \sin 23°10'}{2 \sin 133°40'} \\
&= 39.4 \text{ in}^2
\end{aligned}$$

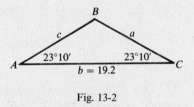

Fig. 13-2

13.13 A painter needs to find the area of the gable end of a house. What is the area of the gable if it is a triangle with two sides of 42.0 ft that meet at a $105°$ angle?

In Fig. 13-3, $a = 42.0$ ft, $b = 42.0$ ft, and $C = 105°$.

$$\begin{aligned}
K &= \tfrac{1}{2}ab \sin C \\
&= \tfrac{1}{2}(42)(42) \sin 105° \\
&= 852 \text{ ft}^2
\end{aligned}$$

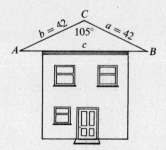

Fig. 13-3

13.14 Three circles with radii 3.0, 5.0, and 9.0 cm are externally tangent. What is the area of the triangle formed by joining their centers?

In Fig. 13-4, $a = 8$ cm, $b = 12$ cm, and $c = 14$ cm.

$$s = \tfrac{1}{2}(a + b + c) = 17 \text{ cm}$$
$$K = \sqrt{s(s - a)(s - b)(s - c)}$$
$$= \sqrt{17(17 - 8)(17 - 12)(17 - 14)}$$
$$= \sqrt{2295}$$
$$= 48 \text{ cm}^2$$

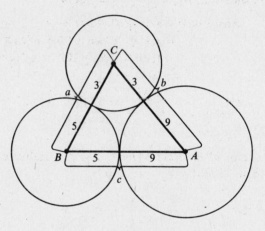

Fig. 13-4

13.15 In a quadrangular field $ABCD$, AB runs N62°10′E 11.4 m, BC runs N22°20′W 19.8 m, and CD runs S40°40′W 15.3 m. DA runs S32°10′E but cannot be measured. Find (*a*) the length of DA and (*b*) the area of the field.

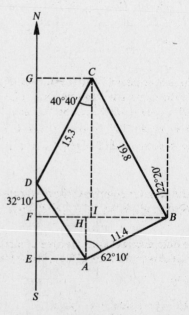

Fig. 13-5

In Fig. 13-5 SN is the north-south line through D, the points E, F, and G are the feet of the perpendiculars to this line through A, B, and C respectively, and the lines AH and CI are perpendicular to BF.

(a)
$$FB = FI + IB = GC + IB$$
$$= 15.3 \sin 40°40' + 19.8 \sin 22°20'$$
$$= 9.97 + 7.52 = 17.49$$

$$FB = FH + HB = EA + HB; \text{ hence}$$
$$EA = FB - HB$$
$$= 17.49 - 11.4 \sin 62°10' = 17.49 - 10.08 = 7.41$$

Since $EA = DA \sin 32°10'$, $DA = \dfrac{7.41}{\sin 32°10'} = 13.9$ m.

(b)
$$\text{Area } ABCD = \text{area } ACD + \text{area } ACB$$
$$= \tfrac{1}{2}(AD)(DC) \sin \angle CDA + \tfrac{1}{2}(AB)(BC) \sin \angle ABC$$
$$= \tfrac{1}{2}(13.9)(15.3) \sin 107°10' + \tfrac{1}{2}(11.4)(19.8) \sin 95°30'$$
$$= 101.6 + 112.3$$
$$= 213.9$$
$$= 214 \text{ m}^2$$

13.16 Prove that the area of a quadrilateral is equal to half the product of its diagonals and the sine of the included angle. See Fig. 13-6(a).

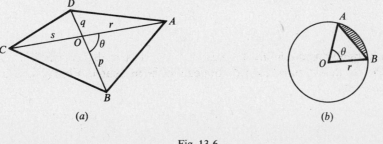

(a)   (b)

Fig. 13-6

Let the diagonals of the quadrilateral $ABCD$ intersect in O, let θ be an angle of intersection of the diagonals, and let O separate the diagonals into segments of length p and q, and r and s, as in the figure.

$\text{Area } ABCD = \text{area } AOB + \text{area } AOD + \text{area } BOC + \text{area } DOC$
$$= \tfrac{1}{2}rp \sin \theta + \tfrac{1}{2}qr \sin (180° - \theta) + \tfrac{1}{2}ps \sin (180° - \theta) + \tfrac{1}{2}qs \sin \theta$$
$$= \tfrac{1}{2}(pr + qr + ps + qs) \sin \theta = \tfrac{1}{2}(p + q)(r + s) \sin \theta = \tfrac{1}{2}(BD)(AC) \sin \theta.$$

13.17 Prove that the area K of the smaller segment (shaded) of a circle of radius r and center O cut off by the chord AB of Fig. 13.6(b) is given by $K = \tfrac{1}{2}r^2(\theta - \sin \theta)$, where θ radians is the central angle intercepted by the chord.

The required area is the difference between the area of sector AOB and triangle AOB.

The area S of the sector AOB is to the area of the circle as the arc AB is to the circumference of the circle:

that is, $\dfrac{S}{\pi r^2} = \dfrac{r\theta}{2\pi r}$ and $S = \tfrac{1}{2}r^2\theta$.

The area of triangle $AOB = \tfrac{1}{2}r \cdot r \sin \theta = \tfrac{1}{2}r^2 \sin \theta$.

Thus, $K = \tfrac{1}{2}r^2\theta - \tfrac{1}{2}r^2 \sin \theta = \tfrac{1}{2}r^2(\theta - \sin \theta)$

13.18 Three circles with centers A, B, and C have respective radii 50, 30, and 20 in and are tangent to each other externally. Find the area of the *curvilinear* triangle formed by the three circles.

Let the points of tangency of the circles be R, S, and T as in Fig. 13-7. The required area is the difference between the area of triangle ABC and the sum of the areas of the three sectors ART, BRS, and SCT.

Since the join of the centers of any two circles passes through their point of tangency, $a = BC = 50$, $b = CA = 70$, and $c = AB = 80$ in. Then

$$s = \tfrac{1}{2}(a + b + c) = 100 \qquad s - a = 50 \qquad s - b = 30 \qquad s - c = 20$$

and
$$K = \text{area } ABC = \sqrt{s(s - a)(s - b)(s - c)} = \sqrt{100(50)(30)(20)} = 1000\sqrt{3} = 1732$$

Since $r = K/s = 17.32$,

$$\tan \tfrac{1}{2}A = \frac{r}{s - a} = \frac{17.32}{50} = 0.3464 \qquad \tfrac{1}{2}A = 19°6' \qquad A = 38°12' = 0.667 \text{ rad}$$

$$\tan \tfrac{1}{2}B = \frac{r}{s - b} = \frac{17.32}{30} = 0.5773 \qquad \tfrac{1}{2}B = 30°0' \qquad B = 60°0' = 1.047 \text{ rad}$$

$$\tan \tfrac{1}{2}C = \frac{r}{s - c} = \frac{17.32}{20} = 0.8660 \qquad \tfrac{1}{2}C = 40°54' \qquad C = 81°48' = 1.428 \text{ rad}$$

Area $ART = \tfrac{1}{2}r^2\theta = \tfrac{1}{2}(50)^2(0.667) = 833.75$, area $BRS = \tfrac{1}{2}(30)^2(1.047) = 471.15$, area $CST = \tfrac{1}{2}(20)^2(1.428) = 285.60$, and their sum is 1590.50.

The required area is $1732 - 1590.50 = 141.50$ or 142 in^2.

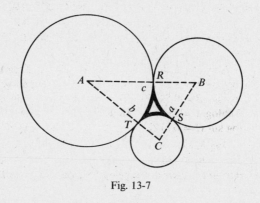

Fig. 13-7

Problems 13.19 to 13.22 demonstrate the use of logarithms to find the area of a triangle. If you are not using logarithms, go directly to the Supplementary Problems.

13.19 Find the area of the triangle ABC, given $A = 37°10'$, $C = 62°30'$, and $b = 34.9$. See Fig. 13-8.

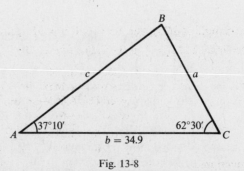

Fig. 13-8

$B = 180° - (A + C) = 80°20'$.

This is a Case II triangle and $K = \dfrac{b^2 \sin C \sin A}{2 \sin B}$.

$$
\begin{aligned}
2 \log b &= 3.0856 \\
\log \sin C &= 9.9479 - 10 \\
\log \sin A &= 9.7811 - 10 \\
\text{colog } 2 &= 9.6990 - 10 \\
\text{colog } \sin B &= 0.0062 \\
\hline
\log K &= 2.5198 \\
K &= 331 \text{ square units}
\end{aligned}
$$

13.20 Find the area of the triangle ABC, given $b = 28.6$, $c = 44.3$, and $B = 23°20'$.

This is a Case III triangle in which there may be two solutions. See Fig. 13-9.

$$\sin C = \frac{c \sin B}{b}$$

$$
\begin{aligned}
\log c &= 1.6464 \\
\log \sin B &= 9.5978 - 10 \\
\text{colog } b &= 8.5436 - 10 \\
\hline
\log \sin C &= 9.7878 - 10
\end{aligned}
$$

$C = 37°50'$ and $C' = 180° - C = 142°10'$
$A = 180° - (B + C) = 118°50'$ and $A' = 180° - (B + C') = 14°30'$

Area of ABC is $K = \dfrac{c^2 \sin A \sin B}{2 \sin C}$.   Area of ABC' is $K = \dfrac{c^2 \sin A' \sin B}{2 \sin C'}$.

$$
\begin{aligned}
2 \log c &= 3.2928 \\
\log \sin A &= 9.9425 - 10 \\
\log \sin B &= 9.5978 - 10 \\
\text{colog } 2 &= 9.6990 - 10 \\
\text{colog } \sin C &= 0.2122 \\
\hline
\log K &= 2.7443 \\
K &= 555
\end{aligned}
\qquad
\begin{aligned}
2 \log c &= 3.2928 \\
\log \sin A' &= 9.3986 - 10 \\
\log \sin B &= 9.5978 - 10 \\
\text{colog } 2 &= 9.6990 - 10 \\
\text{colog } \sin C' &= 0.2122 \\
\hline
\log K &= 2.2004 \\
K &= 159
\end{aligned}
$$

Two triangles are determined, their areas being 555 and 159 square units respectively.

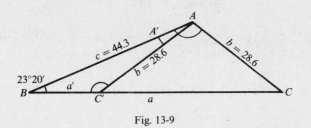

Fig. 13-9

13.21 Find the area of the triangle ABC, given $a = 16.4$, $b = 55.7$, and $C = 27°20'$.

This is a Case IV triangle and $K = \frac{1}{2}ab \sin C$. See Fig. 13-10.

$$
\begin{aligned}
\log a &= 1.2148 \\
\log b &= 1.7458 \\
\log \sin C &= 9.6620 - 10 \\
\text{colog } 2 &= 9.6990 - 10 \\
\hline
\log K &= 2.3216 \\
K &= 210
\end{aligned}
$$

The area is 210 square units.

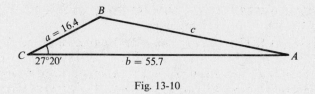

Fig. 13-10

13.22 Find the area of the triangle ABC, given $a = 255$, $b = 290$, and $c = 419$. See Fig. 13-11.

This is a Case V triangle and $K = \sqrt{s(s-a)(s-b)(s-c)}$.

$$s = \tfrac{1}{2}(a+b+c) \qquad K = \sqrt{s(s-a)(s-b)(s-c)}$$

$a = 255$	$s - a = 227$	$\log(s-a) = 2.3560$
$b = 290$	$s - b = 192$	$\log(s-b) = 2.2833$
$c = 419$	$s - c = \underline{\ 63}$	$\log(s-c) = 1.7993$
$2s = \overline{964}$	$s = 482$	$\log s = \underline{2.6830}$
$s = 482$		$2 \log K = 9.1216$
		$\log K = 4.5608$
		$K = 36{,}400$

The area is 36,400 square units.

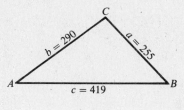

Fig. 13-11

Supplementary Problems

Find the area of the triangle ABC, given:

13.23 $b = 13$ ft, $a = 27$ ft, $C = 85°$ *Ans.*   175 ft^2

13.24 $a = 23.3$ cm, $c = 21.5$ cm, $B = 121.0°$ *Ans.*   215 cm^2

13.25 $a = 4.1$ m, $b = 5.2$ m, $c = 6.7$ m   *Ans.*   11 m^2

13.26 $A = 65°$, $B = 35°$, $c = 12$ yd   *Ans.*   38 yd^2

13.27 $b = 23.84$, $c = 35.26$, $A = 50°32'$ *Ans.*   324.5 square units

13.28 $a = 456.3$, $b = 586.8$, $C = 28°17'$ *Ans.*   63,440 square units

13.29 $a = 512.3$, $B = 52°15'$, $C = 63°46'$ *Ans.*   103,600 square units

13.30 $b = 444.8$, $A = 110°16'$, $B = 30°10'$ *Ans.*   117,600 square units

13.31 $a = 384.2, b = 492.8, c = 677.8$ *Ans.*   93,080 square units

13.32 $a = 28.16, b = 60.15, c = 51.17$ *Ans.*   718.6 square units

13.33   To find the area of a triangular lot, the owner starts at one corner and walks due east 215 m to a second corner. After turning through an angle of 78.4°, the owner walks 314 m to the third corner. What is the area of the lot?

   Ans.   33,100 m^2

13.34   An artist wishes to make a sign in the shape of an isosceles triangle with a 42° vertex angle and a base of 18 m. What is the area of the sign?

   Ans.   211 m^2

13.35   Point C has a bearing of N28°E from point A and a bearing of N12°W from point B. What is the area of triangle ABC if B is due east of A and 23 km from A?

   Ans.   355 km^2

13.36   Three circles have radii of 7.72, 4.84, and 11.4 cm. If they are externally tangent, what is the area of the triangle formed by joining their centers?

   Ans.   101 cm^2

13.37   A woman hikes 503 m, turns and jogs 415 m, turns again, and runs 365 m returning to her starting point. What is the area of the triangle formed by her path?

   Ans.   74,600 m^2

Chapter 14

Inverses of Trigonometric Functions

14.1 INVERSE TRIGONOMETRIC RELATIONS

The equation

$$x = \sin y \tag{1}$$

defines a unique value of x for each given angle y. But when x is given, the equation may have no solution or many solutions. For example: if $x = 2$, there is no solution, since the sine of an angle never exceeds 1; if $x = \frac{1}{2}$, there are many solutions $y = 30°, 150°, 390°, 510°, -210°, -330°, \ldots$.

To express y as a function of x, we will write

$$y = \arcsin x \tag{2}$$

In spite of the use of the word *arc*, (2) is to be interpreted as stating that "y is an angle whose sine is x." Similarly we shall write $y = \arccos x$ if $x = \cos y$, $y = \arctan x$ if $x = \tan y$, etc.

The notation $y = \sin^{-1} x$, $y = \cos^{-1} x$, etc., (to be read "inverse sine of x, inverse cosine of x," etc.) are also used but $\sin^{-1} x$ may be confused with $1/\sin x = (\sin x)^{-1}$, so care in writing negative exponents for trigonometric functions is needed.

14.2 GRAPHS OF THE INVERSE TRIGONOMETRIC RELATIONS

The graph of $y = \arcsin x$ is the graph of $x = \sin y$ and differs from the graph of $y = \sin x$ of Chap. 8 in that the roles of x and y are interchanged. Thus, the graph of $y = \arcsin x$ is a sine curve drawn on the y axis instead of the x axis.

Similarly the graphs of the remaining inverse trigonometric relations are those of the corresponding trigonometric functions except that the roles of x and y are interchanged.

14.3 INVERSE TRIGONOMETRIC FUNCTIONS

It is at times necessary to consider the inverse trigonometric relations as functions (i.e., one value of y corresponding to each admissible value of x). To do this, we agree to select one out of the many angles corresponding to the given value of x. For example, when $x = \frac{1}{2}$, we shall agree to select the value $y = \pi/6$ and when $x = -\frac{1}{2}$, we shall agree to select the value $y = -\pi/6$. This selected value is called the *principal value* of arcsin x. When only the principal value is called for, we shall write arcsin x, arccos x, etc. Alternative notation for the principal value of the inverses of the trigonometric functions is $\sin^{-1} x$, $\text{Cos}^{-1} x$, $\text{Tan}^{-1} x$, etc. The portions of the graphs on which the principal values of each of the inverse trigonometric relations lie are shown in Fig. 14-1(a) to (f) by a heavier line.

When x is positive or zero and the inverse function exists, the principal value is defined as that value of y which lies between 0 and $\frac{1}{2}\pi$ inclusive.

EXAMPLE 14.1 (a) Arcsin $\sqrt{3}/2 = \pi/3$ since $\sin \pi/3 = \sqrt{3}/2$ and $0 < \pi/3 < \pi/2$.

(b) Arccos $\sqrt{3}/2 = \pi/6$ since $\cos \pi/6 = \sqrt{3}/2$ and $0 < \pi/6 < \pi/2$.

(c) Arctan $1 = \pi/4$ since $\tan \pi/4 = 1$ and $0 < \pi/4 < \pi/2$.

When x is negative and the inverse function exists, the principal value is defined as follows:

$$-\tfrac{1}{2}\pi \le \text{Arcsin } x < 0 \qquad \tfrac{1}{2}\pi < \text{Arccot } x < \pi$$

$$\tfrac{1}{2}\pi < \text{Arccos } x \le \pi \qquad \tfrac{1}{2}\pi < \text{Arcsec } x \le \pi$$

$$-\tfrac{1}{2}\pi < \text{Arctan } x < 0 \qquad -\tfrac{1}{2}\pi \le \text{Arccsc } x < 0$$

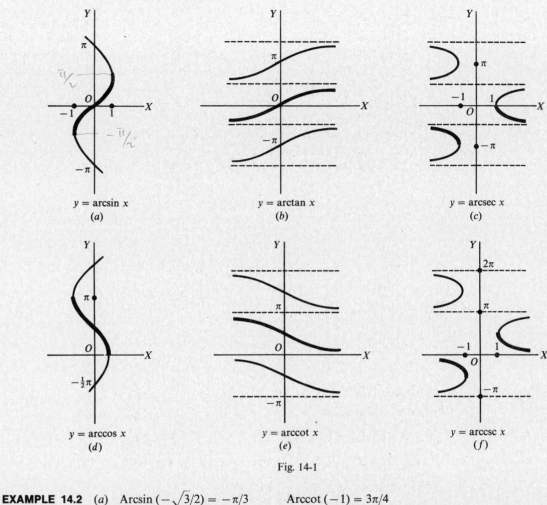

Fig. 14-1

EXAMPLE 14.2   (a)   Arcsin $(-\sqrt{3}/2) = -\pi/3$ Arccot $(-1) = 3\pi/4$

  (b)   Arccos $(-1/2) = 2\pi/3$ Arcsec $(-2/\sqrt{3}) = +5\pi/6$

  (c)   Arctan $(-1/\sqrt{3}) = -\pi/6$ Arccsc $(-\sqrt{2}) = -\pi/4$

14.4 PRINCIPAL-VALUE RANGE

Authors vary in defining the principal values of the inverse functions when x is negative. The definitions given are the most convenient for the calculus. In many calculus textbooks, the inverse of a trigonometric function is defined as the principal-valued inverse and no capital letter is used in their notation. Since only the inverse function is considered, this generally causes no problem in the calculus class.

Inverse Function	Principal-Value Range
$y = $ Arcsin x	$-\frac{1}{2}\pi \leq y \leq \frac{1}{2}\pi$
$y = $ Arccos x	$0 \leq y \leq \pi$
$y = $ Arctan x	$-\frac{1}{2}\pi < y < \frac{1}{2}\pi$
$y = $ Arccot x	$0 < y < \pi$
$y = $ Arcsec x	$0 \leq y \leq \pi, y \neq \frac{1}{2}\pi$
$y = $ Arccsc x	$-\frac{1}{2}\pi \leq y \leq \frac{1}{2}\pi, y \neq 0$

14.5 GENERAL VALUES OF INVERSE TRIGONOMETRIC RELATIONS

Let y be an inverse trigonometric relation of x. Since the value of a trigonometric relation of y is known, two positions are determined in general for the terminal side of the angle y (see Chap. 2). Let y_1 and y_2 respectively be angles determined by the two positions of the terminal side. Then the totality of values of y consist of the angles y_1 and y_2, together with all angles coterminal with them, that is,

$$y_1 + 2n\pi \qquad \text{and} \qquad y_2 + 2n\pi$$

where n is any positive or negative integer or zero.

One of the values y_1 or y_2 may always be taken as the principal value of the inverse trigonometric function.

EXAMPLE 14.3 Write expressions for the general value of (a) arcsin 1/2, (b) arccos (-1), and (c) arctan (-1).

(a) The principal value of arcsin 1/2 is $\pi/6$, and a second value (not coterminal with the principal value) is $5\pi/6$. The general value of arcsin 1/2 is given by

$$\pi/6 + 2n\pi \qquad 5\pi/6 + 2n\pi$$

where n is any positive or negative integer or zero.

(b) The principal value is π and there is no other value not coterminal with it. Thus, the general value is given by $\pi + 2n\pi$, where n is a positive or negative integer or zero.

(c) The principal value is $-\pi/4$, and a second value (not coterminal with the principal value) is $3\pi/4$. Thus, the general value is given by

$$-\pi/4 + 2n\pi \qquad 3\pi/4 + 2n\pi$$

where n is a positive or negative integer or zero.

Solved Problems

14.1 Find the principal value of each of the following.

(a) Arcsin $0 = 0$

(b) Arccos $(-1) = \pi$

(c) Arctan $\sqrt{3} = \pi/3$

(d) Arccot $\sqrt{3} = \pi/6$

(e) Arcsec $2 = \pi/3$

(f) Arccsc $(-\sqrt{2}) = -\pi/4$

(g) Arccos $0 = \pi/2$

(h) Arcsin $(-1) = -\pi/2$

(i) Arctan $(-1) = -\pi/4$

(j) Arccot $0 = \pi/2$

(k) Arcsec $(-\sqrt{2}) = -3\pi/4$

(l) Arccsc $(-2) = -5\pi/6$

14.2 Express the principal value of each of the following to the nearest minute or to the nearest hundredth of a degree.

(a) Arcsin $0.3333 = 19°28'$ or $19.47°$

(b) Arccos $0.4000 = 66°25'$ or $66.42°$

(c) Arctan $1.5000 = 56°19'$ or $56.31°$

(d) Arccot $1.1875 = 40°\ 6'$ or $40.10°$

(e) Arcsec $1.0324 = 14°24'$ or $14.39°$

(f) Arccsc $1.5082 = 41°32'$ or $41.53°$

(g) Arcsin $(-0.6439) = -40°5'$ or $-40.08°$

(h) Arccos $(-0.4519) = 116°52'$ or $116.87°$

(i) Arctan $(-1.4400) = -55°13'$ or $-55.22°$

(j) Arccot $(-0.7340) = 126°17'$ or $126.28°$

(k) Arcsec $(-1.2067) = 145°58'$ or $145.97°$

(l) Arccsc $(-4.1923) = -13°48'$ or $-13.80°$

14.3 Verify each of the following.

(a) $\sin(\text{Arcsin } 1/2) = \sin \pi/6 = 1/2$

(b) $\cos[\text{Arccos}(-1/2)] = \cos 2\pi/3 = -1/2$

(c) $\cos[\text{Arcsin}(-\sqrt{2}/2)] = \cos(-\pi/4) = \sqrt{2}/2$

(d) $\text{Arcsin}(\sin \pi/3) = \text{Arcsin}\sqrt{3}/2 = \pi/3$

(e) $\text{Arccos}[\cos(-\pi/4)] = \text{Arccos}\sqrt{2}/2 = \pi/4$

(f) $\text{Arcsin}(\tan 3\pi/4) = \text{Arcsin}(-1) = -\pi/2$

(g) $\text{Arccos}[\tan(-5\pi/4)] = \text{Arccos}(-1) = \pi$

14.4 Verify each of the following.

(a) $\text{Arcsin}\sqrt{2}/2 - \text{Arcsin } 1/2 = \pi/4 - \pi/6 = \pi/12$

(b) $\text{Arccos } 0 + \text{Arctan}(-1) = \pi/2 + (-\pi/4) = \pi/4 = \text{Arctan } 1$

14.5 Evaluate each of the following:
(a) $\cos(\text{Arcsin } 3/5)$,   (b) $\sin[\text{Arccos}(-2/3)]$,   (c) $\tan[\text{Arcsin}(-3/4)]$

(a) Let $\theta = \text{Arcsin } 3/5$; then $\sin \theta = 3/5$, θ being a first-quadrant angle. From Fig. 14-2(a),

$$\cos(\text{Arcsin } 3/5) = \cos \theta = 4/5$$

(b) Let $\theta = \text{Arccos}(-2/3)$; then $\cos \theta = -2/3$, θ being a second-quadrant angle. From Fig. 14-2(b),

$$\sin[\text{Arccos}(-2/3)] = \sin \theta = \sqrt{5}/3$$

(c) Let $\theta = \text{Arcsin}(-3/4)$; then $\sin \theta = -3/4$, θ being a fourth-quadrant angle. From Fig. 14-2(c),

$$\tan[\text{Arcsin}(-3/4)] = \tan \theta = -3/\sqrt{7} = -3\sqrt{7}/7$$

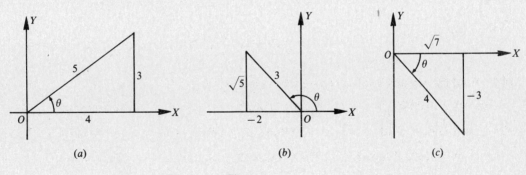

(a)   (b)   (c)

Fig. 14-2

14.6 Evaluate $\sin(\text{Arcsin } 12/13 + \text{Arcsin } 4/5)$.

Let $\theta = \text{Arcsin } 12/13$

and $\phi = \text{Arcsin } 4/5$

Then $\sin \theta = 12/13$ and $\sin \phi = 4/5$, θ and ϕ being first-quadrant angles. From Fig. 14-3(a) and (b),

$$\sin(\text{Arcsin } 12/13 + \text{Arcsin } 4/5) = \sin(\theta + \phi)$$

$$= \sin \theta \cos \phi + \cos \theta \sin \phi$$

$$= \frac{12}{13} \cdot \frac{3}{5} + \frac{5}{13} \cdot \frac{4}{5} = \frac{56}{65}$$

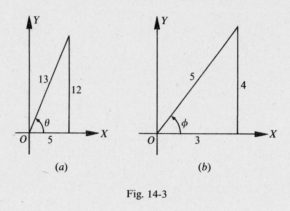

Fig. 14-3

14.7   Evaluate cos (Arctan 15/8 − Arcsin 7/25).

Let $\theta = $ Arctan 15/8

and $\phi = $ Arcsin 7/25

Then tan θ = 15/8 and sin ϕ = 7/25, θ and ϕ being first-quadrant angles. From Fig. 14-4(*a*) and (*b*),

$$\text{cos (Arctan 15/8 − Arcsin 7/25)} = \cos (\theta - \phi)$$

$$= \cos \theta \cos \phi + \sin \theta \sin \phi$$

$$= \frac{8}{17} \cdot \frac{24}{25} + \frac{15}{17} \cdot \frac{7}{25} = \frac{297}{425}$$

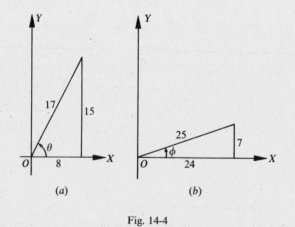

Fig. 14-4

14.8   Evaluate sin (2 Arctan 3).

Let $\theta = $ Arctan 3; then tan θ = 3, θ being a first-quadrant angle. From Fig. 14-5,

$$\text{sin (2 Arctan 3)} = \sin 2\theta$$

$$= 2 \sin \theta \cos \theta$$

$$= 2(3/\sqrt{10})(1/\sqrt{10})$$

$$= 3/5$$

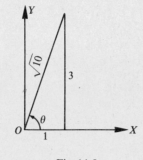

Fig. 14-5

14.9 Show that Arcsin $1/\sqrt{5}$ + Arcsin $2/\sqrt{5}$ = $\pi/2$.

Let θ = Arcsin $1/\sqrt{5}$ and ϕ = Arcsin $2/\sqrt{5}$; then sin θ = $1/\sqrt{5}$ and sin ϕ = $2/\sqrt{5}$, each angle terminating in the first-quadrant. We are to show that $\theta + \phi = \pi/2$ or, taking the sines of both members, that sin $(\theta + \phi)$ = sin $\pi/2$.

From Figs. 14-6(a) and (b),

$$\sin (\theta + \phi) = \sin \theta \cos \phi + \cos \theta \sin \phi$$

$$= \frac{1}{\sqrt{5}} \cdot \frac{1}{\sqrt{5}} + \frac{2}{\sqrt{5}} \cdot \frac{2}{\sqrt{5}} = 1 = \sin \pi/2$$

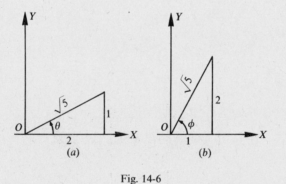

(a)   (b)

Fig. 14-6

14.10 Show that 2 Arctan $1/2$ = Arctan $4/3$.

Let θ = Arctan $1/2$ and ϕ = Arctan $4/3$; then tan θ = $1/2$ and tan ϕ = $4/3$.
We are to show that $2\theta = \phi$ or, taking the tangents of both members, that tan 2θ = tan ϕ.

Now tan $2\theta = \dfrac{2 \tan \theta}{1 - \tan^2 \theta} = \dfrac{2(1/2)}{1 - (1/2)^2} = 4/3 = \tan \phi$.

14.11 Show that Arcsin $77/85$ − Arcsin $3/5$ = Arccos $15/17$.

Let θ = Arcsin $77/85$, ϕ = Arcsin $3/5$, and ψ = Arccos $15/17$; then sin θ = $77/85$, sin ϕ = $3/5$, and cos ψ = $15/17$, each angle terminating in the first-quadrant. Taking the sine of both members of the given relation, we are to show that sin $(\theta - \phi)$ = sin ψ. From Fig. 14-7(a), (b), and (c),

$$\sin (\theta - \phi) = \sin \theta \cos \phi - \cos \theta \sin \phi = \frac{77}{85} \cdot \frac{4}{5} - \frac{36}{85} \cdot \frac{3}{5} = \frac{8}{17} = \sin \psi.$$

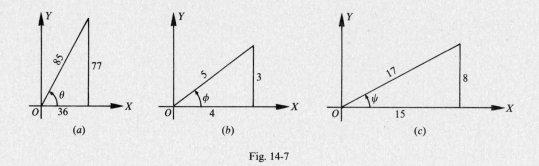

Fig. 14-7

14.12 Show that Arccot 43/32 − Arctan 1/4 = Arccos 12/13.

Let θ = Arccot 43/32, ϕ = Arctan 1/4, and ψ = Arccos 12/13; then cot θ = 43/32, tan ϕ = 1/4, and cos ψ = 12/13, each angle terminating in the first-quadrant. Taking the tangent of both members of the given relation, we are to show that tan $(\theta - \phi)$ = tan ψ. From Fig. 14-8, tan ψ = 5/12.

$$\tan (\theta - \phi) = \frac{\tan \theta - \tan \phi}{1 + \tan \theta \tan \phi} = \frac{32/43 - 1/4}{1 + (32/43)(1/4)} = \frac{5}{12} = \tan \psi$$

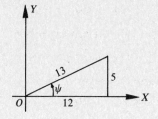

Fig. 14-8

14.13 Show that Arctan 1/2 + Arctan 1/5 + Arctan 1/8 = $\pi/4$.

We shall show that Arctan 1/2 + Arctan 1/5 = $\pi/4$ − Arctan 1/8.

$$\tan (\text{Arctan } 1/2 + \text{Arctan } 1/5) = \frac{1/2 + 1/5}{1 - (1/2)(1/5)} = \frac{7}{9}$$

and

$$\tan (\pi/4 - \text{Arctan } 1/8) = \frac{1 - 1/8}{1 + 1/8} = \frac{7}{9}$$

14.14 Show that 2 Arctan 1/3 + Arctan 1/7 = Arcsec $\sqrt{34}/5$ + Arccsc $\sqrt{17}$.

Let θ = Arctan 1/3, ϕ = Arctan 1/7, λ = Arcsec $\sqrt{34}/5$, and ψ = Arccsc $\sqrt{17}$; then tan θ = 1/3, tan ϕ = 1/7, sec λ = $\sqrt{34}/5$, and csc ψ = $\sqrt{17}$, each angle terminating in the first-quadrant.

Taking the tangent of both members of the given relation, we are to show that

$$\tan (2\theta + \phi) = \tan (\lambda + \psi)$$

Now

$$\tan 2\theta = \frac{2 \tan \theta}{1 - \tan^2 \theta} = \frac{2(1/3)}{1 - (1/3)^2} = 3/4$$

$$\tan (2\theta + \phi) = \frac{\tan 2\theta + \tan \phi}{1 - \tan 2\theta \tan \phi} = \frac{3/4 + 1/7}{1 - (3/4)(1/7)} = 1$$

and, using Fig. 14-9(a) and (b), tan $(\lambda + \psi) = \dfrac{3/5 + 1/4}{1 - (3/5)(1/4)} = 1$.

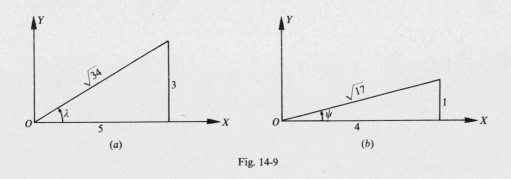

Fig. 14-9

14.15 Find the general value of each of the following.

(a) $\arcsin \sqrt{2}/2 = \pi/4 + 2n\pi,\ 3\pi/4 + 2n\pi$ (d) $\arcsin(-1) = -\pi/2 + 2n\pi$

(b) $\arccos 1/2 = \pi/3 + 2n\pi,\ 5\pi/3 + 2n\pi$ (e) $\arccos 0 = \pi/2 + 2n\pi,\ 3\pi/2 + 2n\pi$

(c) $\arctan 0 = 2n\pi,\ (2n+1)\pi$ (f) $\arctan(-\sqrt{3}) = -\pi/3 + 2n\pi,\ 2\pi/3 + 2n\pi$

where n is a positive or negative integer or zero.

14.16 Show that the general value of   (a) $\arcsin x = n\pi + (-1)^n \operatorname{Arcsin} x$

(b) $\arccos x = 2n\pi \pm \operatorname{Arccos} x$

(c) $\arctan x = n\pi + \operatorname{Arctan} x$

where n is any positive or negative integer or zero.

(a)   Let $\theta = \operatorname{Arcsin} x$. Then since $\sin(\pi - \theta) = \sin \theta$, all values of $\arcsin x$ are given by

$$(1)\ \ \theta + 2m\pi \quad \text{and} \quad (2)\ \ \pi - \theta + 2m\pi = (2m+1)\pi - \theta$$

Now, when $n = 2m$, that is, n is an even integer, (1) may be written as $n\pi + \theta = n\pi + (-1)^n\theta$; and when $n = 2m + 1$, that is, n is an odd integer, (2) may be written as $n\pi - \theta = n\pi + (-1)^n\theta$. Thus, $\arcsin x = n\pi + (-1)^n \operatorname{Arcsin} x$, where n is any positive or negative integer or zero.

(b)   Let $\theta = \operatorname{Arccos} x$. Then since $\cos(-\theta) = \cos \theta$, all values of $\arccos x$ are given by $\theta + 2n\pi$ and $-\theta + 2n\pi$ or $2n\pi \pm \theta = 2n\pi \pm \operatorname{Arccos} x$, where n is any positive or negative integer or zero.

(c)   Let $\theta = \operatorname{Arctan} x$. Then since $\tan(\pi + \theta) = \tan \theta$, all values of $\arctan x$ are given by $\theta + 2m\pi$ and $(\pi + \theta) + 2m\pi = \theta + (2m+1)\pi$ or, as in (a), by $n\pi + \operatorname{Arctan} x$, where n is any positive or negative integer or zero.

14.17 Express the general value of each of the functions of Prob. 14.15, using the form of Prob. 14.16.

(a) $\arcsin \sqrt{2}/2 = n\pi + (-1)^n\pi/4$ (d) $\arcsin(-1) = n\pi + (-1)^n(-\pi/2)$

(b) $\arccos 1/2 = 2n\pi \pm \pi/3$ (e) $\arccos 0 = 2n\pi \pm \pi/2$

(c) $\arctan 0 = n\pi$ (f) $\arctan(-\sqrt{3}) = n\pi - \pi/3$

where n is any positive or negative integer or zero.

Supplementary Problems

14.18 Write the following in inverse-relation notation.
(a) $\sin \theta = 3/4$,   (b) $\cos \alpha = -1$,   (c) $\tan x = -2$,   (d) $\cot \beta = 1/2$

Ans.   (a) $\theta = \arcsin 3/4$,   (b) $\alpha = \arccos(-1)$,   (c) $x = \arctan(-2)$,   (d) $\beta = \operatorname{arccot} 1/2$

14.19 Find the principal value of each of the following.

(a) $\text{Arcsin } \sqrt{3}/2$ (d)   Arccot 1   (g) $\text{Arctan } (-\sqrt{3})$ (j)   Arccsc (-1)

(b) $\text{Aarccos } (-\sqrt{2}/2)$ (e)   Arcsin $(-1/2)$ (h)   Arccot 0

(c) $\text{Arctan } 1/\sqrt{3}$ (f)   Arccos $(-1/2)$ (i) $\text{Arcsec } (-\sqrt{2})$

Ans.   (a) $\pi/3$, (b) $3\pi/4$, (c) $\pi/6$, (d) $\pi/4$, (e) $-\pi/6$, (f) $2\pi/3$, (g) $-\pi/3$, (h) $\pi/2$, (i) $-3\pi/4$, (j) $-\pi/2$

14.20 Evaluate each of the following.

(a)   sin [Arcsin $(-1/2)$]   (f)   sin (Arccos 4/5)   (k)   Arctan (cot 230°)

(b)   cos (Arccos $\sqrt{3}/2$)   (g)   cos [Arcsin $(-12/13)$]   (l)   Arccot (tan 100°)

(c)   tan [Arctan (-1)]   (h)   sin (Arctan 2)   (m)   sin (2 Arcsin 2/3)

(d)   sin [Arccos $(-\sqrt{3}/2)$]   (i)   Arccos (sin 220°)   (n)   cos 2 Arcsin 3/5

(e)   tan (Arcsin 0)   (j)   Arcsin [cos $(-105°)$]   (o)   sin ($\frac{1}{2}$ Arccos 4/5)

Ans.   (a) $-1/2$, (b) $\sqrt{3}/2$, (c) -1, (d) $1/2$, (e) 0, (f) $3/5$, (g) $5/13$, (h) $2/\sqrt{5} = 2\sqrt{5}/5$ (i) $\dfrac{13\pi}{18}$, (j) $-\dfrac{\pi}{12}$, (k) $\dfrac{2\pi}{9}$,

(l) $\dfrac{17\pi}{18}$, (m) $4\sqrt{5}/9$, (n) $7/25$, (o) $1/\sqrt{10} = \sqrt{10}/10$

14.21 Show that

(a) $\sin\left(\text{Arcsin}\dfrac{5}{13} + \text{Arcsin}\dfrac{4}{5}\right) = \dfrac{63}{65}$ (e) $\cos\left(\text{Arctan}\dfrac{-4}{3} + \text{Arcsin}\dfrac{12}{13}\right) = \dfrac{63}{65}$

(b) $\cos\left(\text{Arccos}\dfrac{15}{17} - \text{Arccos}\dfrac{7}{25}\right) = \dfrac{297}{425}$ (f) $\tan\left(\text{Arcsin}\dfrac{-3}{5} - \text{Arccos}\dfrac{5}{13}\right) = \dfrac{63}{16}$

(c) $\sin\left(\text{Arcsin}\dfrac{1}{2} - \text{Arccos}\dfrac{1}{3}\right) = \dfrac{1-2\sqrt{6}}{6}$ (g) $\tan\left(2\,\text{Arcsin}\dfrac{4}{5} + \text{Arccos}\dfrac{12}{13}\right) = -\dfrac{253}{204}$

(d) $\tan\left(\text{Arctan}\dfrac{3}{4} + \text{Arccot}\dfrac{15}{8}\right) = \dfrac{77}{36}$ (h) $\sin\left(2\,\text{Arcsin}\dfrac{4}{5} - \text{Arccos}\dfrac{12}{13}\right) = \dfrac{323}{325}$

14.22 Show that

(a) $\text{Arctan}\dfrac{1}{2} + \text{Arctan}\dfrac{1}{3} = \dfrac{\pi}{4}$ (e) $\text{Arccos}\dfrac{12}{13} + \text{Arctan}\dfrac{1}{4} = \text{Arccot}\dfrac{43}{32}$

(b) $\text{Arcsin}\dfrac{4}{5} + \text{Arctan}\dfrac{3}{4} = \dfrac{\pi}{2}$ (f) $\text{Arcsin}\dfrac{3}{5} + \text{Arcsin}\dfrac{15}{17} = \text{Arccos}\dfrac{-13}{85}$

(c) $\text{Arctan}\dfrac{4}{3} - \text{Arctan}\dfrac{1}{7} = \dfrac{\pi}{4}$ (g) $\text{Arctan } a + \text{Arctan}\dfrac{1}{a} = \dfrac{\pi}{2}$ (a > 0)

(d) $2\,\text{Arctan}\dfrac{1}{3} + \text{Arctan}\dfrac{1}{7} = \dfrac{\pi}{4}$

14.23 Prove: The area of the segment cut from a circle of radius r by a chord at a distance d from the center is given by $K = r^2 \text{ Arccos } d/r - d\sqrt{r^2 - d^2}$.

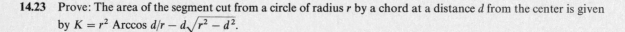

Chapter 15

Trigonometric Equations

15.1 TRIGONOMETRIC EQUATIONS

Trigonometric equations, i.e., equations involving trigonometric functions of unknown angles, are called:

(a) Identical equations or *identities*, if they are satisfied by all values of the unknown angles for which the functions are defined

(b) Conditional equations, or equations, if they are satisfied only by particular values of the unknown angles

EXAMPLE 15.1 (a) $\sin x \csc x = 1$ is an identity, being satisfied by every value of x for which $\csc x$ is defined.

(b) $\sin x = 0$ is a conditional equation since it is not satisfied by $x = \frac{1}{4}\pi$ or $\frac{1}{2}\pi$.

Hereafter in this chapter we shall use the term *equation* instead of *conditional equation*.

A solution of a trigonometric equation, like $\sin x = 0$, is a value of the angle x which satisfies the equation. Two solutions of $\sin x = 0$ are $x = 0$ and $x = \pi$.

If a given equation has one solution, it has in general an unlimited number of solutions. Thus, the complete solution of $\sin x = 0$ is given by

$$x = 0 + 2n\pi \qquad x = \pi + 2n\pi$$

where n is any positive or negative integer or zero.

In this chapter we shall list only the particular solutions for which $0 \leq x < 2\pi$.

15.2 SOLVING TRIGONOMETRIC EQUATIONS

There is no general method for solving trigonometric equations. Several standard procedures are illustrated in the following examples and other procedures are introduced in the Solved Problems. All solutions will be for the interval $0 \leq x < 2\pi$.

(A) The equation may be factorable.

EXAMPLE 15.2 Solve $\sin x - 2 \sin x \cos x = 0$.

Factoring, $\sin x - 2 \sin x \cos x = \sin x (1 - 2 \cos x) = 0$ and setting each factor equal to zero, we have

$$\sin x = 0 \qquad \text{and} \qquad x = 0, \pi$$

$$1 - 2 \cos x = 0 \quad \text{or} \quad \cos x = \tfrac{1}{2} \qquad \text{and} \qquad x = \pi/3, 5\pi/3$$

Check: For $x = 0$, $\quad \sin x - 2 \sin x \cos x = 0 - 2(0)(1) = 0$

For $x = \pi/3$, $\sin x - 2 \sin x \cos x = \frac{1}{2}\sqrt{3} - 2(\frac{1}{2}\sqrt{3})(\frac{1}{2}) = 0$

For $x = \pi$, $\quad \sin x - 2 \sin x \cos x = 0 - 2(0)(-1) = 0$

For $x = 5\pi/3$, $\sin x - 2 \sin x \cos x = -\frac{1}{2}\sqrt{3} - 2(-\frac{1}{2}\sqrt{3})(\frac{1}{2}) = 0$

Thus, the required solutions ($0 \leq x < 2\pi$) are $x = 0, \pi/3, \pi$, and $5\pi/3$.

(B) The various functions occurring in the equation may be expressed in terms of a single function.

EXAMPLE 15.3 Solve $2 \tan^2 x + \sec^2 x = 2$.

Replacing $\sec^2 x$ by $1 + \tan^2 x$, we have

$$2 \tan^2 x + (1 + \tan^2 x) = 2 \qquad 3 \tan^2 x = 1 \qquad \text{and} \qquad \tan x = \pm 1/\sqrt{3}$$

From $\tan x = 1/\sqrt{3}$, $x = \pi/6$ and $7\pi/6$; from $\tan x = -1/\sqrt{3}$, $x = 5\pi/6$ and $11\pi/6$. After checking each of these values in the orginal equation, we find that the required solutions ($0 \leq x < 2\pi$) are $x = \pi/6$, $5\pi/6$, $7\pi/6$, and $11\pi/6$.

The necessity of the check is illustrated in Examples 15.4 and 15.5.

EXAMPLE 15.4   Solve $\sec x + \tan x = 0$.

Multiplying the equation $\sec x + \tan x = \dfrac{1}{\cos x} + \dfrac{\sin x}{\cos x} = 0$ by $\cos x$, we have $1 + \sin x = 0$ or $\sin x = -1$; then $x = 3\pi/2$. However, neither $\sec x$ nor $\tan x$ is defined when $x = 3\pi/2$ and the equation has no solution.

(C)   Both members of the equation are squared.

EXAMPLE 15.5   Solve $\sin x + \cos x = 1$.

If the procedure of *(B)* were used, we would replace $\sin x$ by $\pm\sqrt{1 - \cos^2 x}$ or $\cos x$ by $\pm\sqrt{1 - \sin^2 x}$ and thereby introduce radicals. To avoid this, we write the equation in the form $\sin x = 1 - \cos x$ and square both members. We have

$$\sin^2 x = 1 - 2\cos x + \cos^2 x \tag{1}$$

$$1 - \cos^2 x = 1 - 2\cos x + \cos^2 x$$

$$2\cos^2 x - 2\cos x = 2\cos x \, (\cos x - 1) = 0$$

From $\cos x = 0$, $x = \pi/2$, $3\pi/2$; from $\cos x = 1$, $x = 0$.

Check: For $x = 0$, $\sin x + \cos x =$ $0 + 1 = 1$
          For $x = \pi/2$, $\sin x + \cos x =$ $1 + 0 = 1$
          For $x = 3\pi/2$, $\sin x + \cos x = -1 + 0 \neq 1$

Thus, the required solutions are $x = 0$ and $\pi/2$.

The value $x = 3\pi/2$, called an *extraneous solution*, was introduced by squaring the two members. Note that (1) is also obtained when both members of $\sin x = \cos x - 1$ are squared and that $x = 3\pi/2$ satisfies this latter relation.

(D)   Solutions are approximate values.

(NOTE: Since we will be using real number properties in solving the equation, the approximate values for the angles will be stated in radians which can be found using Table 3 in Appendix 2 or a calculator. These values are not exact and may not yield an exact check when substituted into the given equation.)

EXAMPLE 15.6   Solve $4 \sin \dot{x} = 3$.

$$4 \sin x = 3 \qquad \sin x = 3/4 = 0.75$$

The reference angle is 0.85 and the solutions for x are $x = 0.85$ and $x = \pi - 0.85 = 3.14 - 0.85 = 2.29$. (See Chap. 7 to review the use of reference angles.)

Check: For $x = 0.85$, $4 \sin 0.85 = 4(0.7513) = 3.0052 \approx 3$
          For $x = 2.29$, $4 \sin 2.29 = 4[\sin(3.14 - 2.29)] = 4[\sin 0.85] = 4[0.7513] = 3.0052 \approx 3$

If a calculator is used, $\sin 2.29$ is computed directly, so $4 \sin 2.29 = 4(0.7523) = 3.0092 \approx 3$.
Thus, the solutions to the nearest hundredth radian are 0.85 and 2.29.

(NOTE: Since the checks used approximate numbers, the symbol $\approx$ was used to indicate the result was approximately equal to the value needed.)

EXAMPLE 15.7   Solve $15 \cos^2 x + 7 \cos x - 2 = 0$.

$15 \cos^2 x + 7 \cos x - 2 = 0$, $(3 \cos x + 2)(5 \cos x - 1) = 0$, and $\cos x = -2/3 = -0.6667$ and $\cos x = 1/5 = 0.2$.

From $\cos x = -0.6667$, the reference angle is 0.84 and $x = \pi - 0.84 = 3.14 - 0.84 = 2.3$ and $x = \pi + 0.84 = 3.14 + 0.84 = 3.98$.

From $\cos x = 0.2$, the reference angle is 1.37 and $x = 1.37$ and $x = 2\pi - 1.37 = 6.28 - 1.37 = 4.91$.

Thus, to the nearest hundredth radian the solutions for x are 0.84, 1.37, 3.98, and 4.91.

(E)   Equation contains a multiple angle.

EXAMPLE 15.8   Solve $\cos 2x - 3 \sin x + 1 = 0$.

$\cos 2x - 3 \sin x + 1 = 0$, $(1 - 2 \sin^2 x) - 3 \sin x + 1 = 0$, $-2 \sin^2 x - 3 \sin x + 2 = 0$, $2 \sin^2 x + 3 \sin x - 2 = 0$, $(2 \sin x - 1)(\sin x + 2) = 0$, and $\sin x = \frac{1}{2}$ and $\sin x = -2$.
From $\sin x = \frac{1}{2}$, $x = \pi/6$ and $5\pi/6$.
From $\sin x = -2$, there are no solutions since $-1 \le \sin x \le 1$ for all x.
The solutions for x are $\pi/6$ and $5\pi/6$.

EXAMPLE 15.9   Solve $2 \cos^2 2x = \cos 2x$.

$2 \cos^2 2x = \cos 2x$, $2 \cos^2 2x - \cos 2x = 0$, $\cos 2x (2 \cos 2x - 1) = 0$, and $\cos 2x = 0$ and $\cos 2x = \frac{1}{2}$.
Since we want $0 \le x < 2\pi$, we find all values of $2x$ such that $0 \le 2x < 4\pi$.
From $\cos 2x = 0$, $2x = \pi/2$, $3\pi/2$, $5\pi/2$, and $7\pi/2$ and $x = \pi/4$, $3\pi/4$, $5\pi/4$, and $7\pi/4$.
From $\cos 2x = \frac{1}{2}$, $2x = \pi/3$, $5\pi/3$, $7\pi/3$, and $11\pi/3$ and $x = \pi/6$, $5\pi/6$, $7\pi/6$, and $11\pi/6$.
Thus the required angles x are $\pi/6$, $\pi/4$, $3\pi/4$, $5\pi/6$, $7\pi/6$, $5\pi/4$, $7\pi/4$, and $11\pi/6$.

(F)   Equations containing half angles.

EXAMPLE 15.10   Solve $4 \sin^2 \frac{1}{2}x = 1$.

First Solution. $4 \sin^2 \frac{1}{2}x = 1$, $4(\pm \sqrt{(1 - \cos x)/2})^2 = 1$, $2 - 2 \cos x = 1$, $\cos x = \frac{1}{2}$, and $x = \pi/3$ and $5\pi/3$.
The required solutions are $x = \pi/3$ and $5\pi/3$.

Second Solution. $4 \sin^2 \frac{1}{2}x = 1$, $\sin^2 \frac{1}{2}x = \frac{1}{4}$, and $\sin \frac{1}{2}x = \pm \frac{1}{2}$. Since we want $0 \le x < 2\pi$, we want all solutions for $\frac{1}{2}x$ such that $0 \le \frac{1}{2}x < \pi$.
From $\sin \frac{1}{2}x = \frac{1}{2}$, $\frac{1}{2}x = \pi/6$ and $5\pi/6$, and $x = \pi/3$ and $5\pi/3$.
From $\sin \frac{1}{2}x = -\frac{1}{2}$, there are no solutions since $\sin \frac{1}{2}x \ge 0$ for all x such that $0 \le \frac{1}{2}x < \pi$.
The required solutions are $x = \pi/3$ and $5\pi/3$.

Solved Problems

Solve each of the trigonometric equations in Probs. 15.1 to 15.19 for all x such that $0 \le x < 2\pi$. (If all solutions are required, adjoin $+ 2n\pi$, where n is zero or any positive or negative integer, to each result given.) In a number of the solutions, the details of the check have been omitted.

15.1 $2 \sin x - 1 = 0$.

Here $\sin x = 1/2$ and $x = \pi/6$ and $5\pi/6$.

15.2 $\sin x \cos x = 0$.

From $\sin x = 0$, $x = 0$ and π; from $\cos x = 0$, $x = \pi/2$ and $3\pi/2$.
The required solutions are $x = 0$, $\pi/2$, π, and $3\pi/2$.

15.3 $(\tan x - 1)(4 \sin^2 x - 3) = 0$.

From $\tan x - 1 = 0$, $\tan x = 1$ and $x = \pi/4$ and $5\pi/4$; from $4 \sin^2 x - 3 = 0$, $\sin x = \pm \sqrt{3}/2$ and $x = \pi/3$, $2\pi/3$, $4\pi/3$, and $5\pi/3$.
The required solutions are $x = \pi/4$, $\pi/3$, $2\pi/3$, $5\pi/4$, $4\pi/3$, and $5\pi/3$.

15.4 $\sin^2 x + \sin x - 2 = 0$.

Factoring, $(\sin x + 2)(\sin x - 1) = 0$.

From $\sin x + 2 = 0$, $\sin x = -2$ and there is no solution; from $\sin x - 1 = 0$, $\sin x = 1$ and $x = \pi/2$. The required solution is $x = \pi/2$.

15.5 $3 \cos^2 x = \sin^2 x$.

First Solution. Replacing $\sin^2 x$ by $1 - \cos^2 x$, we have $3 \cos^2 x = 1 - \cos^2 x$ or $4 \cos^2 x = 1$. Then $\cos x = \pm 1/2$ and the required solutions are $x = \pi/3$, $2\pi/3$, $4\pi/3$, and $5\pi/3$.

Second Solution. Dividing the equation by $\cos^2 x$, we have $3 = \tan^2 x$. Then $\tan x = \pm\sqrt{3}$ and the solutions above are obtained.

15.6 $2 \sin x - \csc x = 1$.

Multiplying the equation by $\sin x$, $2 \sin^2 x - 1 = \sin x$, and rearranging, we have

$$2 \sin^2 x - \sin x - 1 = (2 \sin x + 1)(\sin x - 1) = 0$$

From $2 \sin x + 1 = 0$, $\sin x = -1/2$ and $x = 7\pi/6$ and $11\pi/6$; from $\sin x = 1$, $x = \pi/2$.

Check: For $x = \pi/2$, $2 \sin x - \csc x = 2(1) - 1 = 1$
For $x = 7\pi/6$ and $11\pi/6$, $2 \sin x - \csc x = 2(-1/2) - (-2) = 1$

The solutions are $x = \pi/2$, $7\pi/6$, and $11\pi/6$.

15.7 $2 \sec x = \tan x + \cot x$.

Transforming to sines and cosines, and clearing of fractions, we have

$$\frac{2}{\cos x} = \frac{\sin x}{\cos x} + \frac{\cos x}{\sin x} \qquad \text{or} \qquad 2 \sin x = \sin^2 x + \cos^2 x = 1$$

Then $\sin x = 1/2$ and $x = \pi/6$ and $5\pi/6$.

15.8 $\tan x + 3 \cot x = 4$.

Multiplying by $\tan x$ and rearranging, we have $\tan^2 x - 4 \tan x + 3 = (\tan x - 1)(\tan x - 3) = 0$.
From $\tan x - 1 = 0$, $\tan x = 1$ and $x = \pi/4$ and $5\pi/4$; from $\tan x - 3 = 0$, $\tan x = 3$ and $x = 1.25$ and 4.39.

Check: For $x = \pi/4$ and $5\pi/4$, $\tan x + 3 \cot x = 1 + 3(1) = 4$
For $x = 1.25$ and 4.39, $\tan x + 3 \cot x = 3.0096 + 3(0.3323) = 4.0065 \approx 4$

The solutions are $\pi/4$, 1.25, $5\pi/4$, and 4.39.

15.9 $\csc x + \cot x = \sqrt{3}$.

First Solution. Writing the equation in the form $\csc x = \sqrt{3} - \cot x$ and squaring, we have

$$\csc^2 x = 3 - 2\sqrt{3} \cot x + \cot^2 x$$

Replacing $\csc^2 x$ by $1 + \cot^2 x$ and combining, we get $2\sqrt{3} \cot x - 2 = 0$. Then $\cot x = 1/\sqrt{3}$ and $x = \pi/3$ and $4\pi/3$.

Check: For $x = \pi/3$, $\csc x + \cot x = 2/\sqrt{3} + 1/\sqrt{3} = \sqrt{3}$
For $x = 4\pi/3$, $\csc x + \cot x = -2/\sqrt{3} + 1/\sqrt{3} \neq \sqrt{3}$

The required solution is $x = \pi/3$.

Second Solution. Upon making the indicated replacement, the equation becomes

$$\frac{1}{\sin x} + \frac{\cos x}{\sin x} = \sqrt{3} \qquad \text{and, clearing of fractions} \qquad 1 + \cos x = \sqrt{3}\,\sin x$$

Squaring both members, we have $1 + 2\cos x + \cos^2 x = 3\sin^2 x = 3(1 - \cos^2 x)$ or

$$4\cos^2 x + 2\cos x - 2 = 2(2\cos x - 1)(\cos x + 1) = 0$$

From $2\cos x - 1 = 0$, $\cos x = 1/2$ and $x = \pi/3$ and $5\pi/3$; from $\cos x + 1 = 0$, $\cos x = -1$ and $x = \pi$.

Now $x = \pi/3$ is the solution. The values $x = \pi$ and $5\pi/3$ are to be excluded since $\csc \pi$ is not defined while $\csc 5\pi/3$ and $\cot 5\pi/3$ are both negative.

15.10 $\cos x - \sqrt{3}\,\sin x = 1.$

Putting the equation in the form $\cos x - 1 = \sqrt{3}\,\sin x$ and squaring, we have

$$\cos^2 x - 2\cos x + 1 = 3\sin^2 x = 3(1 - \cos^2 x)$$

then, combining and factoring,

$$4\cos^2 x - 2\cos x - 2 = 2(2\cos x + 1)(\cos x - 1) = 0$$

From $2\cos x + 1 = 0$, $\cos x = -1/2$ and $x = 2\pi/3$ and $4\pi/3$; from $\cos x - 1 = 0$, $\cos x = 1$ and $x = 0$.

Check: For $x = 0$, $\cos x - \sqrt{3}\,\sin x = 1 - \sqrt{3}\,(0) = 1$
For $x = 2\pi/3$, $\cos x - \sqrt{3}\,\sin x = -1/2 - \sqrt{3}\,(\sqrt{3}/2) \neq 1$
For $x = 4\pi/3$, $\cos x - \sqrt{3}\,\sin x = -1/2 - \sqrt{3}\,(-\sqrt{3}/2) = 1$

The required solutions are $x = 0$ and $4\pi/3$.

15.11 $2\cos x = 1 - \sin x.$

As in Prob. 15.10, we obtain

$$4\cos^2 x = 1 - 2\sin x + \sin^2 x$$

$$4(1 - \sin^2 x) = 1 - 2\sin x + \sin^2 x$$

$$5\sin^2 x - 2\sin x - 3 = (5\sin x + 3)(\sin x - 1) = 0.$$

From $5\sin x + 3 = 0$, $\sin x = -3/5 = -0.6000$, the reference angle is 0.64, and $x = 3.78$ and 5.64; from $\sin x - 1 = 0$, $\sin x = 1$ and $x = \pi/2$.

Check: For $x = \pi/2$, $2(0) = 1 - 1$
For $x = 3.78$, $2(-0.8021) \neq 1 - (-0.5972)$
For $x = 5.64$, $2(0.8021) \approx 1 - (-0.5972)$

The required solutions are $x = \pi/2$ and 5.64.

Equations Involving Multiple Angles and Half Angles.

15.12 $\sin 3x = -\frac{1}{2}\sqrt{2}.$

Since we require x such that $0 \leq x < 2\pi$, $3x$ must be such that $0 \leq 3x < 6\pi$.

Then $3x = 5\pi/4, \ 7\pi/4, \ 13\pi/4, \ 15\pi/4, \ 21\pi/4, \ 23\pi/4$

and $x = 5\pi/12, \ 7\pi/12, \ 13\pi/12, \ 5\pi/4, \ 7\pi/4, \ 23\pi/12$

Each of these values is a solution.

15.13 $\cos \frac{1}{2}x = \frac{1}{2}$.

Since we require x such that $0 \leqq x < 2\pi$, $\frac{1}{2}x$ must be such that $0 \leqq \frac{1}{2}x < \pi$.
Then $\frac{1}{2}x = \pi/3$ and $x = 2\pi/3$.

15.14 $\sin 2x + \cos x = 0$.

Substituting for $\sin 2x$, we have $2 \sin x \cos x + \cos x = \cos x \, (2 \sin x + 1) = 0$.

From $\cos x = 0$, $x = \pi/2$ and $3\pi/2$; from $\sin x = -1/2$, $x = 7\pi/6$ and $11\pi/6$.

The required solutions are $x = \pi/2$, $7\pi/6$, $3\pi/2$, and $11\pi/6$.

15.15 $2 \cos^2 \frac{1}{2}x = \cos^2 x$.

First Solution. Substituting $1 + \cos x$ for $2 \cos^2 \frac{1}{2}x$, the equation becomes $\cos^2 x - \cos x - 1 = 0$; then
$\cos x = \dfrac{1 \pm \sqrt{5}}{2} = 1.6180,\ -0.6180$. Since $\cos x$ cannot exceed 1, we consider $\cos x = -0.6180$ and obtain
the solutions $x = 2.24$ and 4.04.

Second Solution. To solve $\sqrt{2} \cos \frac{1}{2}x = \cos x$ and $\sqrt{2} \cos \frac{1}{2}x = -\cos x$, we square and obtain the equation of
this problem. The solution of the first of these equations is 4.04 and the solution of the second is 2.24.

15.16 $\cos 2x + \cos x + 1 = 0$.

Substituting $2 \cos^2 x - 1$ for $\cos 2x$, we have $2 \cos^2 x + \cos x \, (2 \cos x + 1) = 0$.

From $\cos x = 0$, $x = \pi/2$ and $3\pi/2$; from $\cos x = -1/2$, $x = 2\pi/3$ and $4\pi/3$.

The required solutions are $x = \pi/2$, $2\pi/3$, $3\pi/2$, and $4\pi/3$.

15.17 $\tan 2x + 2 \sin x = 0$.

Using $\tan 2x = \dfrac{\sin 2x}{\cos 2x} = \dfrac{2 \sin x \cos x}{\cos 2x}$, we have

$$\frac{2 \sin x \cos x}{\cos 2x} + 2 \sin x = 2 \sin x \left(\frac{\cos x}{\cos 2x} + 1 \right) = 2 \sin x \left(\frac{\cos x + \cos 2x}{\cos 2x} \right) = 0$$

From $\sin x = 0$, $x = 0$, π; from $\cos x + \cos 2x = \cos x + 2 \cos^2 x - 1 = (2 \cos x - 1)(\cos x + 1) = 0$,
$x = \pi/3$, $5\pi/3$, and π. The required solutions are $x = 0$, $\pi/3$, π, and $5\pi/3$.

15.18 $\sin 2x = \cos 2x$.

First Solution. Let $2x = \theta$; then we are to solve $\sin \theta = \cos \theta$ for $0 \leqq \theta < 4\pi$. Then $\theta = \pi/4$, $5\pi/4$, $9\pi/4$, and
$13\pi/4$ and $x = \theta/2 = \pi/8$, $5\pi/8$, $9\pi/8$, and $13\pi/8$ are the solutions.
Second Solution. Dividing by $\cos 2x$, the equation becomes $\tan 2x = 1$ for which $2x = \pi/4$, $5\pi/4$, $9\pi/4$, and
$13\pi/4$ as in the first solution.

15.19 $\sin 2x = \cos 4x$.

Since $\cos 4x = \cos 2(2x) = 1 - 2 \sin^2 2x$, the equation becomes

$$2 \sin^2 2x + \sin 2x - 1 = (2 \sin 2x - 1)(\sin 2x + 1) = 0$$

From $2 \sin 2x - 1 = 0$ or $\sin 2x = 1/2$, $2x = \pi/6$, $5\pi/6$, $13\pi/6$, and $17\pi/6$ and $x = \pi/12$, $5\pi/12$, $13\pi/12$, and
$17\pi/12$; from $\sin 2x + 1 = 0$ or $\sin 2x = -1$, $2x = 3\pi/2$ and $7\pi/2$ and $x = 3\pi/4$ and $7\pi/4$. All these values are
solutions.

15.20 Solve the system

$$r \sin \theta = 3 \qquad (1)$$

$$r = 4(1 + \sin \theta) \qquad (2)$$

for $r > 0$ and $0 \le \theta < 2\pi$.

Dividing (2) by (1), $1/\sin \theta = 4(1 + \sin \theta)/3$ or $4 \sin^2 \theta + 4 \sin \theta - 3 = 0$ and

$$(2 \sin \theta + 3)(2 \sin \theta - 1) = 0$$

From $2 \sin \theta - 1 = 0$, $\sin \theta = 1/2$ and $\theta = \pi/6$ and $5\pi/6$; using (1), $r(1/2) = 3$ and $r = 6$. Note that $2 \sin \theta + 3 = 0$ is excluded since when $r > 0$, $\sin \theta > 0$ by (1).

The required solutions are $\theta = \pi/6$ and $r = 6$ and $\theta = 5\pi/6$ and $r = 6$.

15.21 Solve Arccos $2x$ = Arcsin x.

If x is positive, α = Arccos $2x$ and β = Arcsin x terminate in quadrant I; if x is negative, α terminates in quadrant II and β terminates in quadrant IV. Thus, x must be positive.

For x positive, $\sin \beta = x$ and $\cos \beta = \sqrt{1 - x^2}$. Taking the cosine of both members of the given equation, we have

$$\cos (\text{Arccos } 2x) = \cos (\text{Arcsin } x) = \cos \beta \qquad \text{or} \qquad 2x = \sqrt{1 - x^2}$$

Squaring, $4x^2 = 1 - x^2$, $5x^2 = 1$, and $x = \sqrt{5}/5 = 0.4472$.

Check: Arccos $2x$ = Arccos $0.8944 = 0.46 = \arcsin 0.4472$, approximating the angle to the nearest hundredth radian.

15.22 Solve Arccos $(2x^2 - 1) = 2$ Arccos $\frac{1}{2}$.

Let α = Arccos $(2x^2 - 1)$ and β = Arccos $\frac{1}{2}$; then $\cos \alpha = 2x^2 - 1$ and $\cos \beta = \frac{1}{2}$.
Taking the cosine of both members of the given equation,

$$\cos \alpha = 2x^2 - 1 = \cos 2\beta = 2 \cos^2 \beta - 1 = 2(\tfrac{1}{2})^2 - 1 = -\tfrac{1}{2}$$

Then $2x^2 = \frac{1}{2}$ and $x = \pm \frac{1}{2}$.

Check: For $x = \pm \frac{1}{2}$, Arccos $(-\frac{1}{2}) = 2$ Arccos $\frac{1}{2}$ or $2\pi/3 = 2(\pi/3)$.

15.23 Solve Arccos $2x$ − Arccos $x = \pi/3$.

If x is positive, $0 <$ Arccos $2x <$ Arccos x; if x is negative, Arccos $2x >$ Arccos $x > 0$. Thus, x must be negative.

Let α = Arccos $2x$ and β = Arccos x; then $\cos \alpha = 2x$, $\sin \alpha = \sqrt{1 - 4x^2}$, $\cos \beta = x$, and $\sin \beta = \sqrt{1 - x^2}$ since both α and β terminate in quadrant II.

Taking the cosine of both members of the given equation,

$$\cos (\alpha - \beta) = \cos \alpha \cos \beta + \sin \alpha \sin \beta = 2x^2 + \sqrt{1 - 4x^2} \sqrt{1 - x^2} = \cos \pi/3 = \tfrac{1}{2}$$

or

$$\sqrt{1 - 4x^2} \sqrt{1 - x^2} = \tfrac{1}{2} - 2x^2$$

Squaring, $1 - 5x^2 + 4x^4 = \frac{1}{4} - 2x^2 + 4x^4$, $3x^2 = \frac{3}{4}$, and $x = -\frac{1}{2}$.

Check: Arccos (-1) − Arccos $(-\frac{1}{2}) = \pi - 2\pi/3 = \pi/3$.

15.24 Solve Arcsin $2x = \frac{1}{4}\pi$ − Arcsin x.

Let α = Arcsin $2x$ and β = Arcsin x; then $\sin \alpha = 2x$ and $\sin \beta = x$. If x is negative, α and β terminate in quadrant IV; thus, x must be positive and β acute.

Taking the sine of both members of the given equation,

$$\sin \alpha = \sin (\tfrac{1}{4}\pi - \beta) = \sin \tfrac{1}{4}\pi \cos \beta - \cos \tfrac{1}{4}\pi \sin \beta$$

or $\qquad 2x = \tfrac{1}{2}\sqrt{2}\sqrt{1 - x^2} - \tfrac{1}{2}\sqrt{2}x \qquad$ and $\qquad (2\sqrt{2} + 1)x = \sqrt{1 - x^2}$

Squaring, $(8 + 4\sqrt{2} + 1)x^2 = 1 - x^2$, $x^2 = 1/(10 + 4\sqrt{2})$, and $x = 0.2527$.

Check: Arcsin $0.5054 = 0.53$; Arcsin $0.2527 = 0.26$, $\tfrac{1}{4}\pi = \tfrac{1}{4}(3.14) = 0.79$; and $0.53 = 0.79 - 0.26$.

15.25   Solve Arctan x + Arctan $(1 - x)$ = Arctan $\tfrac{4}{3}$.

Let α = Arctan x and β = Arctan $(1 - x)$; then tan $\alpha = x$ and tan $\beta = 1 - x$.

Taking the tangent of both members of the given equation,

$$\tan (\alpha + \beta) = \frac{\tan \alpha + \tan \beta}{1 - \tan \alpha \tan \beta} = \frac{x + (1 - x)}{1 - x(1 - x)} = \frac{1}{1 - x + x^2} = \tan (\text{Arctan} \tfrac{4}{3}) = \tfrac{4}{3}$$

Then $3 = 4 - 4x + 4x^2$, $4x^2 - 4x + 1 = (2x - 1)^2 = 0$, and $x = \tfrac{1}{2}$.

Check: Arctan $\tfrac{1}{2}$ + Arctan $(1 - \tfrac{1}{2})$ = 2 Arctan $0.5000 = 2(0.46) = 0.92$ and Arctan $\tfrac{4}{3}$ = Arctan $1.3333 = 0.93$.

Supplementary Problems

Solve each of the following equations for all x such that $0 \le x < 2\pi$. Use Table 3 in Appendix 2 when finding approximate values for x.

15.26 $\sin x = \sqrt{3}/2$.   *Ans.* $\pi/3, 2\pi/3$

15.27 $\cos^2 x = 1/2$.   *Ans.* $\pi/4, 3\pi/4, 5\pi/4, 7\pi/4$

15.28 $\sin x \cos x = 0$.   *Ans.* $0, \pi/2, \pi, 3\pi/2$

15.29 $(\tan x - 1)(2 \sin x + 1) = 0$.   *Ans.* $\pi/4, 7\pi/6, 5\pi/4, 11\pi/6$

15.30 $2 \sin^2 x - \sin x - 1 = 0$.   *Ans.* $\pi/2, 7\pi/6, 11\pi/6$

15.31 $\sin 2x + \sin x = 0$.   *Ans.* $0, 2\pi/3, \pi, 4\pi/3$

15.32 $\cos x + \cos 2x = 0$.   *Ans.* $\pi/3, \pi, 5\pi/3$

15.33 $2 \tan x \sin x - \tan x = 0$.   *Ans.* $0, \pi/6, 5\pi/6, \pi$

15.34 $2 \cos x + \sec x = 3$.   *Ans.* $0, \pi/3, 5\pi/3$

15.35 $2 \sin x + \csc x = 3$.   *Ans.* $\pi/6, \pi/2, 5\pi/6$

15.36 $\sin x + 1 = \cos x$.   *Ans.* $0, 3\pi/2$

15.37 $\sec x - 1 = \tan x$.   *Ans.* 0

15.38 $2 \cos x + 3 \sin x = 2$.   *Ans.* $0, 1.96$

15.39 $3 \sin x + 5 \cos x + 5 = 0$.   *Ans.* $\pi, 4.22$

15.40 $1 + \sin x = 2 \cos x$.   *Ans.* 0.64, $3\pi/2$

15.41 $3 \sin x + 4 \cos x = 2$.   *Ans.* 1.80, 5.76

15.42 $\sin 2x = -\sqrt{3}/2$.   *Ans.* $2\pi/3$, $5\pi/6$, $5\pi/3$, $11\pi/6$

15.43 $\tan 3x = 1$.   *Ans.* $\pi/12$, $5\pi/12$, $3\pi/4$, $13\pi/12$, $17\pi/12$, $7\pi/4$

15.44 $\cos x/2 = \sqrt{3}/2$.   *Ans.* $\pi/3$

15.45 $\cot x/3 = -1/\sqrt{3}$.   *Ans.* No solution in given interval

15.46 $\sin x \cos x = 1/2$.   *Ans.* $\pi/4$, $5\pi/4$

15.47 $\sin x/2 + \cos x = 1$.   *Ans.* 0, $\pi/3$, $5\pi/3$

Solve each of the following systems for $r \geqq 0$ and $0 \leqq \theta < 2\pi$.

15.48 $r = a \sin \theta$ *Ans.* $\theta = \pi/6$, $r = a/2$
$r = a \cos 2\theta$ $\theta = 5\pi/6$, $r = a/2$; $\theta = 3\pi/2$, $r = -a$

15.49 $r = a \cos \theta$ *Ans.* $\theta = \pi/2$, $r = 0$; $\theta = 3\pi/2$, $r = 0$
$r = a \sin 2\theta$ $\theta = \pi/6$, $r = \sqrt{3}a/2$
$\theta = 5\pi/6$, $r = -\sqrt{3}a/2$

15.50 $r = 4(1 + \cos \theta)$ *Ans.* $\theta = \pi/3$, $r = 6$
$r = 3 \sec \theta$ $\theta = 5\pi/3$, $r = 6$

Solve each of the following equations.

15.51 Arctan $2x$ + Arctan $x = \pi/4$.   *Ans.* $x = 0.2808$

15.52 Arcsin x + Arctan $x = \pi/2$.   *Ans.* $x = 0.7862$

15.53 Arccos x + Arctan $x = \pi/2$.   *Ans.* $x = 0$

Chapter 16

Complex Numbers

16.1 IMAGINARY NUMBERS

The square root of a negative number (e.g., $\sqrt{-1}$, $\sqrt{-5}$, and $\sqrt{-9}$) is called an *imaginary number*. Since by definition $\sqrt{-5} = \sqrt{5} \cdot \sqrt{-1}$ and $\sqrt{-9} = \sqrt{9} \cdot \sqrt{-1} = 3\sqrt{-1}$, it is convenient to introduce the symbol $i = \sqrt{-1}$ and to adopt $\sqrt{-5} = i\sqrt{5}$ and $\sqrt{-9} = 3i$ as the standard form for these numbers.

The symbol i has the property $i^2 = -1$; and for higher integral powers we have $i^3 = i^2 \cdot i = (-1)i = -i$, $i^4 = (i^2)^2 = (-1)^2 = 1$, $i^5 = i^4 \cdot i = i$, etc.

The use of the standard form simplifies the operations on imaginary numbers and eliminates the possibility of certain common errors. Thus $\sqrt{-9} \cdot \sqrt{4} = \sqrt{-36} = 6i$ since $\sqrt{-9} \cdot \sqrt{4} = 3i(2) = 6i$ but $\sqrt{-9} \cdot \sqrt{-4} \neq \sqrt{36}$ since $\sqrt{-9} \cdot \sqrt{-4} = (3i)(2i) = 6i^2 = -6$.

16.2 COMPLEX NUMBERS

A number $a + bi$, where a and b are real numbers, is called a *complex number*. The first term a is called the *real part* of the complex number and the second term bi is called the *imaginary part*.

Complex numbers may be thought of as including all real numbers and all imaginary numbers. For example, $5 = 5 + 0i$ and $3i = 0 + 3i$.

Two complex numbers $a + bi$ and $c + di$ are said to be *equal* if and only if $a = c$ and $b = d$.

The *conjugate* of a complex number $a + bi$ is the complex number $a - bi$. Thus, $2 + 3i$ and $2 - 3i$, and $-3 + 4i$ and $-3 - 4i$ are pairs of conjugate complex numbers.

16.3 ALGEBRAIC OPERATIONS

Addition. To add two complex numbers, add the real parts and add the imaginary parts.

EXAMPLE 16.1 $(2 + 3i) + (4 - 5i) = (2 + 4) + (3 - 5)i = 6 - 2i$.

Subtraction. To subtract two complex numbers, subtract the real parts and subtract the imaginary parts.

EXAMPLE 16.2 $(2 + 3i) - (4 - 5i) = (2 - 4) + [3 - (-5)]i = -2 + 8i$.

Multiplication. To multiply two complex numbers, carry out the multiplication as if the numbers were ordinary binomials and replace i^2 by -1.

EXAMPLE 16.3 $(2 + 3i)(4 - 5i) = 8 + 2i - 15i^2 = 8 + 2i - 15(-1) = 23 + 2i$.

Division. To divide two complex numbers, multiply both numerator and denominator of the fraction by the conjugate of the denominator.

EXAMPLE 16.4 $\dfrac{2 + 3i}{4 - 5i} = \dfrac{(2 + 3i)(4 + 5i)}{(4 - 5i)(4 + 5i)} = \dfrac{(8 - 15) + (10 + 12)i}{16 + 25} = -\dfrac{7}{41} + \dfrac{22}{41}i$.

[Note the form of the result; it is neither $\dfrac{-7 + 22i}{41}$ nor $\dfrac{1}{41}(-7 + 22i)$.]

(See Probs. 16.1 to 16.9.)

16.4 GRAPHIC REPRESENTATION OF COMPLEX NUMBERS

The complex number $x + yi$ may be represented graphically by the point P [(see Fig. 16-1(a)] whose rectangular coordinates are (x, y).

The point O, having coordinates $(0, 0)$ represents the complex number $0 + 0i = 0$. All points on the x axis have coordinates of the form $(x, 0)$ and correspond to real numbers $x + 0i = x$. For this reason, the x axis is called the *axis of reals*. All points on the y axis have coordinates of the form $(0, y)$ and correspond to imaginary numbers $0 + yi = yi$. The y axis is called the *axis of imaginaries*. The plane on which the complex numbers are represented is called the *complex plane*.

In addition to representing a complex number by a point P in the complex plane, the number may be represented [see Fig. 16-1(b)] by the directed line segment or vector OP.

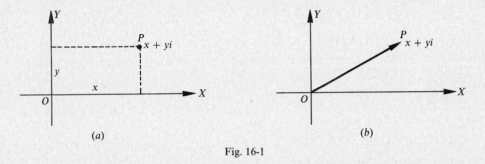

Fig. 16-1

16.5 GRAPHIC REPRESENTATION OF ADDITION AND SUBTRACTION

Let $z_1 = x_1 + iy_1$ and $z_2 = x_2 + iy_2$ be two complex numbers. The vector representation of these numbers [Fig. 16-2(a)] suggests the familiar parallelogram law for determining graphically the sum $z_1 + z_2 = (x_1 + iy_1) + (x_2 + iy_2)$.

Since $z_1 - z_2 = (x_1 + iy_1) - (x_2 + iy_2) = (x_1 + iy_1) + (-x_2 - iy_2)$, the difference $z_1 - z_2$ of the two complex numbers may be obtained graphically by applying the parallelogram law to $x_1 + iy_1$ and $-x_2 - iy_2$. [See Fig. 16-2(b).]

In Fig. 16-2(c) both the sum $OR = z_1 + z_2$ and the difference $OS = z_1 - z_2$ are shown. Note that the segments OS and P_2P_1 (the other diagonal of OP_2RP_1) are equal.

(See Prob. 16.11.)

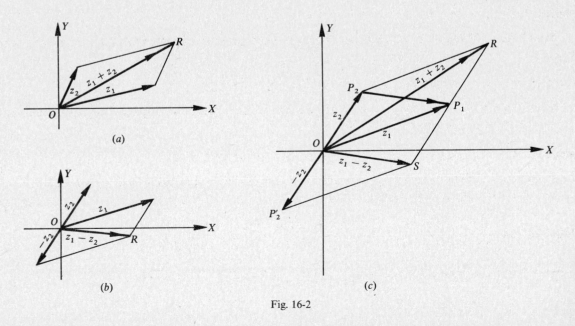

Fig. 16-2

16.6 POLAR OR TRIGONOMETRIC FORM OF COMPLEX NUMBERS

Let the complex number $x + yi$ be represented (Fig. 16-3(a)) by the vector OP. This vector (and hence the complex number) may be described in terms of the length r of the vector and *any* positive angle θ which the vector makes with the positive x axis (axis of positive reals). The number $r = \sqrt{x^2 + y^2}$ is called the *modulus* or *absolute value* of the complex number. The angle θ, called the *amplitude* of the complex number, is usually chosen as the smallest positive angle for which $\tan \theta = y/x$, but at times it will be found more convenient to choose some other angle coterminal with it.

From Fig. 16-3(a), $x = r \cos \theta$ and $y = r \sin \theta$; then $z = x + yi = r \cos \theta + ir \sin \theta = r(\cos \theta + i \sin \theta)$. We call $z = r(\cos \theta + i \sin \theta)$ the *polar* or *trigonometric form* and $z = x + yi$ the *rectangular form* of the complex number z. An abbreviated notation is sometimes used and is written $z = r \operatorname{cis} \theta$.

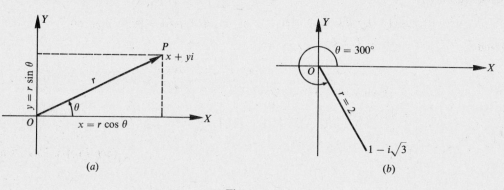

Fig. 16-3

EXAMPLE 16.5 Express $z = 1 - i\sqrt{3}$ in polar form. [See Fig. 16-3(b).]

The modulus is $r = \sqrt{(1)^2 + (-\sqrt{3})^2} = 2$. Since $\tan \theta = y/x = -\sqrt{3}/1 = -\sqrt{3}$, the amplitude θ is either 120° or 300°. Now we know that P lies in quadrant IV; hence, $\theta = 300°$ and the required polar form is $z = r(\cos \theta + i \sin \theta) = 2(\cos 300° + i \sin 300°)$. Note that z may also be represented in polar form by $z = 2[\cos (300° + n360°) + i \sin (300° + n360°)]$, where n is any integer.

EXAMPLE 16.6 Express the complex number $z = 8(\cos 210° + i \sin 210°)$ in rectangular form.

Since $\cos 210° = -\sqrt{3}/2$ and $\sin 210° = -1/2$,

$$z = 8(\cos 210° + i \sin 210°) = 8[-\sqrt{3}/2 + i(-1/2)] = -4\sqrt{3} - 4i$$

is the required rectangular form.

(See Probs. 16.12 to 16.13.)

16.7 MULTIPLICATION AND DIVISION IN POLAR FORM

Multiplication. The modulus of the product of two complex numbers is the product of their moduli, and the amplitude of the product is the sum of their amplitudes.

Division. The modulus of the quotient of two complex numbers is the modulus of the dividend divided by the modulus of the divisor, and the amplitude of the quotient is the amplitude of the dividend minus the amplitude of the divisor. For a proof of these theorems, see Prob. 16.14.

EXAMPLE 16.7 Find (a) the product $z_1 z_2$, (b) the quotient z_1/z_2, and (c) the quotient z_2/z_1 where $z_1 = 2(\cos 300° + i \sin 300°)$ and $z_2 = 8(\cos 210° + i \sin 210°)$.

(a) The modulus of the product is $2(8) = 16$. The amplitude is $300° + 210° = 510°$, but, following the convention, we shall use the smallest positive coterminal angle $510° - 360° = 150°$. Thus $z_1z_2 = 16(\cos 150° + i \sin 150°)$.

(b) The modulus of the quotient z_1/z_2 is $2/8 = \frac{1}{4}$ and the amplitude is $300° - 210° = 90°$. Thus $z_1/z_2 = \frac{1}{4}(\cos 90° + i \sin 90°)$.

(c) The modulus of the quotient z_2/z_1 is $8/2 = 4$. The amplitude is $210° - 300° = -90°$, but we shall use the smallest positive coterminal angle $-90° + 360° = 270°$. Thus

$$z_2/z_1 = 4(\cos 270° + i \sin 270°)$$

[NOTE.   From Examples 16.5 and 16.6 the numbers are

$$z_1 = 1 - i\sqrt{3} \qquad \text{and} \qquad z_2 = -4\sqrt{3} - 4i$$

in rectangular form. Then

$$z_1z_2 = (1 - i\sqrt{3})(-4\sqrt{3} - 4i) = -8\sqrt{3} + 8i = 16(\cos 150° + i \sin 150°)$$

as in (a), and

$$\frac{z_2}{z_1} = \frac{-4\sqrt{3} - 4i}{1 - i\sqrt{3}} = \frac{(-4\sqrt{3} - 4i)(1 + i\sqrt{3})}{(1 - i\sqrt{3})(1 + i\sqrt{3})} = \frac{-16i}{4} = -4i$$

$$= 4(\cos 270° + i \sin 270°)$$

as in (c).]

(See Probs. 16.15 and 16.16.)

16.8   DE MOIVRE'S THEOREM

If n is any rational number,

$$[r(\cos \theta + i \sin \theta)]^n = r^n(\cos n\theta + i \sin n\theta)$$

A proof of this theorem is beyond the scope of this book; a verification for $n = 2$ and $n = 3$ is given in Prob. 16.17.

EXAMPLE 16.8
$$(\sqrt{3} - i)^{10} = [2(\cos 330° + i \sin 330°)]^{10}$$
$$= 2^{10}(\cos 10 \cdot 330° + i \sin 10 \cdot 330°)$$
$$= 1024(\cos 60° + i \sin 60°) = 1024(1/2 + i\sqrt{3}/2)$$
$$= 512 + 512i\sqrt{3}$$

(See Prob. 16.18.)

16.9   ROOTS OF COMPLEX NUMBERS

We state, without proof, the theorem: A complex number $a + bi = r(\cos \theta + i \sin \theta)$ has exactly n distinct nth roots.

The procedure for determining these roots is given in Example 16.9.

EXAMPLE 16.9   Find all fifth roots of $4 - 4i$.

The usual polar form of $4 - 4i$ is $4\sqrt{2}(\cos 315° + i \sin 315°)$ but we shall need the more general form

$$4\sqrt{2}[\cos (315° + k360°) + i \sin (315° + k360°)]$$

where k is any integer, including zero.

Using De Moivre's theorem, a fifth root of $4 - 4i$ is given by

$$\{4\sqrt{2}[\cos(315° + k360°) + i\sin(315° + k360°)]\}^{1/5}$$

$$= (4\sqrt{2})^{1/5}\left(\cos\frac{315° + k360°}{5} + i\sin\frac{315° + k360°}{5}\right)$$

$$= \sqrt{2}[\cos(63° + k72°) + i\sin(63° + k72°)]$$

Assigning in turn the values $k = 0, 1, 2, \ldots$, we find

$$k = 0: \quad \sqrt{2}(\cos 63° + i\sin 63°) = R_1$$
$$k = 1: \quad \sqrt{2}(\cos 135° + i\sin 135°) = R_2$$
$$k = 2: \quad \sqrt{2}(\cos 207° + i\sin 207°) = R_3$$
$$k = 3: \quad \sqrt{2}(\cos 279° + i\sin 279°) = R_4$$
$$k = 4: \quad \sqrt{2}(\cos 351° + i\sin 351°) = R_5$$
$$k = 5: \quad \sqrt{2}(\cos 423° + i\sin 423°)$$
$$= \sqrt{2}(\cos 63° + i\sin 63°) = R_1, \text{ etc.}$$

Thus, the five fifth roots are obtained by assigning the values 0, 1, 2, 3, 4 (i.e., 0, 1, 2, 3, $\ldots$, $n - 1$) to k.

(See also Prob. 16.19.)

The modulus of each of the roots is $\sqrt{2}$; hence these roots lie on a circle of radius $\sqrt{2}$ with center at the origin. The difference in amplitude of two consecutive roots is $72°$; hence the roots are equally spaced on this circle, as shown in Fig. 16-4.

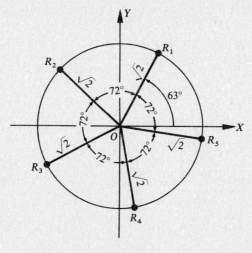

Fig. 16-4

Solved Problems

In Probs. 16.1 to 16.6, perform the indicated operations, simplify, and write the result in the form $a + bi$.

16.1 $(3 - 4i) + (-5 + 7i) = (3 - 5) + (-4 + 7)i = -2 + 3i$

16.2 $(4 + 2i) - (-1 + 3i) = [4 - (-1)] + (2 - 3)i = 5 - i$

16.3 $(2 + i)(3 - 2i) = (6 + 2) + (-4 + 3)i = 8 - i$

16.4 $(3 + 4i)(3 - 4i) = 9 + 16 = 25$

16.5 $\dfrac{1 + 3i}{2 + i} = \dfrac{(1 + 3i)(2 - i)}{(2 + i)(2 - i)} = \dfrac{(2 + 3) + (-1 + 6)i}{4 + 1} = 1 + i$

16.6 $\dfrac{3 - 2i}{2 - 3i} = \dfrac{(3 - 2i)(2 + 3i)}{(2 - 3i)(2 + 3i)} = \dfrac{(6 + 6) + (9 - 4)i}{4 + 9} = \dfrac{12}{13} + \dfrac{5}{13}i$

16.7   Find x and y such that $2x - yi = 4 + 3i$.

   Here $2x = 4$ and $-y = 3$; then $x = 2$ and $y = -3$.

16.8   Show that the conjugate complex numbers $2 + i$ and $2 - i$ are roots of the quadratic equation $x^2 - 4x + 5 = 0$.

   For $x = 2 + i$: $(2 + i)^2 - 4(2 + i) + 5 = 4 + 4i + i^2 - 8 - 4i + 5 = 0$.
   For $x = 2 - i$: $(2 - i)^2 - 4(2 - i) + 5 = 4 - 4i + i^2 - 8 + 4i + 5 = 0$.
   Since each number satisfies the equation, it is a root of the equation.

16.9   Show that the conjugate of the sum of two complex numbers is equal to the sum of their conjugates.

   Let the complex numbers be $a + bi$ and $c + di$. Their sum is $(a + c) + (b + d)i$ and the conjugate of the sum is $(a + c) - (b + d)i$.

   The conjugates of the two given numbers are $a - bi$ and $c - di$, and their sum is

$$(a + c) + (-b - d)i = (a + c) - (b + d)i$$

16.10   Represent graphically (as a vector) the following complex numbers:

   (a) $3 + 2i$,   (b) $2 - i$,   (c) $-2 + i$,   (d) $-1 - 3i$

   We locate, in turn, the points whose coordinates are $(3, 2), (2, -1), (-2, 1), (-1, -3)$ and join each to the origin O.

16.11   Perform graphically the following operations:

   (a) $(3 + 4i) + (2 + 5i)$,   (b) $(3 + 4i) + (2 - 3i)$,   (c) $(4 + 3i) - (2 + i)$,   (d) $(4 + 3i) - (2 - i)$

   For (a) and (b), draw as in Fig. 16-5(a) and (b) the two vectors and apply the parallelogram law.

   For (c) draw the vectors representing $4 + 3i$ and $-2 - i$ and apply the parallelogram law as in Fig. 16-5(c).

   For (d) draw the vectors representing $4 + 3i$ and $-2 + i$ and apply the parallelogram law as in Fig. 16-5(d).

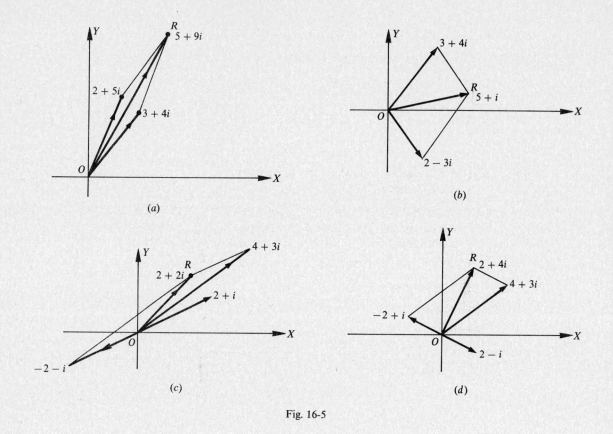

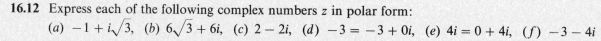

Fig. 16-5

16.12 Express each of the following complex numbers z in polar form:

(a) $-1 + i\sqrt{3}$, (b) $6\sqrt{3} + 6i$, (c) $2 - 2i$, (d) $-3 = -3 + 0i$, (e) $4i = 0 + 4i$, (f) $-3 - 4i$

(a) P lies in the second quadrant; $r = \sqrt{(-1)^2 + (\sqrt{3})^2} = 2$; $\tan \theta = \sqrt{3}/(-1) = -\sqrt{3}$ and $\theta = 120°$. Thus, $z = 2(\cos 120° + i \sin 120°)$.

(b) P lies in the first quadrant; $r\sqrt{(6\sqrt{3})^2 + 6^2} = 12$; $\tan \theta = 6/6\sqrt{3} = 1/\sqrt{3}$ and $\theta = 30°$. Thus, $z = 12(\cos 30° + i \sin 30°)$.

(c) P lies in the fourth quadrant; $r = \sqrt{2^2 + (-2)^2} = 2\sqrt{2}$; $\tan \theta = -2/2 = -1$ and $\theta = 315°$. Thus, $z = 2\sqrt{2}(\cos 315° + i \sin 315°)$.

(d) P lies on the negative x axis and $\theta = 180°$; $r = \sqrt{(-3)^2 + 0^2} = 3$. Thus, $z = 3(\cos 180° + i \sin 180°)$.

(e) P lies on the positive y axis and $\theta = 90°$; $r = \sqrt{0^2 + 4^2} = 4$. Thus, $z = 4(\cos 90° + i \sin 90°)$.

(f) P lies in the third quadrant; $r = \sqrt{(-3)^2 + (-4)^2} = 5$; $\tan \theta = -4/(-3) = 1.3333$ and $\theta = 233°8'$. Thus, $z = 5(\cos 233°8' + i \sin 233°8')$. θ is not a special angle, so it must be approximated to get polar form.

16.13 Express each of the following complex numbers z in rectangular form:

(a) $4(\cos 240° + i \sin 240°)$ (c) $3(\cos 90° + i \sin 90°)$

(b) $2(\cos 315° + i \sin 315°)$ (d) $5(\cos 128° + i \sin 128°)$.

(a) $4(\cos 240° + i \sin 240°) = 4[-1/2 + i(-\sqrt{3}/2)] = -2 - 2i\sqrt{3}$

(b) $2(\cos 315° + i \sin 315°) = 2[1/\sqrt{2} + i(-1/\sqrt{2})] = \sqrt{2} - i\sqrt{2}$

(c) $3(\cos 90° + i \sin 90°) = 3[0 + i(1)] = 3i$

(d) $5(\cos 128° + i \sin 128°) \approx 5[-0.6157 + i(0.7880)] = -3.0785 + 3.9400i$. Since $128°$ is not a special angle its function values are approximated.

16.14 Prove: (a) The modulus of the product of two complex numbers is the product of their moduli, and the amplitude of the product is the sum of their amplitudes.

(b) The modulus of the quotient of two complex numbers is the modulus of the dividend divided by the modulus of the divisor, and the amplitude of the quotient is the amplitude of the dividend minus the amplitude of the divisor.

Let $z_1 = r_1(\cos \theta_1 + i \sin \theta_1)$ and $z_2 = r_2(\cos \theta_2 + i \sin \theta_2)$.

(a) $z_1 z_2 = r_1(\cos \theta_1 + i \sin \theta_1) \cdot r_2(\cos \theta_2 + i \sin \theta_2)$

$= r_1 r_2[(\cos \theta_1 \cos \theta_2 - \sin \theta_1 \sin \theta_2) + i(\sin \theta_1 \cos \theta_2 + \cos \theta_1 \sin \theta_2)]$

$= r_1 r_2[\cos (\theta_1 + \theta_2) + i \sin (\theta_1 + \theta_2)]$

(b) $\dfrac{r_1(\cos \theta_1 + i \sin \theta_1)}{r_2(\cos \theta_2 + i \sin \theta_2)} = \dfrac{r_1(\cos \theta_1 + i \sin \theta_1)(\cos \theta_2 - i \sin \theta_2)}{r_2(\cos \theta_2 + i \sin \theta_2)(\cos \theta_2 - i \sin \theta_2)}$

$= \dfrac{r_1}{r_2} \cdot \dfrac{(\cos \theta_1 \cos \theta_2 + \sin \theta_1 \sin \theta_2) + i(\sin \theta_1 \cos \theta_2 - \cos \theta_1 \sin \theta_2)}{\cos^2 \theta_2 + \sin^2 \theta_2}$

$= \dfrac{r_1}{r_2}[\cos(\theta_1 - \theta_2) + i \sin(\theta_1 - \theta_2)]$

16.15 Perform the indicated operations, giving the result in both polar and rectangular form.

(a) $5(\cos 170° + i \sin 170°) \cdot (\cos 55° + i \sin 55°)$

(b) $2(\cos 50° + i \sin 50°) \cdot 3(\cos 40° + i \sin 40°)$

(c) $6(\cos 110° + i \sin 110°) \cdot \frac{1}{2}(\cos 212° + i \sin 212°)$

(d) $10(\cos 305° + i \sin 305°) \div 2(\cos 65° + i \sin 65°)$

(e) $4(\cos 220° + i \sin 220°) \div 2(\cos 40° + i \sin 40°)$

(f) $6(\cos 230° + i \sin 230°) \div 3(\cos 75° + i \sin 75°)$

(a)   The modulus of the product is $5(1) = 5$ and the amplitude is $170° + 55° = 225°$.
In polar form the product is $5(\cos 225° + i \sin 225°)$, and in rectangular form the product is $5(-\sqrt{2}/2 - i\sqrt{2}/2) = -5\sqrt{2}/2 - 5i\sqrt{2}/2$.

(b)   The modulus of the product is $2(3) = 6$ and the amplitude is $50° + 40° = 90°$.
In polar form the product is $6(\cos 90° + i \sin 90°)$, and in rectangular form it is $6(0 + i) = 6i$.

(c)   The modulus of the product is $6(\frac{1}{2}) = 3$ and the amplitude is $110° + 212° = 322°$.
In polar form the product is $3(\cos 322° + i \sin 322°)$, and in rectangular form it is approximately $3(0.7880 - 0.6157i) = 2.3640 - 1.8471i$.

(d)   The modulus of the quotient is $10/2 = 5$ and the amplitude is $305° - 65° = 240°$.
In polar form the product is $5(\cos 240° + i \sin 240°)$, and in rectangular form it is $5(-1/2 - i\sqrt{3}/2) = -5/2 - 5i\sqrt{3}/2$.

(e)   The modulus of the quotient is $4/2 = 2$ and the amplitude is $220° - 40° = 180°$.
      In polar form the quotient is $2(\cos 180° + i \sin 180°)$, and in rectangular form it is $2(-1 + 0i) = -2$.

(f)   The modulus of the quotient is $6/3 = 2$ and the amplitude is $230° - 75° = 155°$.
      In polar form the quotient is $2(\cos 155° + i \sin 155°)$, and in rectangular form it is approximately
      $2(-0.9063 + 0.4226i) = -1.8126 + 0.8452i$.

16.16 Express each of the numbers in polar form, perform the indicated operation, and give the result in rectangular form.

(a) $(-1 + i\sqrt{3})(\sqrt{3} + i)$ (d) $-2 \div (-\sqrt{3} + i)$

(b) $(3 - 3i\sqrt{3})(-2 - 2i\sqrt{3})$ (e) $6i \div (-3 - 3i)$

(c) $(4 - 4i\sqrt{3}) \div (-2\sqrt{3} + 2i)$ (f) $(1 + i\sqrt{3})(1 + i\sqrt{3})$

(a) $(-1 + i\sqrt{3})(\sqrt{3} + i) = 2(\cos 120° + i \sin 120°) \cdot 2(\cos 30° + i \sin 30°)$
     $= 4(\cos 150° + i \sin 150°) = 4(-\sqrt{3}/2 + \tfrac{1}{2}i) = -2\sqrt{3} + 2i$

(b) $(3 - 3i\sqrt{3})(-2 - 2i\sqrt{3}) = 6(\cos 300° + i \sin 300°) \cdot 4(\cos 240° + i \sin 240°)$
     $= 24(\cos 540° + i \sin 540°) = 24(-1 + 0i) = -24$

(c) $(4 - 4i\sqrt{3}) \div (-2\sqrt{3} + 2i) = 8(\cos 300° + i \sin 300°) \div 4(\cos 150° + i \sin 150°)$
     $= 2(\cos 150° + i \sin 150°) = 2(-\sqrt{3}/2 + \tfrac{1}{2}i) = -\sqrt{3} + i$

(d) $-2 \div (-\sqrt{3} + i) = 2(\cos 180° + i \sin 180°) \div 2(\cos 150° + i \sin 150°)$
     $= \cos 30° + i \sin 30° = \tfrac{1}{2}\sqrt{3} + \tfrac{1}{2}i$

(e) $6i \div (-3 - 3i) = 6(\cos 90° + i \sin 90°) \div 3\sqrt{2}(\cos 225° + i \sin 225°)$
     $= \sqrt{2}(\cos 225° + i \sin 225°) = -1 - i$

(f) $(1 + i\sqrt{3})(1 + i\sqrt{3}) = 2(\cos 60° + i \sin 60°) \cdot 2(\cos 60° + i \sin 60°)$
     $= 4(\cos 120° + i \sin 120°) = 4(-\tfrac{1}{2} + \tfrac{1}{2}i\sqrt{3}) = -2 + 2i\sqrt{3}$

16.17 Verify De Moivre's theorem for $n = 2$ and $n = 3$.

Let $z = r(\cos \theta + i \sin \theta)$.

For $n = 2$: $z^2 = [r(\cos \theta + i \sin \theta)][r(\cos \theta + i \sin \theta)]$
$= r^2[(\cos \theta - \sin \theta) + i(2 \sin \theta \cos \theta)] = r^2(\cos 2\theta + i \sin 2\theta)$

For $n = 3$: $z^3 = z^2 \cdot z = [r^2(\cos 2\theta + i \sin 2\theta)][r(\cos \theta + i \sin \theta)]$
$= r^3[(\cos 2\theta \cos \theta - \sin 2\theta \sin \theta) + i(\sin 2\theta \cos \theta + \cos 2\theta \sin \theta)]$
$= r^3(\cos 3\theta + i \sin 3\theta)$

The theorem may be established for n a positive integer by mathematical induction.

16.18 Evaluate each of the following using De Moivre's theorem and express each result in rectangular form:

(a) $(1 + i\sqrt{3})^4$, (b) $(\sqrt{3} - i)^5$, (c) $(-1 + i)^{10}$

(a) $(1 + i\sqrt{3})^4 = [2(\cos 60° + i \sin 60°)]^4 = 2^4(\cos 4 \cdot 60° + i \sin 4 \cdot 60°)(\cos 210° + i \sin 210°)$
     $= 2^4(\cos 240° + i \sin 240°) = -8 - 8i\sqrt{3}$

(b) $(\sqrt{3} - i)^5 = [2(\cos 330° + i \sin 330°)]^5 = 32(\cos 1650° + i \sin 1650°) = 32 = -16\sqrt{3} - 16i$

(c) $(-1 + i)^{10} = [\sqrt{2}(\cos 135° + i \sin 135°)]^{10} = 32(\cos 270° + i \sin 270°) = -32i$

16.19 Find the indicated roots in rectangular form, except when this would necessitate the use of tables.

(a)	Square roots of $2 - 2i\sqrt{3}$	(e)	Fourth roots of i
(b)	Fourth roots of $-8 - 8i\sqrt{3}$	(f)	Sixth roots of -1
(c)	Cube roots of $-4\sqrt{2} + 4i\sqrt{2}$	(g)	Fourth roots of $-16i$
(d)	Cube roots of 1		

(a)
$$2 - 2i\sqrt{3} = 4[\cos(300° + k\,360°) + i\sin(300° + k\,360°)]$$

and
$$(2 - 2i\sqrt{3})^{1/2} = 2[\cos(150° + k\,180°) + i\sin(150° + k\,180°)]$$

Putting $k = 0$ and 1, the required roots are

$$R_1 = 2(\cos 150° + i\sin 150°) = 2(-\tfrac{1}{2}\sqrt{3} + \tfrac{1}{2}i) = -\sqrt{3} + i$$

$$R_2 = 2(\cos 330° + i\sin 330°) = 2(\tfrac{1}{2}\sqrt{3} - \tfrac{1}{2}i) = \sqrt{3} - i$$

(b)
$$-8 - 8i\sqrt{3} = 16[\cos(240° + k\,360°) + i\sin(240° + k\,360°)]$$

and
$$(-8 - 8i\sqrt{3})^{1/4} = 2[\cos(60° + k\,90°) + i\sin(60° + k\,90°)]$$

Putting $k = 0$, 1, 2, and 3, the required roots are

$$R_1 = 2(\cos 60° + i\sin 60°) = 2(\tfrac{1}{2} + i\tfrac{1}{2}\sqrt{3}) = 1 + i\sqrt{3}$$

$$R_2 = 2(\cos 150° + i\sin 150°) = 2(-\tfrac{1}{2}\sqrt{3} + \tfrac{1}{2}i) = -\sqrt{3} + i$$

$$R_3 = 2(\cos 240° + i\sin 240°) = 2(-\tfrac{1}{2} - i\tfrac{1}{2}\sqrt{3}) = -1 - i\sqrt{3}$$

$$R_4 = 2(\cos 330° + i\sin 330°) = 2(\tfrac{1}{2}\sqrt{3} - \tfrac{1}{2}i) = \sqrt{3} - i$$

(c)
$$-4\sqrt{2} + 4i\sqrt{2} = 8[\cos(135° + k\,360°) + i\sin(135° + k\,360°)]$$

and
$$(-4\sqrt{2} + 4i\sqrt{2})^{1/3} = 2[\cos(45° + k\,120°) + i\sin(45° + k\,120°)]$$

Putting $k = 0$, 1, and 2, the required roots are

$$R_1 = 2(\cos 45° + i\sin 45°) = 2(1/\sqrt{2} + i/\sqrt{2}) = \sqrt{2} + i\sqrt{2}$$

$$R_2 = 2(\cos 165° + i\sin 165°)$$

$$R_3 = 2(\cos 285° + i\sin 285°)$$

(d) $1 = \cos(0° + k\,360°) + i\sin(0° + k\,360°)$ and $1^{1/3} = \cos(k\,120°) + i\sin(k\,120°)$.
Putting $k = 0$, 1, and 2, the required roots are

$$R_1 = \cos 0° + i\sin 0° = 1$$

$$R_2 = \cos 120° + i\sin 120° = -\tfrac{1}{2} + i\tfrac{1}{2}\sqrt{3}$$

$$R_3 = \cos 240° + i\sin 240° = -\tfrac{1}{2} - i\tfrac{1}{2}\sqrt{3}$$

Note that
$$R_2^2 = \cos 2(120°) + i\sin 2(120°) = R^3$$

$$R_3^2 = \cos 2(240°) + i\sin 2(240°) = R_2$$

and
$$R_2 R_3 = (\cos 120° + i\sin 120°)(\cos 240° + i\sin 240°) = \cos 0° + i\sin 0° = R_1$$

(e) $i = \cos(90° + k\,360°) + i\sin(90° + k\,360°)$ and $i^{1/4} = \cos(22\tfrac{1}{2}° + k\,90°) + i\sin(22\tfrac{1}{2}° + k\,90°)$.
Thus, the required roots are

$R_1 = \cos 22\tfrac{1}{2}° + i\sin 22\tfrac{1}{2}°$	$R_3 = \cos 202\tfrac{1}{2}° + i\sin 202\tfrac{1}{2}°$
$R_2 = \cos 112\tfrac{1}{2}° + i\sin 112\tfrac{1}{2}°$	$R_4 = \cos 292\tfrac{1}{2}° + i\sin 292\tfrac{1}{2}°$

(f) $-1 = \cos(180° + k\,360°) + i\sin(180° + k\,360°)$

and $(-1)^{1/6} = \cos(30° + k\,60°) + i\sin(30° + k\,60°)$.

Thus, the required roots are

$$R_1 = \cos 30° + i\sin 30° = \tfrac{1}{2}\sqrt{3} + \tfrac{1}{2}i$$

$$R_2 = \cos 90° + i\sin 90° = i$$

$$R_3 = \cos 150° + i\sin 150° = -\tfrac{1}{2}\sqrt{3} + \tfrac{1}{2}i$$

$$R_4 = \cos 210° + i\sin 210° = -\tfrac{1}{2}\sqrt{3} - \tfrac{1}{2}i$$

$$R_5 = \cos 270° + i\sin 270° = -i$$

$$R_6 = \cos 330° + i\sin 330° = \tfrac{1}{2}\sqrt{3} - \tfrac{1}{2}i$$

Note that $R_2^2 = R_5^2 = \cos 180° + i\sin 180°$ and thus R_2 and R_5 are the square roots of -1; that $R_1^3 = R_3^3 = R_5^3 = \cos 90° + i\sin 90° = i$ and thus R_1, R_3, and R_5 are the cube roots of i; and that $R_2^3 = R_4^3 = R_6^3 = \cos 270° + i\sin 270° = -i$ and thus R_2, R_4, and R_6 are the cube roots of $-i$.

(g) $-16i = 16[\cos(270° + k\,360°) + i\sin(270° + k\,360°)].$

and $(-16i)^{1/4} = 2[\cos(67\tfrac{1}{2}° + k\,90°) + i\sin(67\tfrac{1}{2}° + k\,90°)]$

Thus, the required roots are

$$R_1 = 2(\cos 67\tfrac{1}{2}° + i\sin 67\tfrac{1}{2}°) \qquad\qquad R_3 = 2(\cos 247\tfrac{1}{2}° + i\sin 247\tfrac{1}{2}°)$$

$$R_2 = 2(\cos 157\tfrac{1}{2}° + i\sin 157\tfrac{1}{2}°) \qquad\qquad R_4 = 2(\cos 337\tfrac{1}{2}° + i\sin 337\tfrac{1}{2}°)$$

Supplementary Problems

16.20   Perform the indicated operations, writing the results in the form $a + bi$.

(a) $(6 - 2i) + (2 + 3i) = 8 + i$ (m) $(2 - i)^2 = 3 - 4i$

(b) $(6 - 2i) - (2 + 3i) = 4 - 5i$ (n) $(4 + 2i)^2 = 12 + 16i$

(c) $(3 + 2i) + (-4 - 3i) = -1 - i$ (o) $(1 + i)^2(2 + 3i) = -6 + 4i$

(d) $(3 - 2i) - (4 - 3i) = -1 + i$

(e) $3(2 - i) = 6 - 3i$ (p) $\dfrac{2 + 3i}{1 + i} = \dfrac{5}{2} + \dfrac{1}{2}i$

(f) $2i(3 + 4i) = -8 + 6i$

(g) $(2 + 3i)(1 + 2i) = -4 + 7i$ (q) $\dfrac{3 - 2i}{3 - 4i} = \dfrac{17}{25} + \dfrac{6}{25}i$

(h) $(2 - 3i)(5 + 2i) = 16 - 11i$ (r) $\dfrac{3 - 2i}{2 + 3i} = -i$

(i) $(3 - 2i)(-4 + i) = -10 + 11i$

(j) $(2 - 3i)(3 + 2i) = 13i$

(k) $(2 + \sqrt{-5})(3 - 2\sqrt{-4}) = (6 + 4\sqrt{5}) + (3\sqrt{5} - 8)i$

(l) $(1 + 2\sqrt{-3})(2 - \sqrt{-3}) = 8 + 3\sqrt{3}i$

16.21   Show that $3 + 2i$ and $3 - 2i$ are roots of $x^2 - 6x + 13 = 0$.

16.22   Perform graphically the following operations.

     (a) $(2 + 3i) + (1 + 4i)$ (c) $(2 + 3i) - (1 + 4i)$

     (b) $(4 - 2i) + (2 + 3i)$ (d) $(4 - 2i) - (2 + 3i)$

16.23   Express each of the following complex numbers in polar form.

     (a) $3 + 3i = 3\sqrt{2}(\cos 45° + i \sin 45°)$ (e) $-8 = 8(\cos 180° + i \sin 180°)$

     (b) $1 + \sqrt{3}i = 2(\cos 60° + i \sin 60°)$ (f) $-2i = 2(\cos 270° + i \sin 270°)$

     (c) $-2\sqrt{3} - 2i = 4(\cos 210° + i \sin 210°)$ (g) $-12 + 5i \approx 13(\cos 157°23' + i \sin 157°23')$

     (d) $\sqrt{2} - i\sqrt{2} = 2(\cos 315° + i \sin 315°)$ (h) $-4 - 3i \approx 5(\cos 216°52' + i \sin 216°52')$

16.24   Perform the indicated operation and express the results in the form $a + bi$.

     (a) $3(\cos 25° + i \sin 25°)8(\cos 200° + i \sin 200°) = -12\sqrt{2} - 12\sqrt{2}i$

     (b) $4(\cos 50° + i \sin 50°)2(\cos 100° + i \sin 100°) = -4\sqrt{3} + 4i$

     (c) $\dfrac{4(\cos 190° + i \sin 190°)}{2(\cos 70° + i \sin 70°)} = -1 + i\sqrt{3}$

     (d) $\dfrac{12(\cos 200° + i \sin 200°)}{3(\cos 350° + i \sin 350°)} = -2\sqrt{3} - 2i$

16.25   Use the polar form in finding each of the following products and quotients, and express each result in the form $a + bi$.

     (a) $(1 + i)(\sqrt{2} - i\sqrt{2}) = 2\sqrt{2}$ (c) $\dfrac{1 - i}{1 + i} = -i$

     (b) $(-1 - i\sqrt{3})(-4\sqrt{3} + 4i) = 8\sqrt{3} + 8i$ (d) $\dfrac{4 + 4\sqrt{3}i}{\sqrt{3} + i} = 2\sqrt{3} + 2i$

16.26   Use De Moivre's theorem to evaluate each of the following and express each result in the form $a + bi$.

     (a) $[2(\cos 6° + i \sin 6°)]^5 = 16\sqrt{3} + 16i$ (f) $(\sqrt{3}/2 + i/2)^9 = -i$

     (b) $[\sqrt{2}(\cos 75° + i \sin 75°)]^4 = 2 - 2\sqrt{3}i$ (g) $\dfrac{(1 - i\sqrt{3})^3}{(-2 + 2i)^4} = \dfrac{1}{8}$

     (c) $(1 + i)^8 = 16$

     (d) $(1 - i)^6 = 8i$ (h) $\dfrac{(1 + i)(\sqrt{3} + i)^3}{(1 - i\sqrt{3})^3} = 1 - i$

     (e) $(1/2 - i\sqrt{3}/2)^{20} = -1/2 - i\sqrt{3}/2$

16.27   Find all the indicated roots, expressing the results in the form $a + bi$ unless tables would be needed to do so.

     (a) The square roots of i *Ans.* $\sqrt{2}/2 + i\sqrt{2}/2, -\sqrt{2}/2 - i\sqrt{2}/2$

     (b) The square roots of $1 + i\sqrt{3}$ *Ans.* $\sqrt{6}/2 + i\sqrt{2}/2, -\sqrt{6}/2 - i\sqrt{2}/2$

     (c) The cube roots of -8 *Ans.* $1 + i\sqrt{3}, -2, 1 - i\sqrt{3}$

     (d) The cube roots of $27i$ *Ans.* $3\sqrt{3}/2 + 3i/2, -3\sqrt{3}/2 + 3i/2, -3i$

     (e) The cube roots of $-4\sqrt{3} + 4i$ *Ans.* $2(\cos 50° + i \sin 50°), 2(\cos 170° + i \sin 170°),$
                                                $2(\cos 290° + i \sin 290°)$

(f) The fifth roots of $1 + i$ *Ans.* $\sqrt[10]{2}(\cos 9° + i \sin 9°)$, $\sqrt[10]{2}(\cos 81° + i \sin 81°)$, etc.

(g) The sixth roots of $-\sqrt{3} + i$ *Ans.* $\sqrt[6]{2}(\cos 25° + i \sin 25°)$, $\sqrt[6]{2}(\cos 85° + i \sin 85°)$, etc.

16.28 Find the tenth roots of 1 and show that the product of any two of them is again one of the tenth roots of 1.

16.29 Show that the reciprocal of any one of the tenth roots of 1 is again a tenth root of 1.

16.30 Denote either of the complex cube roots of 1 (Prob. 16.19d) by ω_1 and the other by ω_2. Show that $\omega_1^2 \omega_2 = \omega_1$ and $\omega_1 \omega_2^2 = \omega_2$.

16.31 Show that $(\cos \theta + i \sin \theta)^{-n} = \cos n\theta - i \sin n\theta$.

16.32 Use the fact that the segments OS and $P_2 P_1$ in Fig. 16-2(c) are equal to devise a second procedure for constructing the difference $OS = z_1 - z_2$ of two complex numbers z_1 and z_2.

Appendix 1

Geometry

A1.1 INTRODUCTION

Appendix 1 is a summary of basic geometry definitions, relations, and theorems. The purpose of this material is to provide information useful in solving problems in trigonometry.

A1.2 ANGLES

An *angle* is a figure determined by two rays having a common endpoint. An *acute* angle is an angle with a measure between 0 and 90°. A *right* angle is an angle with a measure of 90°, while an *obtuse* angle has a measure between 90 and 180°. When the sum of the measures of two angles is 90°, the angles are *complementary*. When the sum of the measures of two angles is 180°, the angles are *supplementary*. Two angles are *equal* when they have the same measure.

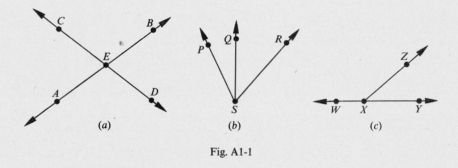

Fig. A1-1

If two lines intersect, the opposite angles are *vertical* angles. In Fig. A1-1(*a*), $\angle AED$ and $\angle BEC$ are vertical angles and $\angle CEA$ and $\angle BED$ are also a pair of vertical angles. When two angles have a common vertex and a common side between them, the angles are *adjacent* angles. In Fig. A1-1(*b*), $\angle PSQ$ and $\angle QSR$ are a pair of adjacent angles. If the exterior sides of two adjacent angles form a straight line, the angles form a *linear pair*. In Fig. A1-1(*c*), $\angle WXZ$ and $\angle ZXY$ are a linear pair.

Properties and Theorems

▲ The measures of vertical angles are equal.

▲ The angles in a linear pair are supplementary.

▲ If the angles in a linear pair are equal, the angles are right angles.

▲ Angles complementary to the same or to equal angles are equal to each other.

▲ Angles supplementary to the same or to equal angles are equal to each other.

A1.3 LINES

Two lines in a plane either intersect or they are parallel. If two lines *intersect*, they have exactly one point in common. Two lines in a plane are *parallel* if they have no common point.

When two lines intersect to form equal adjacent angles, the lines are *perpendicular*. Each of the angles formed by two perpendicular lines is a right angle. The sides of a right angle are perpendicular.

183

A *transversal* is a line that intersects two or more coplanar lines in distinct points. In Fig. A1-2, lines *m* and *n* are cut by transversal *t*. When two lines are cut by a transversal, the angles formed are classified by their location. Angles between the two lines are called *interior* angles and the angles not between the two lines are called *exterior* angles. Interior or exterior angles are said to *alternate* if the two angles have different vertices and lie on opposite sides of the transversal. A pair of *corresponding* angles are two angles, one an interior angle and one an exterior angle, that have different vertices and lie on the same side of the transversal. In Fig. A1-2, the interior angles are numbered 3, 4, 5, and 6 while the exterior angles are numbered 1, 2, 7, and 8. The angles numbered 3 and 6 and the angles numbered 4 and 5 are pairs of alternate interior angles. The angles numbered 1 and 8 and those numbered 2 and 7 are pairs of alternate exterior angles. The pairs of corresponding angles are numbered 1 and 5, 2 and 6, 3 and 7, and 4 and 8.

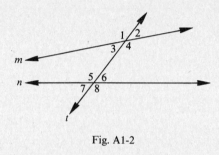

Fig. A1-2

Properties and Theorems

▲ In a plane, if two lines are perpendicular to the same line, then the lines are parallel.

▲ Two lines parallel to a third line are parallel to each other.

▲ If two parallel lines are cut by a transversal, then the alternate interior angles are equal.

▲ If two parallel lines are cut by a transversal, then the alternate exterior angles are equal.

▲ If two parallel lines are cut by a transversal, then the corresponding angles are equal.

▲ If two parallel lines are cut by a transversal, then the interior angles on the same side of the transversal are supplementary.

▲ Any pair of horizontal lines are parallel.

▲ Any pair of vertical lines are parallel.

▲ Any vertical line is perpendicular to any horizontal line.

▲ If two angles have their sides parallel, right side to right side and left side to left side, then the angles are equal.

▲ If two angles have their sides perpendicular, right side to right side and left side to left side, then the angles are equal.

▲ If two lines are cut by a transversal so that the alternate interior angles formed are equal, then the lines are parallel.

▲ If two lines are cut by a transversal so that the alternate exterior angles formed are equal, then the lines are parallel.

▲ If two lines are cut by a transversal so that the corresponding angles formed are equal, then the lines are parallel.

▲ If two lines are cut by a transversal so that the interior angles on the same side of the transversal are supplementary, then the lines are parallel.

▲ If a transversal is perpendicular to one of two parallel lines, it is perpendicular to the other one as well.

▲ Through a point not on a line there is exactly one line parallel to a given line.

▲ Through a point not on a line there is exactly one line perpendicular to a given line.

▲ In a plane, there is exactly one line perpendicular to a given line at any point on the line.

A1.4   TRIANGLES

A *triangle* is a plane closed figure formed by three line segments that intersect each other at their endpoints. Triangles that have no two sides with the same length are called *scalene* triangles, those with at least two sides having the same length are called *isosceles* triangles, and those with all three sides having the same length are called *equilateral* triangles. If a triangle contains a right angle, it is a *right* triangle. A triangle that does not contain a right angle is called an *oblique* triangle.

Two triangles are said to be *congruent* when they have the same size and the same shape. When two triangles are congruent, the pairs of corresponding sides have the same length and the pairs of corresponding angles are equal. Triangles that have the same shape are said to be *similar*. Similar triangles have corresponding sides that are proportional in length and corresponding angles that are equal.

A *median* of a triangle is a line segment from a vertex to the midpoint of the opposite side. An *altitude* of a triangle is a line segment from one vertex perpendicular to the opposite side.

Properties and Theorems

▲ If three sides of one triangle are equal to the three sides of another triangle, the triangles are congruent.

▲ If two sides and the included angle of one triangle are equal to the corresponding two sides and included angle of another triangle, the triangles are congruent.

▲ If two angles and the included side of one triangle are equal to the corresponding two angles and included side of another triangle, the triangles are congruent.

▲ If two angles and a nonincluded side of one triangle are equal to the corresponding two angles and nonincluded side of another triangle, the triangles are congruent.

▲ If the hypotenuse and leg of one right triangle are equal to the corresponding sides of another right triangle, then the two right triangles are congruent.

▲ If two angles of one triangle are equal to the corresponding two angles of another triangle, the triangles are similar.

▲ The sum of the measures of the angles of a triangle is 180°.

▲ An exterior angle of a triangle is equal to the sum of the measures of the two nonadjacent interior angles of the triangle.

▲ The acute angles of a right triangle are complementary.

▲ The measure of each angle of an equiangular triangle is 60°.

▲ If two sides of a triangle are equal, then the angles opposite these sides are equal.

▲ If a triangle is equilateral, then it is also equiangular.

▲ If two angles of a triangle are equal, then the sides opposite these angles are equal.

▲ If a triangle is equiangular, then it is also equilateral.

▲ The altitude to the base of an isosceles triangle bisects the base and the vertex angle.

▲ The median to the base of an isosceles triangle bisects the vertex angle and is perpendicular to the base.

▲ The bisector of the vertex angle of an isosceles triangle is the perpendicular bisector of the base.

▲ In a right triangle, the square of the length of the hypotenuse c is equal to the sum of the squares of the lengths of the two legs a and b; i.e., $c^2 = a^2 + b^2$. (Pythagorean Theorem)

▲ In a 45°–45° right triangle, the length of the hypotenuse c is equal to $\sqrt{2}$ times the length of a leg a; i.e., $c = \sqrt{2}a$.

▲ In a 30°–60° right triangle, the length of the hypotenuse c is equal to 2 times the length of the leg a opposite the 30° angle; i.e., $c = 2a$. Also, the length of the leg b opposite the 60° angle is equal to $\sqrt{3}$ times the length of the leg a opposite the 30° angle; i.e., $b = \sqrt{3}a$.

▲ The midpoint of the hypotenuse of a right triangle is equidistant from all three vertices of the triangle.

▲ If the square of the length of one side c of a triangle is equal to the sum of the squares of the lengths of other two sides a and b of the triangle, i.e., $c^2 = a^2 + b^2$, then the triangle is a right triangle.

▲ The area K of a triangle is one-half the product of its altitude a and base b; i.e., $K = \frac{1}{2}ab$.

▲ The area of an equilateral triangle is equal to one-fourth of the square of a side s times $\sqrt{3}$; i.e., $K = \frac{1}{4}s^2\sqrt{3}$.

A1.5 POLYGONS

A *polygon* is a plane closed figure whose sides are line segments that are noncollinear and each side intersects exactly two other line segments at their endpoints. A *quadrilateral* is a polygon having four sides. A *regular* polygon is a polygon that is both equilateral and equiangular. A *diagonal* of a polygon is a line segment that joins two nonadjacent vertices of the polygon.

A *parallelogram* is a quadrilateral with opposite sides parallel. A *rectangle* is a parallelogram with one right angle. A *rhombus* is a parallelogram with two adjacent sides equal. A *square* is a rectangle with two adjacent sides equal.

A *trapezoid* is a quadrilateral with exactly one pair of parallel sides. An *isosceles trapezoid* is a trapezoid with the nonparallel sides having equal lengths.

Properties and Theorems

▲ The opposite angles of a parallelogram are equal.

▲ The opposite sides of a parallelogram are equal.

▲ The diagonals of a parallelogram bisect each other.

▲ The consecutive interior angles of a parallelogram are supplementary.

▲ The sum of the interior angles of a quadrilateral is 360°.

▲ If both pairs of opposite angles of a quadrilateral are equal, the quadrilateral is a parallelogram.

▲ If both pairs of opposite sides of a quadrilateral are equal, the quadrilateral is a parallelogram.

▲ If the diagonals of a quadrilateral bisect each other, the quadrilateral is a parallelogram.

▲ A rectangle is an equiangular parallelogram.

▲ A rhombus is an equilateral parallelogram.

▲ The diagonals of a rectangle are equal.

▲ The diagonals of a rhombus are perpendicular.

▲ If the diagonals of a parallelogram are equal, the parallelogram is a rectangle.

▲ If the diagonals of a parallelogram are perpendicular, the parallelogram is a rhombus.

▲ The diagonals of a square are the perpendicular bisectors of each other.

▲ A rhombus with a right angle is a square.

▲ A square is a regular polygon.

▲ The diagonals of an isosceles trapezoid are equal.

▲ The area K of a parallelogram is equal to the product of its altitude a and base b; i.e., $K = ab$.

▲ The area K of a rectangle is equal to the product of its length l and width w; i.e., $K = lw$.

▲ The area K of a rhombus is equal to one-half the product of its diagonals d and d'; i.e., $K = \frac{1}{2}dd'$.

▲ The area K of a square is equal to the square of its side s; i.e., $K = s^2$.

▲ The area K of a square is equal to one-half the square of its diagonal d; i.e., $K = \frac{1}{2}d^2$.

▲ The area K of a trapezoid is equal to one-half the product of the altitude h and the sum of the bases b and b'; i.e., $K = \frac{1}{2}h(b + b')$.

A1.6 CIRCLES

A *circle* is the set of all points in a plane that are at a given distance from a given point. Any line segment which has its endpoints on a circle is a *chord* of the circle. If a chord of a circle passes through the center of the circle, then it is a *diameter*. A *radius* is a line segment from the center of a circle to a point on the circle. A *secant* is a line that intersects a circle in two points. A *tangent* is a line that intersects a circle in one point.

An *arc* of a circle is a part of the circle from one point on the circle to another. A *semicircle* is an arc of a circle joining the endpoints of a diameter of the circle. An *inscribed angle* is an angle with sides that are chords of a circle. A *central angle* is an angle with sides that are radii of the circle.

Properties and Theorems

▲ A circle contains 360°.

▲ A semicircle contains 180°.

▲ A central angle is equal in degrees to its intercepted arc.

▲ An inscribed angle is equal in degrees to one-half its intercepted arc.

▲ An angle inscribed in a semicircle is a right angle.

▲ In a circle, if a diameter is perpendicular to a chord, it bisects the chord and its arc.

▲ In a circle two chords that are equal are equidistant from the center of the circle.

▲ In a circle two chords that are equidistant from the center of the circle are equal.

▲ A tangent to a circle is perpendicular to the radius drawn to the point of tangency.

▲ If a triangle is inscribed in a semicircle, then the triangle is a right triangle.

▲ In a plane, if a line is perpendicular to a radius of a circle at its endpoint on the circle, then the line is tangent to the circle.

▲ The line segment joining an external point to the center of a circle bisects the angle formed by the two tangents to the circle from that point.

▲ The lengths of the two tangent segments from an external point to a circle are equal.

▲ If two secants intersect in the interior of a circle, then the angle formed is equal in degrees to one-half the sum of the arcs intercepted by it and its vertical angle.

▲ If a line through the center of a circle bisects a chord that is not a diameter, then it is perpendicular to the chord.

▲ The circumference C of a circle is equal to π times the diameter d; i.e. $C = \pi d$.

▲ The radius r of a circle is equal to one-half the diameter d; i.e., $r = \frac{1}{2}d$.

▲ The area K of a circle is equal to π times the radius r squared; i.e., $K = \pi r^2$.

Appendix 2

Tables

Table 1   Trigonometric Functions—Angle in 10 Minute Intervals

A	sin A	cos A	tan A	cot A	sec A	csc A	
0° 0′	0.0000	1.0000	0.0000	Undefined	1.0000	Undefined	90° 0′
0°10′	0.0029	1.0000	0.0029	343.7730	1.0000	343.7740	89°50′
0°20′	0.0058	1.0000	0.0058	171.8850	1.0000	171.8880	89°40′
0°30′	0.0087	1.0000	0.0087	114.5880	1.0000	114.5930	89°30′
0°40′	0.0116	0.9999	0.0116	85.9396	1.0001	85.9454	89°20′
0°50′	0.0145	0.9999	0.0145	68.7499	1.0001	68.7572	89°10′
1° 0′	0.0175	0.9998	0.0175	57.2901	1.0002	57.2989	89° 0′
1°10′	0.0204	0.9998	0.0204	49.1040	1.0002	49.1142	88°50′
1°20′	0.0233	0.9997	0.0233	42.9641	1.0003	42.9758	88°40′
1°30′	0.0262	0.9997	0.0262	38.1885	1.0003	38.2016	88°30′
1°40′	0.0291	0.9996	0.0291	34.3678	1.0004	34.3823	88°20′
1°50′	0.0320	0.9995	0.0320	31.2416	1.0005	31.2576	88°10′
2° 0′	0.0349	0.9994	0.0349	28.6363	1.0006	28.6537	88° 0′
2°10′	0.0378	0.9993	0.0378	26.4316	1.0007	26.4505	87°50′
2°20′	0.0407	0.9992	0.0407	24.5418	1.0008	24.5621	87°40′
2°30′	0.0436	0.9990	0.0437	22.9038	1.0010	22.9256	87°30′
2°40′	0.0465	0.9989	0.0466	21.4704	1.0011	21.4937	87°20′
2°50′	0.0494	0.9988	0.0495	20.2055	1.0012	20.2303	87°10′
3° 0′	0.0523	0.9986	0.0524	19.0811	1.0014	19.1073	87° 0′
3°10′	0.0552	0.9985	0.0553	18.0750	1.0015	18.1026	86°50′
3°20′	0.0581	0.9983	0.0582	17.1694	1.0017	17.1984	86°40′
3°30′	0.0610	0.9981	0.0612	16.3499	1.0019	16.3804	86°30′
3°40′	0.0640	0.9980	0.0641	15.6048	1.0021	15.6368	86°20′
3°50′	0.0669	0.9978	0.0670	14.9244	1.0022	14.9579	86°10′
4° 0′	0.0698	0.9976	0.0699	14.3007	1.0024	14.3356	86° 0′
4°10′	0.0727	0.9974	0.0729	13.7267	1.0027	13.7631	85°50′
4°20′	0.0756	0.9971	0.0758	13.1969	1.0029	13.2347	85°40′
4°30′	0.0785	0.9969	0.0787	12.7062	1.0031	12.7455	85°30′
4°40′	0.0814	0.9967	0.0816	12.2505	1.0033	12.2912	85°20′
4°50′	0.0843	0.9964	0.0846	11.8262	1.0036	11.8684	85°10′
5° 0′	0.0872	0.9962	0.0875	11.4301	1.0038	11.4737	85° 0′
5°10′	0.0901	0.9959	0.0904	11.0594	1.0041	11.1046	84°50′
5°20′	0.0929	0.9957	0.0934	10.7119	1.0043	10.7585	84°40′
5°30′	0.0958	0.9954	0.0963	10.3854	1.0046	10.4334	84°30′
5°40′	0.0987	0.9951	0.0992	10.0780	1.0049	10.1275	84°20′
5°50′	0.1016	0.9948	0.1022	9.7882	1.0052	9.8391	84°10′
6° 0′	0.1045	0.9945	0.1051	9.5144	1.0055	9.5668	84° 0′
6°10′	0.1074	0.9942	0.1080	9.2553	1.0058	9.3092	83°50′
6°20′	0.1103	0.9939	0.1110	9.0098	1.0061	9.0652	83°40′
6°30′	0.1132	0.9936	0.1139	8.7769	1.0065	8.8337	83°30′
6°40′	0.1161	0.9932	0.1169	8.5555	1.0068	8.6138	83°20′
6°50′	0.1190	0.9929	0.1198	8.3450	1.0072	8.4047	83°10′
7° 0′	0.1219	0.9925	0.1228	8.1444	1.0075	8.2055	83° 0′
7°10′	0.1248	0.9922	0.1257	7.9530	1.0079	8.0156	82°50′
7°20′	0.1276	0.9918	0.1287	7.7704	1.0082	7.8344	82°40′
7°30′	0.1305	0.9914	0.1317	7.5958	1.0086	7.6613	82°30′
7°40′	0.1334	0.9911	0.1346	7.4287	1.0090	7.4957	82°20′
7°50′	0.1363	0.9907	0.1376	7.2687	1.0094	7.3372	82°10′
	cos A	sin A	cot A	tan A	csc A	sec A	A

A	$\sin A$	$\cos A$	$\tan A$	$\cot A$	$\sec A$	$\csc A$	
8° 0′	0.1392	0.9903	0.1405	7.1154	1.0098	7.1853	82° 0′
8°10′	0.1421	0.9899	0.1435	6.9682	1.0102	7.0396	81°50′
8°20′	0.1449	0.9894	0.1465	6.8269	1.0107	6.8998	81°40′
8°30′	0.1478	0.9890	0.1495	6.6912	1.0111	6.7655	81°30′
8°40′	0.1507	0.9886	0.1524	6.5606	1.0116	6.6363	81°20′
8°50′	0.1536	0.9881	0.1554	6.4348	1.0120	6.5121	81°10′
9° 0′	0.1564	0.9877	0.1584	6.3138	1.0125	6.3925	81° 0′
9°10′	0.1593	0.9872	0.1614	6.1970	1.0129	6.2772	80°50′
9°20′	0.1622	0.9868	0.1644	6.0844	1.0134	6.1661	80°40′
9°30′	0.1650	0.9863	0.1673	5.9758	1.0139	6.0589	80°30′
9°40′	0.1679	0.9858	0.1703	5.8708	1.0144	5.9554	80°20′
9°50′	0.1708	0.9853	0.1733	5.7694	1.0149	5.8554	80°10′
10° 0′	0.1736	0.9848	0.1763	5.6713	1.0154	5.7588	80° 0′
10°10′	0.1765	0.9843	0.1793	5.5764	1.0160	5.6653	79°50′
10°20′	0.1794	0.9838	0.1823	5.4845	1.0165	5.5749	79°40′
10°30′	0.1822	0.9833	0.1853	5.3955	1.0170	5.4874	79°30′
10°40′	0.1851	0.9827	0.1883	5.3093	1.0176	5.4026	79°20′
10°50′	0.1880	0.9822	0.1914	5.2257	1.0181	5.3205	79°10′
11° 0′	0.1908	0.9816	0.1944	5.1446	1.0187	5.2408	79° 0′
11°10′	0.1937	0.9811	0.1974	5.0658	1.0193	5.1636	78°50′
11°20′	0.1965	0.9805	0.2004	4.9894	1.0199	5.0886	78°40′
11°30′	0.1994	0.9799	0.2035	4.9152	1.0205	5.0159	78°30′
11°40′	0.2022	0.9793	0.2065	4.8430	1.0211	4.9452	78°20′
11°50′	0.2051	0.9787	0.2095	4.7729	1.0217	4.8765	78°10′
12° 0′	0.2079	0.9781	0.2126	4.7046	1.0223	4.8097	78° 0′
12°10′	0.2108	0.9775	0.2156	4.6382	1.0230	4.7448	77°50′
12°20′	0.2136	0.9769	0.2186	4.5736	1.0236	4.6817	77°40′
12°30′	0.2164	0.9763	0.2217	4.5107	1.0243	4.6202	77°30′
12°40′	0.2193	0.9757	0.2247	4.4494	1.0249	4.5604	77°20′
12°50′	0.2221	0.9750	0.2278	4.3897	1.0256	4.5022	77°10′
13° 0′	0.2250	0.9744	0.2309	4.3315	1.0263	4.4454	77° 0′
13°10′	0.2278	0.9737	0.2339	4.2747	1.0270	4.3901	76°50′
13°20′	0.2306	0.9730	0.2370	4.2193	1.0277	4.3362	76°40′
13°30′	0.2334	0.9724	0.2401	4.1653	1.0284	4.2837	76°30′
13°40′	0.2363	0.9717	0.2432	4.1126	1.0291	4.2324	76°20′
13°50′	0.2391	0.9710	0.2462	4.0611	1.0299	4.1824	76°10′
14° 0′	0.2419	0.9703	0.2493	4.0108	1.0306	4.1336	76° 0′
14°10′	0.2447	0.9696	0.2524	3.9617	1.0314	4.0859	75°50′
14°20′	0.2476	0.9689	0.2555	3.9136	1.0321	4.0394	75°40′
14°30′	0.2504	0.9681	0.2586	3.8667	1.0329	3.9939	75°30′
14°40′	0.2532	0.9674	0.2617	3.8208	1.0337	3.9495	75°20′
14°50′	0.2560	0.9667	0.2648	3.7760	1.0345	3.9061	75°10′
15° 0′	0.2588	0.9659	0.2679	3.7321	1.0353	3.8637	75° 0′
15°10′	0.2616	0.9652	0.2711	3.6891	1.0361	3.8222	74°50′
15°20′	0.2644	0.9644	0.2742	3.6470	1.0369	3.7817	74°40′
15°30′	0.2672	0.9636	0.2773	3.6059	1.0377	3.7420	74°30′
15°40′	0.2700	0.9628	0.2805	3.5656	1.0386	3.7032	74°20′
15°50′	0.2728	0.9621	0.2836	3.5261	1.0394	3.6652	74°10′
	$\cos A$	$\sin A$	$\cot A$	$\tan A$	$\csc A$	$\sec A$	A

A	sin A	cos A	tan A	cot A	sec A	csc A	
16° 0′	0.2756	0.9613	0.2867	3.4874	1.0403	3.6280	74° 0′
16°10′	0.2784	0.9605	0.2899	3.4495	1.0412	3.5915	73°50′
16°20′	0.2812	0.9596	0.2931	3.4124	1.0421	3.5559	73°40′
16°30′	0.2840	0.9588	0.2962	3.3759	1.0429	3.5209	73°30′
16°40′	0.2868	0.9580	0.2994	3.3402	1.0439	3.4867	73°20′
16°50′	0.2896	0.9572	0.3026	3.3052	1.0448	3.4532	73°10′
17° 0′	0.2924	0.9563	0.3057	3.2709	1.0457	3.4203	73° 0′
17°10′	0.2952	0.9555	0.3089	3.2371	1.0466	3.3881	72°50′
17°20′	0.2979	0.9546	0.3121	3.2041	1.0476	3.3565	72°40′
17°30′	0.3007	0.9537	0.3153	3.1716	1.0485	3.3255	72°30′
17°40′	0.3035	0.9528	0.3185	3.1397	1.0495	3.2951	72°20′
17°50′	0.3062	0.9520	0.3217	3.1084	1.0505	3.2653	72°10′
18° 0′	0.3090	0.9511	0.3249	3.0777	1.0515	3.2361	72° 0′
18°10′	0.3118	0.9502	0.3281	3.0475	1.0525	3.2074	71°50′
18°20′	0.3145	0.9492	0.3314	3.0178	1.0535	3.1792	71°40′
18°30′	0.3173	0.9483	0.3346	2.9887	1.0545	3.1515	71°30′
18°40′	0.3201	0.9474	0.3378	2.9600	1.0555	3.1244	71°20′
18°50′	0.3228	0.9465	0.3411	2.9319	1.0566	3.0977	71°10′
19° 0′	0.3256	0.9455	0.3443	2.9042	1.0576	3.0716	71° 0′
19°10′	0.3283	0.9446	0.3476	2.8770	1.0587	3.0458	70°50′
19°20′	0.3311	0.9436	0.3508	2.8502	1.0598	3.0206	70°40′
19°30′	0.3338	0.9426	0.3541	2.8239	1.0608	2.9957	70°30′
19°40′	0.3365	0.9417	0.3574	2.7980	1.0619	2.9713	70°20′
19°50′	0.3393	0.9407	0.3607	2.7725	1.0631	2.9474	70°10′
20° 0′	0.3420	0.9397	0.3640	2.7475	1.0642	2.9238	70° 0′
20°10′	0.3448	0.9387	0.3673	2.7228	1.0653	2.9006	69°50′
20°20′	0.3475	0.9377	0.3706	2.6985	1.0665	2.8779	69°40′
20°30′	0.3502	0.9367	0.3739	2.6746	1.0676	2.8555	69°30′
20°40′	0.3529	0.9356	0.3772	2.6511	1.0688	2.8334	69°20′
20°50′	0.3557	0.9346	0.3805	2.6279	1.0700	2.8117	69°10′
21° 0′	0.3584	0.9336	0.3839	2.6051	1.0711	2.7904	69° 0′
21°10′	0.3611	0.9325	0.3872	2.5826	1.0723	2.7695	68°50′
21°20′	0.3638	0.9315	0.3906	2.5605	1.0736	2.7488	68°40′
21°30′	0.3665	0.9304	0.3939	2.5386	1.0748	2.7285	68°30′
21°40′	0.3692	0.9293	0.3973	2.5172	1.0760	2.7085	68°20′
21°50′	0.3719	0.9283	0.4006	2.4960	1.0773	2.6888	68°10′
22° 0′	0.3746	0.9272	0.4040	2.4751	1.0785	2.6695	68° 0′
22°10′	0.3773	0.9261	0.4074	2.4545	1.0798	2.6504	67°50′
22°20′	0.3800	0.9250	0.4108	2.4342	1.0811	2.6316	67°40′
22°30′	0.3827	0.9239	0.4142	2.4142	1.0824	2.6131	67°30′
22°40′	0.3854	0.9228	0.4176	2.3945	1.0837	2.5949	67°20′
22°50′	0.3881	0.9216	0.4210	2.3750	1.0850	2.5770	67°10′
23° 0′	0.3907	0.9205	0.4245	2.3559	1.0864	2.5593	67° 0′
23°10′	0.3934	0.9194	0.4279	2.3369	1.0877	2.5419	66°50′
23°20′	0.3961	0.9182	0.4314	2.3183	1.0891	2.5247	66°40′
23°30′	0.3987	0.9171	0.4348	2.2998	1.0904	2.5078	66°30′
23°40′	0.4014	0.9159	0.4383	2.2817	1.0918	2.4912	66°20′
23°50′	0.4041	0.9147	0.4417	2.2637	1.0932	2.4748	66°10′
	cos A	sin A	cot A	tan A	csc A	sec A	A

A	$\sin A$	$\cos A$	$\tan A$	$\cot A$	$\sec A$	$\csc A$	
24° 0′	0.4067	0.9135	0.4452	2.2460	1.0946	2.4586	66° 0′
24°10′	0.4094	0.9124	0.4487	2.2286	1.0961	2.4426	65°50′
24°20′	0.4120	0.9112	0.4522	2.2113	1.0975	2.4269	65°40′
24°30′	0.4147	0.9100	0.4557	2.1943	1.0989	2.4114	65°30′
24°40′	0.4173	0.9088	0.4592	2.1775	1.1004	2.3961	65°20′
24°50′	0.4200	0.9075	0.4628	2.1609	1.1019	2.3811	65°10′
25° 0′	0.4226	0.9063	0.4663	2.1445	1.1034	2.3662	65° 0′
25°10′	0.4253	0.9051	0.4699	2.1283	1.1049	2.3515	64°50′
25°20′	0.4279	0.9038	0.4734	2.1123	1.1064	2.3371	64°40′
25°30′	0.4305	0.9026	0.4770	2.0965	1.1079	2.3228	64°30′
25°40′	0.4331	0.9013	0.4806	2.0809	1.1095	2.3088	64°20′
25°50′	0.4358	0.9001	0.4841	2.0655	1.1110	2.2949	64°10′
26° 0′	0.4384	0.8988	0.4877	2.0503	1.1126	2.2812	64° 0′
26°10′	0.4410	0.8975	0.4913	2.0353	1.1142	2.2677	63°50′
26°20′	0.4436	0.8962	0.4950	2.0204	1.1158	2.2543	63°40′
26°30′	0.4462	0.8949	0.4986	2.0057	1.1174	2.2412	63°30′
26°40′	0.4488	0.8936	0.5022	1.9912	1.1190	2.2282	63°20′
26°50′	0.4514	0.8923	0.5059	1.9768	1.1207	2.2153	63°10′
27° 0′	0.4540	0.8910	0.5095	1.9626	1.1223	2.2027	63° 0′
27°10′	0.4566	0.8897	0.5132	1.9486	1.1240	2.1902	62°50′
27°20′	0.4592	0.8884	0.5169	1.9347	1.1257	2.1779	62°40′
27°30′	0.4617	0.8870	0.5206	1.9210	1.1274	2.1657	62°30′
27°40′	0.4643	0.8857	0.5243	1.9074	1.1291	2.1537	62°20′
27°50′	0.4669	0.8843	0.5280	1.8940	1.1308	2.1418	62°10′
28° 0′	0.4695	0.8829	0.5317	1.8807	1.1326	2.1301	62° 0′
28°10′	0.4720	0.8816	0.5354	1.8676	1.1343	2.1185	61°50′
28°20′	0.4746	0.8802	0.5392	1.8546	1.1361	2.1070	61°40′
28°30′	0.4772	0.8788	0.5430	1.8418	1.1379	2.0957	61°30′
28°40′	0.4797	0.8774	0.5467	1.8291	1.1397	2.0846	61°20′
28°50′	0.4823	0.8760	0.5505	1.8165	1.1415	2.0736	61°10′
29° 0′	0.4848	0.8746	0.5543	1.8040	1.1434	2.0627	61° 0′
29°10′	0.4874	0.8732	0.5581	1.7917	1.1452	2.0519	60°50′
29°20′	0.4899	0.8718	0.5619	1.7796	1.1471	2.0413	60°40′
29°30′	0.4924	0.8704	0.5658	1.7675	1.1490	2.0308	60°30′
29°40′	0.4950	0.8689	0.5696	1.7556	1.1509	2.0204	60°20′
29°50′	0.4975	0.8675	0.5735	1.7437	1.1528	2.0101	60°10′
30° 0′	0.5000	0.8660	0.5774	1.7321	1.1547	2.0000	60° 0′
30°10′	0.5025	0.8646	0.5812	1.7205	1.1566	1.9900	59°50′
30°20′	0.5050	0.8631	0.5851	1.7090	1.1586	1.9801	59°40′
30°30′	0.5075	0.8616	0.5890	1.6977	1.1606	1.9703	59°30′
30°40′	0.5100	0.8601	0.5930	1.6864	1.1626	1.9606	59°20′
30°50′	0.5125	0.8587	0.5969	1.6753	1.1646	1.9511	59°10′
31° 0′	0.5150	0.8572	0.6009	1.6643	1.1666	1.9416	59° 0′
31°10′	0.5175	0.8557	0.6048	1.6534	1.1687	1.9323	58°50′
31°20′	0.5200	0.8542	0.6088	1.6426	1.1707	1.9230	58°40′
31°30′	0.5225	0.8526	0.6128	1.6319	1.1728	1.9139	58°30′
31°40′	0.5250	0.8511	0.6168	1.6212	1.1749	1.9048	58°20′
31°50′	0.5275	0.8496	0.6208	1.6107	1.1770	1.8959	58°10′
	$\cos A$	$\sin A$	$\cot A$	$\tan A$	$\csc A$	$\sec A$	A

A	$\sin A$	$\cos A$	$\tan A$	$\cot A$	$\sec A$	$\csc A$	
32° 0′	0.5299	0.8480	0.6249	1.6003	1.1792	1.8871	58° 0′
32°10′	0.5324	0.8465	0.6289	1.5900	1.1813	1.8783	57°50′
32°20′	0.5348	0.8450	0.6330	1.5798	1.1835	1.8697	57°40′
32°30′	0.5373	0.8434	0.6371	1.5697	1.1857	1.8612	57°30′
32°40′	0.5398	0.8418	0.6412	1.5597	1.1879	1.8527	57°20′
32°50′	0.5422	0.8403	0.6453	1.5497	1.1901	1.8443	57°10′
33° 0′	0.5446	0.8387	0.6494	1.5399	1.1924	1.8361	57° 0′
33°10′	0.5471	0.8371	0.6536	1.5301	1.1946	1.8279	56°50′
33°20′	0.5495	0.8355	0.6577	1.5204	1.1969	1.8198	56°40′
33°30′	0.5519	0.8339	0.6619	1.5108	1.1992	1.8118	56°30′
33°40′	0.5544	0.8323	0.6661	1.5013	1.2015	1.8039	56°20′
33°50′	0.5568	0.8307	0.6703	1.4919	1.2039	1.7960	56°10′
34° 0′	0.5592	0.8290	0.6745	1.4826	1.2062	1.7883	56° 0′
34°10′	0.5616	0.8274	0.6787	1.4733	1.2086	1.7806	55°50′
34°20′	0.5640	0.8258	0.6830	1.4641	1.2110	1.7730	55°40′
34°30′	0.5664	0.8241	0.6873	1.4550	1.2134	1.7655	55°30′
34°40′	0.5688	0.8225	0.6916	1.4460	1.2158	1.7581	55°20′
34°50′	0.5712	0.8208	0.6959	1.4370	1.2183	1.7507	55°10′
35° 0′	0.5736	0.8192	0.7002	1.4281	1.2208	1.7434	55° 0′
35°10′	0.5760	0.8175	0.7046	1.4193	1.2233	1.7362	54°50′
35°20′	0.5783	0.8158	0.7089	1.4106	1.2258	1.7291	54°40′
35°30′	0.5807	0.8141	0.7133	1.4019	1.2283	1.7221	54°30′
35°40′	0.5831	0.8124	0.7177	1.3934	1.2309	1.7151	54°20′
35°50′	0.5854	0.8107	0.7221	1.3848	1.2335	1.7081	54°10′
36° 0′	0.5878	0.8090	0.7265	1.3764	1.2361	1.7013	54° 0′
36°10′	0.5901	0.8073	0.7310	1.3680	1.2387	1.6945	53°50′
36°20′	0.5925	0.8056	0.7355	1.3597	1.2413	1.6878	53°40′
36°30′	0.5948	0.8039	0.7400	1.3514	1.2440	1.6812	53°30′
36°40′	0.5972	0.8021	0.7445	1.3432	1.2467	1.6746	53°20′
36°50′	0.5995	0.8004	0.7490	1.3351	1.2494	1.6681	53°10′
37° 0′	0.6018	0.7986	0.7536	1.3270	1.2521	1.6616	53° 0′
37°10′	0.6041	0.7969	0.7581	1.3190	1.2549	1.6553	52°50′
37°20′	0.6065	0.7951	0.7627	1.3111	1.2577	1.6489	52°40′
37°30′	0.6088	0.7934	0.7673	1.3032	1.2605	1.6427	52°30′
37°40′	0.6111	0.7916	0.7720	1.2954	1.2633	1.6365	52°20′
37°50′	0.6134	0.7898	0.7766	1.2876	1.2661	1.6303	52°10′
38° 0′	0.6157	0.7880	0.7813	1.2799	1.2690	1.6243	52° 0′
38°10′	0.6180	0.7862	0.7860	1.2723	1.2719	1.6183	51°50′
38°20′	0.6202	0.7844	0.7907	1.2647	1.2748	1.6123	51°40′
38°30′	0.6225	0.7826	0.7954	1.2572	1.2778	1.6064	51°30′
38°40′	0.6248	0.7808	0.8002	1.2497	1.2807	1.6005	51°20′
38°50′	0.6271	0.7790	0.8050	1.2423	1.2837	1.5948	51°10′
39° 0′	0.6293	0.7771	0.8098	1.2349	1.2868	1.5890	51° 0′
39°10′	0.6316	0.7753	0.8146	1.2276	1.2898	1.5833	50°50′
39°20′	0.6338	0.7735	0.8195	1.2203	1.2929	1.5777	50°40′
39°30′	0.6361	0.7716	0.8243	1.2131	1.2960	1.5721	50°30′
39°40′	0.6383	0.7698	0.8292	1.2059	1.2991	1.5666	50°20′
39°50′	0.6406	0.7679	0.8342	1.1988	1.3022	1.5611	50°10′
	$\cos A$	$\sin A$	$\cot A$	$\tan A$	$\csc A$	$\sec A$	A

A	$\sin A$	$\cos A$	$\tan A$	$\cot A$	$\sec A$	$\csc A$	
40° 0′	0.6428	0.7660	0.8391	1.1918	1.3054	1.5557	50° 0′
40°10′	0.6450	0.7642	0.8441	1.1847	1.3086	1.5504	49°50′
40°20′	0.6472	0.7623	0.8491	1.1778	1.3118	1.5450	49°40′
40°30′	0.6494	0.7604	0.8541	1.1708	1.3151	1.5398	49°30′
40°40′	0.6517	0.7585	0.8591	1.1640	1.3184	1.5345	49°20′
40°50′	0.6539	0.7566	0.8642	1.1571	1.3217	1.5294	49°10′
41° 0′	0.6561	0.7547	0.8693	1.1504	1.3250	1.5243	49° 0′
41°10′	0.6583	0.7528	0.8744	1.1436	1.3284	1.5192	48°50′
41°20′	0.6604	0.7509	0.8796	1.1369	1.3318	1.5141	48°40′
41°30′	0.6626	0.7490	0.8847	1.1303	1.3352	1.5092	48°30′
41°40′	0.6648	0.7470	0.8899	1.1237	1.3386	1.5042	48°20′
41°50′	0.6670	0.7451	0.8952	1.1171	1.3421	1.4993	48°10′
42° 0′	0.6691	0.7431	0.9004	1.1106	1.3456	1.4945	48° 0′
42°10′	0.6713	0.7412	0.9057	1.1041	1.3492	1.4897	47°50′
42°20′	0.6734	0.7392	0.9110	1.0977	1.3527	1.4849	47°40′
42°30′	0.6756	0.7373	0.9163	1.0913	1.3563	1.4802	47°30′
42°40′	0.6777	0.7353	0.9217	1.0850	1.3600	1.4755	47°20′
42°50′	0.6799	0.7333	0.9271	1.0786	1.3636	1.4709	47°10′
43° 0′	0.6820	0.7314	0.9325	1.0724	1.3673	1.4663	47° 0′
43°10′	0.6841	0.7294	0.9380	1.0661	1.3711	1.4617	46°50′
43°20′	0.6862	0.7274	0.9435	1.0599	1.3748	1.4572	46°40′
43°30′	0.6884	0.7254	0.9490	1.0538	1.3786	1.4527	46°30′
43°40′	0.6905	0.7234	0.9545	1.0477	1.3824	1.4483	46°20′
43°50′	0.6926	0.7214	0.9601	1.0416	1.3863	1.4439	46°10′
44° 0′	0.6947	0.7193	0.9657	1.0355	1.3902	1.4396	46° 0′
44°10′	0.6967	0.7173	0.9713	1.0295	1.3941	1.4352	45°50′
44°20′	0.6988	0.7153	0.9770	1.0235	1.3980	1.4310	45°40′
44°30′	0.7009	0.7133	0.9827	1.0176	1.4020	1.4267	45°30′
44°40′	0.7030	0.7112	0.9884	1.0117	1.4061	1.4225	45°20′
44°50′	0.7050	0.7092	0.9942	1.0058	1.4101	1.4183	45°10′
45° 0′	0.7071	0.7071	1.0000	1.0000	1.4142	1.4142	45° 0
	$\cos A$	$\sin A$	$\cot A$	$\tan A$	$\csc A$	$\sec A$	A

Table 2   Trigonometric Functions—Angle in Tenth of Degree Intervals

A	$\sin A$	$\cos A$	$\tan A$	$\cot A$	$\sec A$	$\csc A$	
0.0°	0.0000	1.0000	0.0000	Undefined	1.0000	Undefined	90.0°
0.1°	0.0017	1.0000	0.0017	572.9680	1.0000	572.9590	89.9°
0.2°	0.0035	1.0000	0.0035	286.4750	1.0000	286.4770	89.8°
0.3°	0.0052	1.0000	0.0052	190.9840	1.0000	190.9870	89.7°
0.4°	0.0070	1.0000	0.0070	143.2380	1.0000	143.2410	89.6°
0.5°	0.0087	1.0000	0.0087	114.5880	1.0000	114.5930	89.5°
0.6°	0.0105	0.9999	0.0105	95.4896	1.0001	95.4948	89.4°
0.7°	0.0122	0.9999	0.0122	81.8473	1.0001	81.8534	89.3°
0.8°	0.0140	0.9999	0.0140	71.6150	1.0001	71.6220	89.2°
0.9°	0.0157	0.9999	0.0157	63.6568	1.0001	63.6647	89.1°
1.0°	0.0175	0.9998	0.0175	57.2898	1.0002	57.2986	89.0°
1.1°	0.0192	0.9998	0.0192	52.0806	1.0002	52.0902	88.9°
1.2°	0.0209	0.9998	0.0209	47.7396	1.0002	47.7500	88.8°
1.3°	0.0227	0.9997	0.0227	44.0660	1.0003	44.0774	88.7°
1.4°	0.0244	0.9997	0.0244	40.9174	1.0003	40.9296	88.6°
1.5°	0.0262	0.9997	0.0262	38.1885	1.0003	38.2016	88.5°
1.6°	0.0279	0.9996	0.0279	35.8005	1.0004	35.8145	88.4°
1.7°	0.0297	0.9996	0.0297	33.6935	1.0004	33.7083	88.3°
1.8°	0.0314	0.9995	0.0314	31.8205	1.0005	31.8363	88.2°
1.9°	0.0332	0.9995	0.0332	30.1446	1.0006	30.1612	88.1°
2.0°	0.0349	0.9994	0.0349	28.6363	1.0006	28.6537	88.0°
2.1°	0.0366	0.9993	0.0367	27.2715	1.0007	27.2898	87.9°
2.2°	0.0384	0.9993	0.0384	26.0307	1.0007	26.0499	87.8°
2.3°	0.0401	0.9992	0.0402	24.8978	1.0008	24.9179	87.7°
2.4°	0.0419	0.9991	0.0419	23.8593	1.0009	23.8802	87.6°
2.5°	0.0436	0.9990	0.0437	22.9038	1.0010	22.9256	87.5°
2.6°	0.0454	0.9990	0.0454	22.0217	1.0010	22.0444	87.4°
2.7°	0.0471	0.9989	0.0472	21.2050	1.0011	21.2285	87.3°
2.8°	0.0488	0.9988	0.0489	20.4465	1.0012	20.4709	87.2°
2.9°	0.0506	0.9987	0.0507	19.7403	1.0013	19.7656	87.1°
3.0°	0.0523	0.9986	0.0524	19.0812	1.0014	19.1073	87.0°
3.1°	0.0541	0.9985	0.0542	18.4645	1.0015	18.4915	86.9°
3.2°	0.0558	0.9984	0.0559	17.8863	1.0016	17.9143	86.8°
3.3°	0.0576	0.9983	0.0577	17.3432	1.0017	17.3720	86.7°
3.4°	0.0593	0.9982	0.0594	16.8319	1.0018	16.8616	86.6°
3.5°	0.0610	0.9981	0.0612	16.3499	1.0019	16.3804	86.5°
3.6°	0.0628	0.9980	0.0629	15.8946	1.0020	15.9260	86.4°
3.7°	0.0645	0.9979	0.0647	15.4638	1.0021	15.4961	86.3°
3.8°	0.0663	0.9978	0.0664	15.0557	1.0022	15.0889	86.2°
3.9°	0.0680	0.9977	0.0682	14.6685	1.0023	14.7026	86.1°
	$\cos A$	$\sin A$	$\cot A$	$\tan A$	$\csc A$	$\sec A$	A

A	$\sin A$	$\cos A$	$\tan A$	$\cot A$	$\sec A$	$\csc A$	
4.0°	0.0698	0.9976	0.0699	14.3007	1.0024	14.3356	86.0°
4.1°	0.0715	0.9974	0.0717	13.9507	1.0026	13.9865	85.9°
4.2°	0.0732	0.9973	0.0734	13.6174	1.0027	13.6541	85.8°
4.3°	0.0750	0.9972	0.0752	13.2996	1.0028	13.3371	85.7°
4.4°	0.0767	0.9971	0.0769	12.9962	1.0030	13.0346	85.6°
4.5°	0.0785	0.9969	0.0787	12.7062	1.0031	12.7455	85.5°
4.6°	0.0802	0.9968	0.0805	12.4288	1.0032	12.4690	85.4°
4.7°	0.0819	0.9966	0.0822	12.1632	1.0034	12.2043	85.3°
4.8°	0.0837	0.9965	0.0840	11.9087	1.0035	11.9506	85.2°
4.9°	0.0854	0.9963	0.0857	11.6645	1.0037	11.7073	85.1°
5.0°	0.0872	0.9962	0.0875	11.4301	1.0038	11.4737	85.0°
5.1°	0.0889	0.9960	0.0892	11.2048	1.0040	11.2493	84.9°
5.2°	0.0906	0.9959	0.0910	10.9882	1.0041	11.0336	84.8°
5.3°	0.0924	0.9957	0.0928	10.7797	1.0043	10.8260	84.7°
5.4°	0.0941	0.9956	0.0945	10.5789	1.0045	10.6261	84.6°
5.5°	0.0958	0.9954	0.0963	10.3854	1.0046	10.4334	84.5°
5.6°	0.0976	0.9952	0.0981	10.1988	1.0048	10.2477	84.4°
5.7°	0.0993	0.9951	0.0998	10.0187	1.0050	10.0685	84.3°
5.8°	0.1011	0.9949	0.1016	9.8448	1.0051	9.8955	84.2°
5.9°	0.1028	0.9947	0.1033	9.6768	1.0053	9.7283	84.1°
6.0°	0.1045	0.9945	0.1051	9.5144	1.0055	9.5668	84.0°
6.1°	0.1063	0.9943	0.1069	9.3572	1.0057	9.4105	83.9°
6.2°	0.1080	0.9942	0.1086	9.2052	1.0059	9.2593	83.8°
6.3°	0.1097	0.9940	0.1104	9.0579	1.0061	9.1129	83.7°
6.4°	0.1115	0.9938	0.1122	8.9152	1.0063	8.9711	83.6°
6.5°	0.1132	0.9936	0.1139	8.7769	1.0065	8.8337	83.5°
6.6°	0.1149	0.9934	0.1157	8.6428	1.0067	8.7004	83.4°
6.7°	0.1167	0.9932	0.1175	8.5126	1.0069	8.5711	83.3°
6.8°	0.1184	0.9930	0.1192	8.3863	1.0071	8.4457	83.2°
6.9°	0.1201	0.9928	0.1210	8.2636	1.0073	8.3239	83.1°
7.0°	0.1219	0.9925	0.1228	8.1444	1.0075	8.2055	83.0°
7.1°	0.1236	0.9923	0.1246	8.0285	1.0077	8.0905	82.9°
7.2°	0.1253	0.9921	0.1263	7.9158	1.0079	7.9787	82.8°
7.3°	0.1271	0.9919	0.1281	7.8062	1.0082	7.8700	82.7°
7.4°	0.1288	0.9917	0.1299	7.6996	1.0084	7.7642	82.6°
7.5°	0.1305	0.9914	0.1317	7.5958	1.0086	7.6613	82.5°
7.6°	0.1323	0.9912	0.1334	7.4947	1.0089	7.5611	82.4°
7.7°	0.1340	0.9910	0.1352	7.3962	1.0091	7.4635	82.3°
7.8°	0.1357	0.9907	0.1370	7.3002	1.0093	7.3684	82.2°
7.9°	0.1374	0.9905	0.1388	7.2066	1.0096	7.2757	82.1°
	$\cos A$	$\sin A$	$\cot A$	$\tan A$	$\csc A$	$\sec A$	A

A	sin A	cos A	tan A	cot A	sec A	csc A	
8.0°	0.1392	0.9903	0.1405	7.1154	1.0098	7.1853	82.0°
8.1°	0.1409	0.9900	0.1423	7.0264	1.0101	7.0972	81.9°
8.2°	0.1426	0.9898	0.1441	6.9395	1.0103	7.0112	81.8°
8.3°	0.1444	0.9895	0.1459	6.8548	1.0106	6.9273	81.7°
8.4°	0.1461	0.9893	0.1477	6.7720	1.0108	6.8454	81.6°
8.5°	0.1478	0.9890	0.1495	6.6912	1.0111	6.7655	81.5°
8.6°	0.1495	0.9888	0.1512	6.6122	1.0114	6.6874	81.4°
8.7°	0.1513	0.9885	0.1530	6.5350	1.0116	6.6111	81.3°
8.8°	0.1530	0.9882	0.1548	6.4596	1.0119	6.5366	81.2°
8.9°	0.1547	0.9880	0.1566	6.3859	1.0122	6.4637	81.1°
9.0°	0.1564	0.9877	0.1584	6.3138	1.0125	6.3925	81.0°
9.1°	0.1582	0.9874	0.1602	6.2432	1.0127	6.3228	80.9°
9.2°	0.1599	0.9871	0.1620	6.1742	1.0130	6.2546	80.8°
9.3°	0.1616	0.9869	0.1638	6.1066	1.0133	6.1880	80.7°
9.4°	0.1633	0.9866	0.1655	6.0405	1.0136	6.1227	80.6°
9.5°	0.1650	0.9863	0.1673	5.9758	1.0139	6.0589	80.5°
9.6°	0.1668	0.9860	0.1691	5.9124	1.0142	5.9963	80.4°
9.7°	0.1685	0.9857	0.1709	5.8502	1.0145	5.9351	80.3°
9.8°	0.1702	0.9854	0.1727	5.7894	1.0148	5.8751	80.2°
9.9°	0.1719	0.9851	0.1745	5.7297	1.0151	5.8164	80.1°
10.0°	0.1736	0.9848	0.1763	5.6713	1.0154	5.7588	80.0°
10.1°	0.1754	0.9845	0.1781	5.6140	1.0157	5.7023	79.9°
10.2°	0.1771	0.9842	0.1799	5.5578	1.0161	5.6470	79.8°
10.3°	0.1788	0.9839	0.1817	5.5026	1.0164	5.5928	79.7°
10.4°	0.1805	0.9836	0.1835	5.4486	1.0167	5.5396	79.6°
10.5°	0.1822	0.9833	0.1853	5.3955	1.0170	5.4874	79.5°
10.6°	0.1840	0.9829	0.1871	5.3435	1.0174	5.4362	79.4°
10.7°	0.1857	0.9826	0.1890	5.2923	1.0177	5.3860	79.3°
10.8°	0.1874	0.9823	0.1908	5.2422	1.0180	5.3367	79.2°
10.9°	0.1891	0.9820	0.1926	5.1929	1.0184	5.2883	79.1°
11.0°	0.1908	0.9816	0.1944	5.1446	1.0187	5.2408	79.0°
11.1°	0.1925	0.9813	0.1962	5.0970	1.0191	5.1942	78.9°
11.2°	0.1942	0.9810	0.1980	5.0504	1.0194	5.1484	78.8°
11.3°	0.1959	0.9806	0.1998	5.0045	1.0198	5.1034	78.7°
11.4°	0.1977	0.9803	0.2016	4.9594	1.0201	5.0593	78.6°
11.5°	0.1994	0.9799	0.2035	4.9152	1.0205	5.0158	78.5°
11.6°	0.2011	0.9796	0.2053	4.8716	1.0209	4.9732	78.4°
11.7°	0.2028	0.9792	0.2071	4.8288	1.0212	4.9313	78.3°
11.8°	0.2045	0.9789	0.2089	4.7867	1.0216	4.8901	78.2°
11.9°	0.2062	0.9785	0.2107	4.7453	1.0220	4.8496	78.1°
	cos A	sin A	cot A	tan A	csc A	sec A	A

A	$\sin A$	$\cos A$	$\tan A$	$\cot A$	$\sec A$	$\csc A$	
12.0°	0.2079	0.9781	0.2126	4.7046	1.0223	4.8097	78.0°
12.1°	0.2096	0.9778	0.2144	4.6646	1.0227	4.7706	77.9°
12.2°	0.2113	0.9774	0.2162	4.6252	1.0231	4.7320	77.8°
12.3°	0.2130	0.9770	0.2180	4.5864	1.0235	4.6942	77.7°
12.4°	0.2147	0.9767	0.2199	4.5483	1.0239	4.6569	77.6°
12.5°	0.2164	0.9763	0.2217	4.5107	1.0243	4.6202	77.5°
12.6°	0.2181	0.9759	0.2235	4.4737	1.0247	4.5841	77.4°
12.7°	0.2198	0.9755	0.2254	4.4373	1.0251	4.5486	77.3°
12.8°	0.2215	0.9751	0.2272	4.4015	1.0255	4.5137	77.2°
12.9°	0.2233	0.9748	0.2290	4.3662	1.0259	4.4793	77.1°
13.0°	0.2250	0.9744	0.2309	4.3315	1.0263	4.4454	77.0°
13.1°	0.2267	0.9740	0.2327	4.2972	1.0267	4.4121	76.9°
13.2°	0.2284	0.9736	0.2345	4.2635	1.0271	4.3792	76.8°
13.3°	0.2300	0.9732	0.2364	4.2303	1.0276	4.3469	76.7°
13.4°	0.2317	0.9728	0.2382	4.1976	1.0280	4.3150	76.6°
13.5°	0.2334	0.9724	0.2401	4.1653	1.0284	4.2837	76.5°
13.6°	0.2351	0.9720	0.2419	4.1335	1.0288	4.2527	76.4°
13.7°	0.2368	0.9715	0.2438	4.1022	1.0293	4.2223	76.3°
13.8°	0.2385	0.9711	0.2456	4.0713	1.0297	4.1923	76.2°
13.9°	0.2402	0.9707	0.2475	4.0408	1.0302	4.1627	76.1°
14.0°	0.2419	0.9703	0.2493	4.0108	1.0306	4.1336	76.0°
14.1°	0.2436	0.9699	0.2512	3.9812	1.0311	4.1048	75.9°
14.2°	0.2453	0.9694	0.2530	3.9520	1.0315	4.0765	75.8°
14.3°	0.2470	0.9690	0.2549	3.9232	1.0320	4.0486	75.7°
14.4°	0.2487	0.9686	0.2568	3.8947	1.0324	4.0211	75.6°
14.5°	0.2504	0.9681	0.2586	3.8667	1.0329	3.9939	75.5°
14.6°	0.2521	0.9677	0.2605	3.8391	1.0334	3.9672	75.4°
14.7°	0.2538	0.9673	0.2623	3.8118	1.0338	3.9408	75.3°
14.8°	0.2554	0.9668	0.2642	3.7848	1.0343	3.9147	75.2°
14.9°	0.2571	0.9664	0.2661	3.7583	1.0348	3.8890	75.1°
15.0°	0.2588	0.9659	0.2679	3.7320	1.0353	3.8637	75.0°
15.1°	0.2605	0.9655	0.2698	3.7062	1.0358	3.8387	74.9°
15.2°	0.2622	0.9650	0.2717	3.6806	1.0363	3.8140	74.8°
15.3°	0.2639	0.9646	0.2736	3.6554	1.0367	3.7897	74.7°
15.4°	0.2656	0.9641	0.2754	3.6305	1.0372	3.7657	74.6°
15.5°	0.2672	0.9636	0.2773	3.6059	1.0377	3.7420	74.5°
15.6°	0.2689	0.9632	0.2792	3.5816	1.0382	3.7186	74.4°
15.7°	0.2706	0.9627	0.2811	3.5576	1.0388	3.6955	74.3°
15.8°	0.2723	0.9622	0.2830	3.5339	1.0393	3.6727	74.2°
15.9°	0.2740	0.9617	0.2849	3.5105	1.0398	3.6502	74.1°
	$\cos A$	$\sin A$	$\cot A$	$\tan A$	$\csc A$	$\sec A$	A

A	$\sin A$	$\cos A$	$\tan A$	$\cot A$	$\sec A$	$\csc A$	
16.0°	0.2756	0.9613	0.2867	3.4874	1.0403	3.6280	74.0°
16.1°	0.2773	0.9608	0.2886	3.4646	1.0408	3.6060	73.9°
16.2°	0.2790	0.9603	0.2905	3.4420	1.0413	3.5843	73.8°
16.3°	0.2807	0.9598	0.2924	3.4197	1.0419	3.5629	73.7°
16.4°	0.2823	0.9593	0.2943	3.3977	1.0424	3.5418	73.6°
16.5°	0.2840	0.9588	0.2962	3.3759	1.0429	3.5209	73.5°
16.6°	0.2857	0.9583	0.2981	3.3544	1.0435	3.5003	73.4°
16.7°	0.2874	0.9578	0.3000	3.3332	1.0440	3.4799	73.3°
16.8°	0.2890	0.9573	0.3019	3.3122	1.0446	3.4598	73.2°
16.9°	0.2907	0.9568	0.3038	3.2914	1.0451	3.4399	73.1°
17.0°	0.2924	0.9563	0.3057	3.2708	1.0457	3.4203	73.0°
17.1°	0.2940	0.9558	0.3076	3.2505	1.0463	3.4009	72.9°
17.2°	0.2957	0.9553	0.3096	3.2305	1.0468	3.3817	72.8°
17.3°	0.2974	0.9548	0.3115	3.2106	1.0474	3.3628	72.7°
17.4°	0.2990	0.9542	0.3134	3.1910	1.0480	3.3440	72.6°
17.5°	0.3007	0.9537	0.3153	3.1716	1.0485	3.3255	72.5°
17.6°	0.3024	0.9532	0.3172	3.1524	1.0491	3.3072	72.4°
17.7°	0.3040	0.9527	0.3191	3.1334	1.0497	3.2891	72.3°
17.8°	0.3057	0.9521	0.3211	3.1146	1.0503	3.2712	72.2°
17.9°	0.3074	0.9516	0.3230	3.0961	1.0509	3.2535	72.1°
18.0°	0.3090	0.9511	0.3249	3.0777	1.0515	3.2361	72.0°
18.1°	0.3107	0.9505	0.3269	3.0595	1.0521	3.2188	71.9°
18.2°	0.3123	0.9500	0.3288	3.0415	1.0527	3.2017	71.8°
18.3°	0.3140	0.9494	0.3307	3.0237	1.0533	3.1848	71.7°
18.4°	0.3156	0.9489	0.3327	3.0061	1.0539	3.1681	71.6°
18.5°	0.3173	0.9483	0.3346	2.9887	1.0545	3.1515	71.5°
18.6°	0.3190	0.9478	0.3365	2.9714	1.0551	3.1352	71.4°
18.7°	0.3206	0.9472	0.3385	2.9544	1.0557	3.1190	71.3°
18.8°	0.3223	0.9466	0.3404	2.9375	1.0564	3.1030	71.2°
18.9°	0.3239	0.9461	0.3424	2.9208	1.0570	3.0872	71.1°
19.0°	0.3256	0.9455	0.3443	2.9042	1.0576	3.0715	71.0°
19.1°	0.3272	0.9449	0.3463	2.8878	1.0583	3.0561	70.9°
19.2°	0.3289	0.9444	0.3482	2.8716	1.0589	3.0407	70.8°
19.3°	0.3305	0.9438	0.3502	2.8555	1.0595	3.0256	70.7°
19.4°	0.3322	0.9432	0.3522	2.8396	1.0602	3.0106	70.6°
19.5°	0.3338	0.9426	0.3541	2.8239	1.0608	2.9957	70.5°
19.6°	0.3355	0.9421	0.3561	2.8083	1.0615	2.9811	70.4°
19.7°	0.3371	0.9415	0.3581	2.7929	1.0622	2.9665	70.3°
19.8°	0.3387	0.9409	0.3600	2.7776	1.0628	2.9521	70.2°
19.9°	0.3404	0.9403	0.3620	2.7625	1.0635	2.9379	70.1°
	$\cos A$	$\sin A$	$\cot A$	$\tan A$	$\csc A$	$\sec A$	A

A	$\sin A$	$\cos A$	$\tan A$	$\cot A$	$\sec A$	$\csc A$	
20.0°	0.3420	0.9397	0.3640	2.7475	1.0642	2.9238	70.0°
20.1°	0.3437	0.9391	0.3659	2.7326	1.0649	2.9098	69.9°
20.2°	0.3453	0.9385	0.3679	2.7179	1.0655	2.8960	69.8°
20.3°	0.3469	0.9379	0.3699	2.7033	1.0662	2.8824	69.7°
20.4°	0.3486	0.9373	0.3719	2.6889	1.0669	2.8688	69.6°
20.5°	0.3502	0.9367	0.3739	2.6746	1.0676	2.8554	69.5°
20.6°	0.3518	0.9361	0.3759	2.6605	1.0683	2.8422	69.4°
20.7°	0.3535	0.9354	0.3779	2.6464	1.0690	2.8291	69.3°
20.8°	0.3551	0.9348	0.3799	2.6325	1.0697	2.8160	69.2°
20.9°	0.3567	0.9342	0.3819	2.6187	1.0704	2.8032	69.1°
21.0°	0.3584	0.9336	0.3839	2.6051	1.0711	2.7904	69.0°
21.1°	0.3600	0.9330	0.3859	2.5916	1.0719	2.7778	68.9°
21.2°	0.3616	0.9323	0.3879	2.5781	1.0726	2.7653	68.8°
21.3°	0.3633	0.9317	0.3899	2.5649	1.0733	2.7529	68.7°
21.4°	0.3649	0.9311	0.3919	2.5517	1.0740	2.7406	68.6°
21.5°	0.3665	0.9304	0.3939	2.5386	1.0748	2.7285	68.5°
21.6°	0.3681	0.9298	0.3959	2.5257	1.0755	2.7165	68.4°
21.7°	0.3697	0.9291	0.3979	2.5129	1.0763	2.7045	68.3°
21.8°	0.3714	0.9285	0.4000	2.5002	1.0770	2.6927	68.2°
21.9°	0.3730	0.9278	0.4020	2.4876	1.0778	2.6810	68.1°
22.0°	0.3746	0.9272	0.4040	2.4751	1.0785	2.6695	68.0°
22.1°	0.3762	0.9265	0.4061	2.4627	1.0793	2.6580	67.9°
22.2°	0.3778	0.9259	0.4081	2.4504	1.0801	2.6466	67.8°
22.3°	0.3795	0.9252	0.4101	2.4382	1.0808	2.6353	67.7°
22.4°	0.3811	0.9245	0.4122	2.4262	1.0816	2.6242	67.6°
22.5°	0.3827	0.9239	0.4142	2.4142	1.0824	2.6131	67.5°
22.6°	0.3843	0.9232	0.4163	2.4023	1.0832	2.6022	67.4°
22.7°	0.3859	0.9225	0.4183	2.3906	1.0840	2.5913	67.3°
22.8°	0.3875	0.9219	0.4204	2.3789	1.0848	2.5805	67.2°
22.9°	0.3891	0.9212	0.4224	2.3673	1.0856	2.5699	67.1°
23.0°	0.3907	0.9205	0.4245	2.3558	1.0864	2.5593	67.0°
23.1°	0.3923	0.9198	0.4265	2.3445	1.0872	2.5488	66.9°
23.2°	0.3939	0.9191	0.4286	2.3332	1.0880	2.5384	66.8°
23.3°	0.3955	0.9184	0.4307	2.3220	1.0888	2.5281	66.7°
23.4°	0.3971	0.9178	0.4327	2.3109	1.0896	2.5179	66.6°
23.5°	0.3987	0.9171	0.4348	2.2998	1.0904	2.5078	66.5°
23.6°	0.4003	0.9164	0.4369	2.2889	1.0913	2.4978	66.4°
23.7°	0.4019	0.9157	0.4390	2.2781	1.0921	2.4879	66.3°
23.8°	0.4035	0.9150	0.4411	2.2673	1.0929	2.4780	66.2°
23.9°	0.4051	0.9143	0.4431	2.2566	1.0938	2.4683	66.1°
	$\cos A$	$\sin A$	$\cot A$	$\tan A$	$\csc A$	$\sec A$	A

A	$\sin A$	$\cos A$	$\tan A$	$\cot A$	$\sec A$	$\csc A$	
24.0°	0.4067	0.9135	0.4452	2.2460	1.0946	2.4586	66.0°
24.1°	0.4083	0.9128	0.4473	2.2355	1.0955	2.4490	65.9°
24.2°	0.4099	0.9121	0.4494	2.2251	1.0963	2.4395	65.8°
24.3°	0.4115	0.9114	0.4515	2.2147	1.0972	2.4300	65.7°
24.4°	0.4131	0.9107	0.4536	2.2045	1.0981	2.4207	65.6°
24.5°	0.4147	0.9100	0.4557	2.1943	1.0989	2.4114	65.5°
24.6°	0.4163	0.9092	0.4578	2.1842	1.0998	2.4022	65.4°
24.7°	0.4179	0.9085	0.4599	2.1742	1.1007	2.3931	65.3°
24.8°	0.4195	0.9078	0.4621	2.1642	1.1016	2.3841	65.2°
24.9°	0.4210	0.9070	0.4642	2.1543	1.1025	2.3751	65.1°
25.0°	0.4226	0.9063	0.4663	2.1445	1.1034	2.3662	65.0°
25.1°	0.4242	0.9056	0.4684	2.1348	1.1043	2.3574	64.9°
25.2°	0.4258	0.9048	0.4706	2.1251	1.1052	2.3486	64.8°
25.3°	0.4274	0.9041	0.4727	2.1155	1.1061	2.3400	64.7°
25.4°	0.4289	0.9033	0.4748	2.1060	1.1070	2.3313	64.6°
25.5°	0.4305	0.9026	0.4770	2.0965	1.1079	2.3228	64.5°
25.6°	0.4321	0.9018	0.4791	2.0872	1.1089	2.3144	64.4°
25.7°	0.4337	0.9011	0.4813	2.0778	1.1098	2.3060	64.3°
25.8°	0.4352	0.9003	0.4834	2.0686	1.1107	2.2976	64.2°
25.9°	0.4368	0.8996	0.4856	2.0594	1.1117	2.2894	64.1°
26.0°	0.4384	0.8988	0.4877	2.0503	1.1126	2.2812	64.0°
26.1°	0.4399	0.8980	0.4899	2.0412	1.1136	2.2730	63.9°
26.2°	0.4415	0.8973	0.4921	2.0323	1.1145	2.2650	63.8°
26.3°	0.4431	0.8965	0.4942	2.0233	1.1155	2.2570	63.7°
26.4°	0.4446	0.8957	0.4964	2.0145	1.1164	2.2490	63.6°
26.5°	0.4462	0.8949	0.4986	2.0057	1.1174	2.2412	63.5°
26.6°	0.4478	0.8942	0.5008	1.9969	1.1184	2.2333	63.4°
26.7°	0.4493	0.8934	0.5029	1.9883	1.1194	2.2256	63.3°
26.8°	0.4509	0.8926	0.5051	1.9797	1.1203	2.2179	63.2°
26.9°	0.4524	0.8918	0.5073	1.9711	1.1213	2.2103	63.1°
27.0°	0.4540	0.8910	0.5095	1.9626	1.1223	2.2027	63.0°
27.1°	0.4555	0.8902	0.5117	1.9542	1.1233	2.1952	62.9°
27.2°	0.4571	0.8894	0.5139	1.9458	1.1243	2.1877	62.8°
27.3°	0.4587	0.8886	0.5161	1.9375	1.1253	2.1803	62.7°
27.4°	0.4602	0.8878	0.5184	1.9292	1.1264	2.1730	62.6°
27.5°	0.4617	0.8870	0.5206	1.9210	1.1274	2.1657	62.5°
27.6°	0.4633	0.8862	0.5228	1.9128	1.1284	2.1584	62.4°
27.7°	0.4648	0.8854	0.5250	1.9047	1.1294	2.1513	62.3°
27.8°	0.4664	0.8846	0.5272	1.8967	1.1305	2.1441	62.2°
27.9°	0.4679	0.8838	0.5295	1.8887	1.1315	2.1371	62.1°
	$\cos A$	$\sin A$	$\cot A$	$\tan A$	$\csc A$	$\sec A$	A

A	$\sin A$	$\cos A$	$\tan A$	$\cot A$	$\sec A$	$\csc A$	
28.0°	0.4695	0.8829	0.5317	1.8807	1.1326	2.1300	62.0°
28.1°	0.4710	0.8821	0.5340	1.8728	1.1336	2.1231	61.9°
28.2°	0.4726	0.8813	0.5362	1.8650	1.1347	2.1162	61.8°
28.3°	0.4741	0.8805	0.5384	1.8572	1.1357	2.1093	61.7°
28.4°	0.4756	0.8796	0.5407	1.8495	1.1368	2.1025	61.6°
28.5°	0.4772	0.8788	0.5430	1.8418	1.1379	2.0957	61.5°
28.6°	0.4787	0.8780	0.5452	1.8341	1.1390	2.0890	61.4°
28.7°	0.4802	0.8771	0.5475	1.8265	1.1401	2.0824	61.3°
28.8°	0.4818	0.8763	0.5498	1.8190	1.1412	2.0757	61.2°
28.9°	0.4833	0.8755	0.5520	1.8115	1.1423	2.0692	61.1°
29.0°	0.4848	0.8746	0.5543	1.8040	1.1434	2.0627	61.0°
29.1°	0.4863	0.8738	0.5566	1.7966	1.1445	2.0562	60.9°
29.2°	0.4879	0.8729	0.5589	1.7893	1.1456	2.0498	60.8°
29.3°	0.4894	0.8721	0.5612	1.7820	1.1467	2.0434	60.7°
29.4°	0.4909	0.8712	0.5635	1.7747	1.1478	2.0371	60.6°
29.5°	0.4924	0.8704	0.5658	1.7675	1.1490	2.0308	60.5°
29.6°	0.4939	0.8695	0.5681	1.7603	1.1501	2.0245	60.4°
29.7°	0.4955	0.8686	0.5704	1.7532	1.1512	2.0183	60.3°
29.8°	0.4970	0.8678	0.5727	1.7461	1.1524	2.0122	60.2°
29.9°	0.4985	0.8669	0.5750	1.7390	1.1535	2.0061	60.1°
30.0°	0.5000	0.8660	0.5774	1.7320	1.1547	2.0000	60.0°
30.1°	0.5015	0.8652	0.5797	1.7251	1.1559	1.9940	59.9°
30.2°	0.5030	0.8643	0.5820	1.7182	1.1570	1.9880	59.8°
30.3°	0.5045	0.8634	0.5844	1.7113	1.1582	1.9820	59.7°
30.4°	0.5060	0.8625	0.5867	1.7045	1.1594	1.9761	59.6°
30.5°	0.5075	0.8616	0.5890	1.6977	1.1606	1.9703	59.5°
30.6°	0.5090	0.8607	0.5914	1.6909	1.1618	1.9645	59.4°
30.7°	0.5105	0.8599	0.5938	1.6842	1.1630	1.9587	59.3°
30.8°	0.5120	0.8590	0.5961	1.6775	1.1642	1.9530	59.2°
30.9°	0.5135	0.8581	0.5985	1.6709	1.1654	1.9473	59.1°
31.0°	0.5150	0.8572	0.6009	1.6643	1.1666	1.9416	59.0°
31.1°	0.5165	0.8563	0.6032	1.6577	1.1679	1.9360	58.9°
31.2°	0.5180	0.8554	0.6056	1.6512	1.1691	1.9304	58.8°
31.3°	0.5195	0.8545	0.6080	1.6447	1.1703	1.9249	58.7°
31.4°	0.5210	0.8536	0.6104	1.6383	1.1716	1.9193	58.6°
31.5°	0.5225	0.8526	0.6128	1.6318	1.1728	1.9139	58.5°
31.6°	0.5240	0.8517	0.6152	1.6255	1.1741	1.9084	58.4°
31.7°	0.5255	0.8508	0.6176	1.6191	1.1754	1.9030	58.3°
31.8°	0.5270	0.8499	0.6200	1.6128	1.1766	1.8977	58.2°
31.9°	0.5284	0.8490	0.6224	1.6066	1.1779	1.8924	58.1°
	$\cos A$	$\sin A$	$\cot A$	$\tan A$	$\csc A$	$\sec A$	A

A	$\sin A$	$\cos A$	$\tan A$	$\cot A$	$\sec A$	$\csc A$	
32.0°	0.5299	0.8480	0.6249	1.6003	1.1792	1.8871	58.0°
32.1°	0.5314	0.8471	0.6273	1.5941	1.1805	1.8818	57.9°
32.2°	0.5329	0.8462	0.6297	1.5880	1.1818	1.8766	57.8°
32.3°	0.5344	0.8453	0.6322	1.5818	1.1831	1.8714	57.7°
32.4°	0.5358	0.8443	0.6346	1.5757	1.1844	1.8663	57.6°
32.5°	0.5373	0.8434	0.6371	1.5697	1.1857	1.8612	57.5°
32.6°	0.5388	0.8425	0.6395	1.5637	1.1870	1.8561	57.4°
32.7°	0.5402	0.8415	0.6420	1.5577	1.1883	1.8510	57.3°
32.8°	0.5417	0.8406	0.6445	1.5517	1.1897	1.8460	57.2°
32.9°	0.5432	0.8396	0.6469	1.5458	1.1910	1.8410	57.1°
33.0°	0.5446	0.8387	0.6494	1.5399	1.1924	1.8361	57.0°
33.1°	0.5461	0.8377	0.6519	1.5340	1.1937	1.8312	56.9°
33.2°	0.5476	0.8368	0.6544	1.5282	1.1951	1.8263	56.8°
33.3°	0.5490	0.8358	0.6569	1.5224	1.1964	1.8214	56.7°
33.4°	0.5505	0.8348	0.6594	1.5166	1.1978	1.8166	56.6°
33.5°	0.5519	0.8339	0.6619	1.5108	1.1992	1.8118	56.5°
33.6°	0.5534	0.8329	0.6644	1.5051	1.2006	1.8070	56.4°
33.7°	0.5548	0.8320	0.6669	1.4994	1.2020	1.8023	56.3°
33.8°	0.5563	0.8310	0.6694	1.4938	1.2034	1.7976	56.2°
33.9°	0.5577	0.8300	0.6720	1.4882	1.2048	1.7929	56.1°
34.0°	0.5592	0.8290	0.6745	1.4826	1.2062	1.7883	56.0°
34.1°	0.5606	0.8281	0.6771	1.4770	1.2076	1.7837	55.9°
34.2°	0.5621	0.8271	0.6796	1.4715	1.2091	1.7791	55.8°
34.3°	0.5635	0.8261	0.6822	1.4659	1.2105	1.7745	55.7°
34.4°	0.5650	0.8251	0.6847	1.4605	1.2120	1.7700	55.6°
34.5°	0.5664	0.8241	0.6873	1.4550	1.2134	1.7655	55.5°
34.6°	0.5678	0.8231	0.6899	1.4496	1.2149	1.7610	55.4°
34.7°	0.5693	0.8221	0.6924	1.4442	1.2163	1.7566	55.3°
34.8°	0.5707	0.8211	0.6950	1.4388	1.2178	1.7522	55.2°
34.9°	0.5721	0.8202	0.6976	1.4335	1.2193	1.7478	55.1°
35.0°	0.5736	0.8192	0.7002	1.4281	1.2208	1.7434	55.0°
35.1°	0.5750	0.8181	0.7028	1.4229	1.2223	1.7391	54.9°
35.2°	0.5764	0.8171	0.7054	1.4176	1.2238	1.7348	54.8°
35.3°	0.5779	0.8161	0.7080	1.4123	1.2253	1.7305	54.7°
35.4°	0.5793	0.8151	0.7107	1.4071	1.2268	1.7263	54.6°
35.5°	0.5807	0.8141	0.7133	1.4019	1.2283	1.7220	54.5°
35.6°	0.5821	0.8131	0.7159	1.3968	1.2299	1.7178	54.4°
35.7°	0.5835	0.8121	0.7186	1.3916	1.2314	1.7137	54.3°
35.8°	0.5850	0.8111	0.7212	1.3865	1.2329	1.7095	54.2°
35.9°	0.5864	0.8100	0.7239	1.3814	1.2345	1.7054	54.1°
	$\cos A$	$\sin A$	$\cot A$	$\tan A$	$\csc A$	$\sec A$	A

A	$\sin A$	$\cos A$	$\tan A$	$\cot A$	$\sec A$	$\csc A$	
36.0°	0.5878	0.8090	0.7265	1.3764	1.2361	1.7013	54.0°
36.1°	0.5892	0.8080	0.7292	1.3713	1.2376	1.6972	53.9°
36.2°	0.5906	0.8070	0.7319	1.3663	1.2392	1.6932	53.8°
36.3°	0.5920	0.8059	0.7346	1.3613	1.2408	1.6892	53.7°
36.4°	0.5934	0.8049	0.7373	1.3564	1.2424	1.6851	53.6°
36.5°	0.5948	0.8039	0.7400	1.3514	1.2440	1.6812	53.5°
36.6°	0.5962	0.8028	0.7427	1.3465	1.2456	1.6772	53.4°
36.7°	0.5976	0.8018	0.7454	1.3416	1.2472	1.6733	53.3°
36.8°	0.5990	0.8007	0.7481	1.3367	1.2489	1.6694	53.2°
36.9°	0.6004	0.7997	0.7508	1.3319	1.2505	1.6655	53.1°
37.0°	0.6018	0.7986	0.7536	1.3270	1.2521	1.6616	53.0°
37.1°	0.6032	0.7976	0.7563	1.3222	1.2538	1.6578	52.9°
37.2°	0.6046	0.7965	0.7590	1.3175	1.2554	1.6540	52.8°
37.3°	0.6060	0.7955	0.7618	1.3127	1.2571	1.6502	52.7°
37.4°	0.6074	0.7944	0.7646	1.3079	1.2588	1.6464	52.6°
37.5°	0.6088	0.7934	0.7673	1.3032	1.2605	1.6427	52.5°
37.6°	0.6101	0.7923	0.7701	1.2985	1.2622	1.6390	52.4°
37.7°	0.6115	0.7912	0.7729	1.2938	1.2639	1.6353	52.3°
37.8°	0.6129	0.7902	0.7757	1.2892	1.2656	1.6316	52.2°
37.9°	0.6143	0.7891	0.7785	1.2846	1.2673	1.6279	52.1°
38.0°	0.6157	0.7880	0.7813	1.2799	1.2690	1.6243	52.0°
38.1°	0.6170	0.7869	0.7841	1.2753	1.2708	1.6207	51.9°
38.2°	0.6184	0.7859	0.7869	1.2708	1.2725	1.6171	51.8°
38.3°	0.6198	0.7848	0.7898	1.2662	1.2742	1.6135	51.7°
38.4°	0.6211	0.7837	0.7926	1.2617	1.2760	1.6099	51.6°
38.5°	0.6225	0.7826	0.7954	1.2572	1.2778	1.6064	51.5°
38.6°	0.6239	0.7815	0.7983	1.2527	1.2796	1.6029	51.4°
38.7°	0.6252	0.7804	0.8012	1.2482	1.2813	1.5994	51.3°
38.8°	0.6266	0.7793	0.8040	1.2438	1.2831	1.5959	51.2°
38.9°	0.6280	0.7782	0.8069	1.2393	1.2849	1.5925	51.1°
39.0°	0.6293	0.7771	0.8098	1.2349	1.2868	1.5890	51.0°
39.1°	0.6307	0.7760	0.8127	1.2305	1.2886	1.5856	50.9°
39.2°	0.6320	0.7749	0.8156	1.2261	1.2904	1.5822	50.8°
39.3°	0.6334	0.7738	0.8185	1.2218	1.2923	1.5788	50.7°
39.4°	0.6347	0.7727	0.8214	1.2174	1.2941	1.5755	50.6°
39.5°	0.6361	0.7716	0.8243	1.2131	1.2960	1.5721	50.5°
39.6°	0.6374	0.7705	0.8273	1.2088	1.2978	1.5688	50.4°
39.7°	0.6388	0.7694	0.8302	1.2045	1.2997	1.5655	50.3°
39.8°	0.6401	0.7683	0.8332	1.2002	1.3016	1.5622	50.2°
39.9°	0.6414	0.7672	0.8361	1.1960	1.3035	1.5590	50.1°
	$\cos A$	$\sin A$	$\cot A$	$\tan A$	$\csc A$	$\sec A$	A

A	$\sin A$	$\cos A$	$\tan A$	$\cot A$	$\sec A$	$\csc A$	
40.0°	0.6428	0.7660	0.8391	1.1918	1.3054	1.5557	50.0°
40.1°	0.6441	0.7649	0.8421	1.1875	1.3073	1.5525	49.9°
40.2°	0.6455	0.7638	0.8451	1.1833	1.3092	1.5493	49.8°
40.3°	0.6468	0.7627	0.8481	1.1792	1.3112	1.5461	49.7°
40.4°	0.6481	0.7615	0.8511	1.1750	1.3131	1.5429	49.6°
40.5°	0.6494	0.7604	0.8541	1.1709	1.3151	1.5398	49.5°
40.6°	0.6508	0.7593	0.8571	1.1667	1.3171	1.5366	49.4°
40.7°	0.6521	0.7581	0.8601	1.1626	1.3190	1.5335	49.3°
40.8°	0.6534	0.7570	0.8632	1.1585	1.3210	1.5304	49.2°
40.9°	0.6547	0.7559	0.8662	1.1544	1.3230	1.5273	49.1°
41.0°	0.6561	0.7547	0.8693	1.1504	1.3250	1.5243	49.0°
41.1°	0.6574	0.7536	0.8724	1.1463	1.3270	1.5212	48.9°
41.2°	0.6587	0.7524	0.8754	1.1423	1.3291	1.5182	48.8°
41.3°	0.6600	0.7513	0.8785	1.1383	1.3311	1.5151	48.7°
41.4°	0.6613	0.7501	0.8816	1.1343	1.3331	1.5121	48.6°
41.5°	0.6626	0.7490	0.8847	1.1303	1.3352	1.5092	48.5°
41.6°	0.6639	0.7478	0.8878	1.1263	1.3373	1.5062	48.4°
41.7°	0.6652	0.7466	0.8910	1.1224	1.3393	1.5032	48.3°
41.8°	0.6665	0.7455	0.8941	1.1184	1.3414	1.5003	48.2°
41.9°	0.6678	0.7443	0.8972	1.1145	1.3435	1.4974	48.1°
42.0°	0.6691	0.7431	0.9004	1.1106	1.3456	1.4945	48.0°
42.1°	0.6704	0.7420	0.9036	1.1067	1.3478	1.4916	47.9°
42.2°	0.6717	0.7408	0.9067	1.1028	1.3499	1.4887	47.8°
42.3°	0.6730	0.7396	0.9099	1.0990	1.3520	1.4859	47.7°
42.4°	0.6743	0.7385	0.9131	1.0951	1.3542	1.4830	47.6°
42.5°	0.6756	0.7373	0.9163	1.0913	1.3563	1.4802	47.5°
42.6°	0.6769	0.7361	0.9195	1.0875	1.3585	1.4774	47.4°
42.7°	0.6782	0.7349	0.9228	1.0837	1.3607	1.4746	47.3°
42.8°	0.6794	0.7337	0.9260	1.0799	1.3629	1.4718	47.2°
42.9°	0.6807	0.7325	0.9293	1.0761	1.3651	1.4690	47.1°
43.0°	0.6820	0.7314	0.9325	1.0724	1.3673	1.4663	47.0°
43.1°	0.6833	0.7302	0.9358	1.0686	1.3696	1.4635	46.9°
43.2°	0.6845	0.7290	0.9391	1.0649	1.3718	1.4608	46.8°
43.3°	0.6858	0.7278	0.9423	1.0612	1.3741	1.4581	46.7°
43.4°	0.6871	0.7266	0.9457	1.0575	1.3763	1.4554	46.6°
43.5°	0.6884	0.7254	0.9490	1.0538	1.3786	1.4527	46.5°
43.6°	0.6896	0.7242	0.9523	1.0501	1.3809	1.4501	46.4°
43.7°	0.6909	0.7230	0.9556	1.0464	1.3832	1.4474	46.3°
43.8°	0.6921	0.7218	0.9590	1.0428	1.3855	1.4448	46.2°
43.9°	0.6934	0.7206	0.9623	1.0392	1.3878	1.4422	46.1°
	$\cos A$	$\sin A$	$\cot A$	$\tan A$	$\csc A$	$\sec A$	A

A	$\sin A$	$\cos A$	$\tan A$	$\cot A$	$\sec A$	$\csc A$	
44.0°	0.6947	0.7193	0.9657	1.0355	1.3902	1.4396	46.0°
44.1°	0.6959	0.7181	0.9691	1.0319	1.3925	1.4370	45.9°
44.2°	0.6972	0.7169	0.9725	1.0283	1.3949	1.4344	45.8°
44.3°	0.6984	0.7157	0.9759	1.0247	1.3972	1.4318	45.7°
44.4°	0.6997	0.7145	0.9793	1.0212	1.3996	1.4293	45.6°
44.5°	0.7009	0.7133	0.9827	1.0176	1.4020	1.4267	45.5°
44.6°	0.7022	0.7120	0.9861	1.0141	1.4044	1.4242	45.4°
44.7°	0.7034	0.7108	0.9896	1.0105	1.4069	1.4217	45.3°
44.8°	0.7046	0.7096	0.9930	1.0070	1.4093	1.4192	45.2°
44.9°	0.7059	0.7083	0.9965	1.0035	1.4117	1.4167	45.1°
45.0°	0.7071	0.7071	1.0000	1.0000	1.4142	1.4142	45.0°
	$\cos A$	$\sin A$	$\cot A$	$\tan A$	$\csc A$	$\sec A$	A

Table 3   Trigonometric Functions—Angle in Hundredth Radian Intervals

X	sin X	cos X	tan X	cot X	sec X	csc X
0.00	0.0000	1.0000	0.0000	Undefined	1.0000	Undefined
0.01	0.0100	1.0000	0.0100	99.9967	1.0000	100.0020
0.02	0.0200	0.9998	0.0200	49.9933	1.0002	50.0033
0.03	0.0300	0.9996	0.0300	33.3233	1.0005	33.3383
0.04	0.0400	0.9992	0.0400	24.9867	1.0008	25.0067
0.05	0.0500	0.9988	0.0500	19.9833	1.0013	20.0083
0.06	0.0600	0.9982	0.0601	16.6467	1.0018	16.6767
0.07	0.0699	0.9976	0.0701	14.2624	1.0025	14.2974
0.08	0.0799	0.9968	0.0802	12.4733	1.0032	12.5133
0.09	0.0899	0.9960	0.0902	11.0811	1.0041	11.1261
0.10	0.0998	0.9950	0.1003	9.9666	1.0050	10.0167
0.11	0.1098	0.9940	0.1104	9.0542	1.0061	9.1093
0.12	0.1197	0.9928	0.1206	8.2933	1.0072	8.3534
0.13	0.1296	0.9916	0.1307	7.6489	1.0085	7.7140
0.14	0.1395	0.9902	0.1409	7.0961	1.0099	7.1662
0.15	0.1494	0.9888	0.1511	6.6166	1.0114	6.6917
0.16	0.1593	0.9872	0.1614	6.1966	1.0129	6.2767
0.17	0.1692	0.9856	0.1717	5.8256	1.0146	5.9108
0.18	0.1790	0.9838	0.1820	5.4954	1.0164	5.5857
0.19	0.1889	0.9820	0.1923	5.1997	1.0183	5.2950
0.20	0.1987	0.9801	0.2027	4.9332	1.0203	5.0335
0.21	0.2085	0.9780	0.2131	4.6917	1.0225	4.7971
0.22	0.2182	0.9759	0.2236	4.4719	1.0247	4.5823
0.23	0.2280	0.9737	0.2341	4.2709	1.0270	4.3864
0.24	0.2377	0.9713	0.2447	4.0864	1.0295	4.2069
0.25	0.2474	0.9689	0.2553	3.9163	1.0321	4.0420
0.26	0.2571	0.9664	0.2660	3.7591	1.0348	3.8898
0.27	0.2667	0.9638	0.2768	3.6133	1.0376	3.7491
0.28	0.2764	0.9611	0.2876	3.4776	1.0405	3.6185
0.29	0.2860	0.9582	0.2984	3.3511	1.0436	3.4971
0.30	0.2955	0.9553	0.3093	3.2327	1.0468	3.3839
0.31	0.3051	0.9523	0.3203	3.1218	1.0501	3.2781
0.32	0.3146	0.9492	0.3314	3.0176	1.0535	3.1790
0.33	0.3240	0.9460	0.3425	2.9195	1.0570	3.0860
0.34	0.3335	0.9428	0.3537	2.8270	1.0607	2.9986
0.35	0.3429	0.9394	0.3650	2.7395	1.0645	2.9163
0.36	0.3523	0.9359	0.3764	2.6567	1.0685	2.8387
0.37	0.3616	0.9323	0.3879	2.5782	1.0726	2.7654
0.38	0.3709	0.9287	0.3994	2.5037	1.0768	2.6960
0.39	0.3802	0.9249	0.4111	2.4328	1.0812	2.6303
0.40	0.3894	0.9211	0.4228	2.3652	1.0857	2.5679
0.41	0.3986	0.9171	0.4346	2.3008	1.0904	2.5087
0.42	0.4078	0.9131	0.4466	2.2393	1.0952	2.4524
0.43	0.4169	0.9090	0.4586	2.1804	1.1002	2.3988
0.44	0.4259	0.9048	0.4708	2.1241	1.1053	2.3478

| X | sin X | cos X | tan X | cot X | sec X | csc X |

X	$\sin X$	$\cos X$	$\tan X$	$\cot X$	$\sec X$	$\csc X$
0.45	0.4350	0.9004	0.4831	2.0702	1.1106	2.2990
0.46	0.4439	0.8961	0.4954	2.0184	1.1160	2.2525
0.47	0.4529	0.8916	0.5080	1.9686	1.1216	2.2081
0.48	0.4618	0.8870	0.5206	1.9208	1.1274	2.1655
0.49	0.4706	0.8823	0.5334	1.8748	1.1334	2.1248
0.50	0.4794	0.8776	0.5463	1.8305	1.1395	2.0858
0.51	0.4882	0.8727	0.5594	1.7878	1.1458	2.0484
0.52	0.4969	0.8678	0.5726	1.7465	1.1523	2.0126
0.53	0.5055	0.8628	0.5859	1.7067	1.1590	1.9781
0.54	0.5141	0.8577	0.5994	1.6683	1.1659	1.9450
0.55	0.5227	0.8525	0.6131	1.6310	1.1730	1.9132
0.56	0.5312	0.8473	0.6269	1.5950	1.1803	1.8826
0.57	0.5396	0.8419	0.6410	1.5601	1.1878	1.8531
0.58	0.5480	0.8365	0.6552	1.5263	1.1955	1.8247
0.59	0.5564	0.8309	0.6696	1.4935	1.2035	1.7974
0.60	0.5646	0.8253	0.6841	1.4617	1.2116	1.7710
0.61	0.5729	0.8196	0.6989	1.4308	1.2200	1.7456
0.62	0.5810	0.8139	0.7139	1.4007	1.2287	1.7211
0.63	0.5891	0.8080	0.7291	1.3715	1.2376	1.6974
0.64	0.5972	0.8021	0.7445	1.3431	1.2467	1.6745
0.65	0.6052	0.7961	0.7602	1.3154	1.2561	1.6524
0.66	0.6131	0.7900	0.7761	1.2885	1.2658	1.6310
0.67	0.6210	0.7838	0.7923	1.2622	1.2758	1.6103
0.68	0.6288	0.7776	0.8087	1.2366	1.2861	1.5903
0.69	0.6365	0.7712	0.8253	1.2116	1.2966	1.5710
0.70	0.6442	0.7648	0.8423	1.1872	1.3075	1.5523
0.71	0.6518	0.7584	0.8595	1.1634	1.3186	1.5341
0.72	0.6594	0.7518	0.8771	1.1402	1.3301	1.5166
0.73	0.6669	0.7452	0.8949	1.1174	1.3420	1.4995
0.74	0.6743	0.7385	0.9131	1.0952	1.3542	1.4830
0.75	0.6816	0.7317	0.9316	1.0734	1.3667	1.4671
0.76	0.6889	0.7248	0.9505	1.0521	1.3796	1.4515
0.77	0.6961	0.7179	0.9697	1.0313	1.3929	1.4365
0.78	0.7033	0.7109	0.9893	1.0109	1.4066	1.4219
0.79	0.7104	0.7038	1.0092	0.9908	1.4208	1.4078
0.80	0.7174	0.6967	1.0296	0.9712	1.4353	1.3940
0.81	0.7243	0.6895	1.0505	0.9520	1.4503	1.3807
0.82	0.7311	0.6822	1.0717	0.9331	1.4658	1.3677
0.83	0.7379	0.6749	1.0934	0.9146	1.4818	1.3551
0.84	0.7446	0.6675	1.1156	0.8964	1.4982	1.3429
0.85	0.7513	0.6600	1.1383	0.8785	1.5152	1.3311
0.86	0.7578	0.6524	1.1616	0.8609	1.5327	1.3195
0.87	0.7643	0.6448	1.1853	0.8437	1.5508	1.3083
0.88	0.7707	0.6372	1.2097	0.8267	1.5695	1.2975
0.89	0.7771	0.6294	1.2346	0.8100	1.5888	1.2869
X	$\sin X$	$\cos X$	$\tan X$	$\cot X$	$\sec X$	$\csc X$

X	sin X	cos X	tan X	cot X	sec X	csc X
0.90	0.7833	0.6216	1.2602	0.7936	1.6087	1.2766
0.91	0.7895	0.6137	1.2864	0.7774	1.6293	1.2666
0.92	0.7956	0.6058	1.3133	0.7615	1.6507	1.2569
0.93	0.8016	0.5978	1.3409	0.7458	1.6727	1.2475
0.94	0.8076	0.5898	1.3692	0.7303	1.6955	1.2383
0.95	0.8134	0.5817	1.3984	0.7151	1.7191	1.2294
0.96	0.8192	0.5735	1.4284	0.7001	1.7436	1.2207
0.97	0.8249	0.5653	1.4592	0.6853	1.7690	1.2123
0.98	0.8305	0.5570	1.4910	0.6707	1.7953	1.2041
0.99	0.8360	0.5487	1.5237	0.6563	1.8225	1.1961
1.00	0.8415	0.5403	1.5574	0.6421	1.8508	1.1884
1.01	0.8468	0.5319	1.5922	0.6281	1.8802	1.1809
1.02	0.8521	0.5234	1.6281	0.6142	1.9107	1.1736
1.03	0.8573	0.5148	1.6652	0.6005	1.9424	1.1665
1.04	0.8624	0.5062	1.7036	0.5870	1.9754	1.1595
1.05	0.8674	0.4976	1.7433	0.5736	2.0098	1.1528
1.06	0.8724	0.4889	1.7844	0.5604	2.0455	1.1463
1.07	0.8772	0.4801	1.8270	0.5473	2.0828	1.1400
1.08	0.8820	0.4713	1.8712	0.5344	2.1217	1.1338
1.09	0.8866	0.4625	1.9171	0.5216	2.1622	1.1279
1.10	0.8912	0.4536	1.9648	0.5090	2.2046	1.1221
1.11	0.8957	0.4447	2.0143	0.4964	2.2489	1.1164
1.12	0.9001	0.4357	2.0660	0.4840	2.2952	1.1110
1.13	0.9044	0.4267	2.1197	0.4718	2.3438	1.1057
1.14	0.9086	0.4176	2.1759	0.4596	2.3947	1.1006
1.15	0.9128	0.4085	2.2345	0.4475	2.4481	1.0956
1.16	0.9168	0.3993	2.2958	0.4356	2.5041	1.0907
1.17	0.9208	0.3902	2.3600	0.4237	2.5631	1.0861
1.18	0.9246	0.3809	2.4273	0.4120	2.6252	1.0815
1.19	0.9284	0.3717	2.4979	0.4003	2.6906	1.0772
1.20	0.9320	0.3624	2.5722	0.3888	2.7597	1.0729
1.21	0.9356	0.3530	2.6503	0.3773	2.8327	1.0688
1.22	0.9391	0.3436	2.7328	0.3659	2.9100	1.0648
1.23	0.9425	0.3342	2.8198	0.3546	2.9919	1.0610
1.24	0.9458	0.3248	2.9119	0.3434	3.0789	1.0573
1.25	0.9490	0.3153	3.0096	0.3323	3.1714	1.0538
1.26	0.9521	0.3058	3.1133	0.3212	3.2699	1.0503
1.27	0.9551	0.2963	3.2236	0.3102	3.3752	1.0470
1.28	0.9580	0.2867	3.3414	0.2993	3.4878	1.0438
1.29	0.9608	0.2771	3.4672	0.2884	3.6085	1.0408
1.30	0.9636	0.2675	3.6021	0.2776	3.7383	1.0378
1.31	0.9662	0.2579	3.7471	0.2669	3.8782	1.0350
1.32	0.9687	0.2482	3.9033	0.2562	4.0294	1.0323
1.33	0.9711	0.2385	4.0723	0.2456	4.1933	1.0297
1.34	0.9735	0.2288	4.2556	0.2350	4.3715	1.0272
X	sin X	cos X	tan X	cot X	sec X	csc X

X	$\sin X$	$\cos X$	$\tan X$	$\cot X$	$\sec X$	$\csc X$
1.35	0.9757	0.2190	4.4552	0.2245	4.5661	1.0249
1.36	0.9779	0.2092	4.6734	0.2140	4.7792	1.0226
1.37	0.9799	0.1994	4.9131	0.2035	5.0138	1.0205
1.38	0.9819	0.1896	5.1774	0.1931	5.2731	1.0185
1.39	0.9837	0.1798	5.4707	0.1828	5.5613	1.0166
1.40	0.9854	0.1700	5.7979	0.1725	5.8835	1.0148
1.41	0.9871	0.1601	6.1654	0.1622	6.2459	1.0131
1.42	0.9887	0.1502	6.5811	0.1519	6.6567	1.0115
1.43	0.9901	0.1403	7.0555	0.1417	7.1260	1.0100
1.44	0.9915	0.1304	7.6018	0.1315	7.6673	1.0086
1.45	0.9927	0.1205	8.2381	0.1214	8.2986	1.0073
1.46	0.9939	0.1106	8.9886	0.1113	9.0441	1.0062
1.47	0.9949	0.1006	9.8874	0.1011	9.9378	1.0051
1.48	0.9959	0.0907	10.9834	0.0910	11.0288	1.0041
1.49	0.9967	0.0807	12.3499	0.0810	12.3903	1.0033
1.50	0.9975	0.0707	14.1014	0.0709	14.1368	1.0025
1.51	0.9982	0.0608	16.4281	0.0609	16.4585	1.0019
1.52	0.9987	0.0508	19.6696	0.0508	19.6950	1.0013
1.53	0.9992	0.0408	24.4984	0.0408	24.5188	1.0008
1.54	0.9995	0.0308	32.4612	0.0308	32.4766	1.0005
1.55	0.9998	0.0208	48.0784	0.0208	48.0888	1.0002
1.56	0.9999	0.0108	92.6208	0.0108	92.6262	1.0001
1.57	1.0000	0.0008	1255.6700	0.0008	1255.6700	1.0000
1.58	1.0000	−0.0092	−108.6510	−0.0092	−108.6560	1.0000
1.59	0.9998	−0.0192	−52.0672	−0.0192	−52.0768	1.0002

X	$\sin X$	$\cos X$	$\tan X$	$\cot X$	$\sec X$	$\csc X$

Table 4　Four-Place Table of Common Logarithms

N	0	1	2	3	4	5	6	7	8	9
10	0000	0043	0086	0128	0170	0212	0253	0294	0334	0374
11	0414	0453	0492	0531	0569	0607	0645	0682	0719	0755
12	0792	0828	0864	0899	0934	0969	1004	1038	1072	1106
13	1139	1173	1206	1239	1271	1303	1335	1367	1399	1430
14	1461	1492	1523	1553	1584	1614	1644	1673	1703	1732
15	1761	1790	1818	1847	1875	1903	1931	1959	1987	2014
16	2041	2068	2095	2122	2148	2175	2201	2227	2253	2279
17	2304	2330	2355	2380	2405	2430	2455	2480	2504	2529
18	2553	2577	2601	2625	2648	2672	2695	2718	2742	2765
19	2788	2810	2833	2856	2878	2900	2923	2945	2967	2989
20	3010	3032	3054	3075	3096	3118	3139	3160	3181	3201
21	3222	3243	3263	3284	3304	3324	3345	3365	3385	3404
22	3424	3444	3464	3483	3502	3522	3541	3560	3579	3598
23	3617	3636	3655	3674	3692	3711	3729	3747	3766	3784
24	3802	3820	3838	3856	3874	3892	3909	3927	3945	3962
25	3979	3997	4014	4031	4048	4065	4082	4099	4116	4133
26	4150	4166	4183	4200	4216	4232	4249	4265	4281	4298
27	4314	4330	4346	4362	4378	4393	4409	4425	4440	4456
28	4472	4487	4502	4518	4533	4548	4564	4579	4594	4609
29	4624	4639	4654	4669	4683	4698	4713	4728	4742	4757
30	4771	4786	4800	4814	4829	4843	4857	4871	4886	4900
31	4914	4928	4942	4955	4969	4983	4997	5011	5024	5038
32	5052	5065	5079	5092	5105	5119	5132	5145	5159	5172
33	5185	5198	5211	5224	5237	5250	5263	5276	5289	5302
34	5315	5328	5340	5353	5366	5378	5391	5403	5416	5428
35	5441	5453	5465	5478	5490	5502	5515	5527	5539	5551
36	5563	5575	5587	5599	5611	5623	5635	5647	5658	5670
37	5682	5694	5705	5717	5729	5740	5752	5763	5775	5786
38	5798	5809	5821	5832	5843	5855	5866	5877	5888	5899
39	5911	5922	5933	5944	5955	5966	5977	5988	5999	6010
40	6021	6031	6042	6053	6064	6075	6085	6096	6107	6117
41	6128	6138	6149	6160	6170	6180	6191	6201	6212	6222
42	6232	6243	6253	6263	6274	6284	6294	6304	6314	6325
43	6335	6345	6355	6365	6375	6385	6395	6405	6415	6425
44	6435	6444	6454	6464	6474	6484	6493	6503	6513	6522
45	6532	6542	6551	6561	6571	6580	6590	6599	6609	6618
46	6628	6637	6646	6656	6665	6675	6684	6693	6702	6712
47	6721	6730	6739	6749	6758	6767	6776	6785	6794	6803
48	6812	6821	6830	6839	6848	6857	6866	6875	6884	6893
49	6902	6911	6920	6928	6937	6946	6955	6964	6972	6981
50	6990	6998	7007	7016	7024	7033	7042	7050	7059	7067
51	7076	7084	7093	7101	7110	7118	7126	7135	7143	7152
52	7160	7168	7177	7185	7193	7202	7210	7218	7226	7235
53	7243	7251	7259	7267	7275	7284	7292	7300	7308	7316
54	7324	7332	7340	7348	7356	7364	7372	7380	7388	7396
N	0	1	2	3	4	5	6	7	8	9

N	0	1	2	3	4	5	6	7	8	9
55	7404	7412	7419	7427	7435	7443	7451	7459	7466	7474
56	7482	7490	7497	7505	7513	7520	7528	7536	7543	7551
57	7559	7566	7574	7582	7589	7597	7604	7612	7619	7627
58	7634	7642	7649	7657	7664	7672	7679	7686	7694	7701
59	7709	7716	7723	7731	7738	7745	7752	7760	7767	7774
60	7782	7789	7796	7803	7810	7818	7825	7832	7839	7846
61	7853	7860	7868	7875	7882	7889	7896	7903	7910	7917
62	7924	7931	7938	7945	7952	7959	7966	7973	7980	7987
63	7993	8000	8007	8014	8021	8028	8035	8041	8048	8055
64	8062	8069	8075	8082	8089	8096	8102	8109	8116	8122
65	8129	8136	8142	8149	8156	8162	8169	8176	8182	8189
66	8195	8202	8209	8215	8222	8228	8235	8241	8248	8254
67	8261	8267	8274	8280	8287	8293	8299	8306	8312	8319
68	8325	8331	8338	8344	8351	8357	8363	8370	8376	8382
69	8388	8395	8401	8407	8414	8420	8426	8432	8439	8445
70	8451	8457	8463	8470	8476	8482	8488	8494	8500	8506
71	8513	8519	8525	8531	8537	8543	8549	8555	8561	8567
72	8573	8579	8585	8591	8597	8603	8609	8615	8621	8627
73	8633	8639	8645	8651	8657	8663	8669	8675	8681	8686
74	8692	8698	8704	8710	8716	8722	8727	8733	8739	8745
75	8751	8756	8762	8768	8774	8779	8785	8791	8797	8802
76	8808	8814	8820	8825	8831	8837	8842	8848	8854	8859
77	8865	8871	8876	8882	8887	8893	8899	8904	8910	8915
78	8921	8927	8932	8938	8943	8949	8954	8960	8965	8971
79	8976	8982	8987	8993	8998	9004	9009	9015	9020	9025
80	9031	9036	9042	9047	9053	9058	9063	9069	9074	9079
81	9085	9090	9096	9101	9106	9112	9117	9122	9128	9133
82	9138	9143	9149	9154	9159	9165	9170	9175	9180	9186
83	9191	9196	9201	9206	9212	9217	9222	9227	9232	9238
84	9243	9248	9253	9258	9263	9269	9274	9279	9284	9289
85	9294	9299	9304	9309	9315	9320	9325	9330	9335	9340
86	9345	9350	9355	9360	9365	9370	9375	9380	9385	9390
87	9395	9400	9405	9410	9415	9420	9425	9430	9435	9440
88	9445	9450	9455	9460	9465	9469	9474	9479	9484	9489
89	9494	9499	9504	9509	9513	9518	9523	9528	9533	9538
90	9542	9547	9552	9557	9562	9566	9571	9576	9581	9586
91	9590	9595	9600	9605	9609	9614	9619	9624	9628	9633
92	9638	9643	9647	9652	9657	9661	9666	9671	9675	9680
93	9685	9689	9694	9699	9703	9708	9713	9717	9722	9727
94	9731	9736	9741	9745	9750	9754	9759	9764	9768	9773
95	9777	9782	9786	9791	9795	9800	9805	9809	9814	9818
96	9823	9827	9832	9836	9841	9845	9850	9854	9859	9863
97	9868	9872	9877	9881	9886	9890	9894	9899	9903	9908
98	9912	9917	9921	9926	9930	9934	9939	9943	9948	9952
99	9956	9961	9965	9969	9974	9978	9983	9987	9991	9996
N	0	1	2	3	4	5	6	7	8	9

Appendix 3

Logarithms

A3.1 COMMON LOGARITHMS

The common logarithm of a given positive number N (written $\log N$) is the exponent of the power of 10 which will produce the given number. For example,

$$\log 1 = 0 \quad \text{since} \quad 10^0 = 1 \qquad\qquad \log 100 = 2 \quad \text{since} \quad 10^2 = 100$$

$$\log 10 = 1 \quad \text{since} \quad 10^1 = 10 \qquad \log 0.001 = -3 \quad \text{since} \quad 10^{-3} = 0.001$$

while

$$\log P = p \quad \text{if} \quad 10^p = P$$

A3.2 FUNDAMENTAL LAWS OF LOGARITHMS

I. The logarithm of a product of two or more positive numbers is equal to the sum of the logarithms of the several numbers; i.e.,

$$\log P \cdot Q = \log P + \log Q$$

$$\log P \cdot Q \cdot R = \log P + \log Q + \log E, \text{etc}$$

II. The logarithm of the quotient of two positive numbers is equal to the logarithm of the dividend minus the logarithm of the divisor; i.e.,

$$\log \frac{P}{Q} = \log P - \log Q$$

III. The logarithm of a power of a positive number is equal to the logarithm of the number multiplied by the exponent of the power; i.e.,

$$\log (P^n) = n \log P$$

IV. The logarithm of a root of a positive number is equal to the logarithm of the number divided by the index of the root; i.e.,

$$\log \sqrt[n]{P} = \frac{1}{n} \log P$$

For proofs of these laws see Prob. A3.1.

The logarithm of an expression involving two or more of the operations in laws I to IV is obtained by combining the results of the several laws; e.g.,

$$\log \frac{P \cdot Q}{R} = \log (P \cdot Q) - \log R = \log P + \log Q - \log R$$

For other examples see Prob. A3.2 to A3.4.

A3.3 CHARACTERISTIC AND MANTISSA

The common logarithm of a positive number (e.g., $\log 300 = 2.4771$ and $\log 0.2 = 9.3010 - 10$) consists of two parts: an integral part called the *characteristic* and a pure decimal part called the *mantissa*.

From Probs. A3.3 and A3.4 it is seen that the characteristic depends only upon the position of the decimal point in the number. For example,

$$\log 2 = 0.3010 \qquad \text{and} \qquad \log 200 = 2.3010$$

$$\log 25 = 1.3979 \qquad \text{and} \qquad \log 2.5 = 0.3979$$

The characteristic of the common logarithm of any number greater than 1 is one less than the number of digits to the left of the decimal point in the given number.

The characteristic of the common logarithm of any positive number smaller than 1 is obtained by subtracting the number of zeros immediately following the decimal point from 9 and affixing -10. Thus the characteristic of the common logarithm of 0.2 is $9 - 10$, of 0.04 is $8 - 10$, of 0.0005 is $6 - 10$.

(See also Prob. A3.5.)

The mantissa of the common logarithm of a positive number is usually a continuous decimal. All references here are to Table 4, Appendix 2, giving the mantissas to four decimal places.

A3.4 TO FIND THE LOGARITHM OF A GIVEN POSITIVE NUMBER

(*a*) Write down the characteristic in accordance with the above rules.

(b_1) When the given number contains three or fewer significant digits, read the mantissa from the table.

EXAMPLE A3.1 Find log 32.8.
The characteristic is 1. To find the mantissa locate the entry 5159 in the row opposite 32 and the column headed 8. Then log 32.8 = 1.5159.

EXAMPLE A3.2 Find log 5.2.
The characteristic is 0. Since 5.2 = 5.20, we find the mantissa by locating the entry 7160 in the row opposite 52 and the column headed 0. Then log 5.2 = 0.7160.

(b_2) When the given number contains four digits, interpolate using the method of proportional parts.

EXAMPLE A3.3 Find log 654.8.
The characteristic is 2. For the mantissa, we have

$$\begin{aligned}
\text{Mantissa of log } 654.0 &= .8156 \\
\text{Mantissa of log } 655.0 &= \underline{.8162} \\
\text{Tabular difference} &= .0006
\end{aligned}$$

$$0.8 \times \text{tabular difference} = .00048 \text{ or } .0005 \text{ to four decimal places}$$

$$\text{Mantissa of log } 654.8 = .8156 + .0005 = .8161$$

Then log 654.8 = 2.8161.

Note that the essential calculation here is $8156 + 0.8(6) = 8166.8$ or 8161.

(b_3) When the given number contains more than four digits, round it to four digits, then interpolate using the method of proportional parts.

EXAMPLE A3.4 Find log 1.1917.
To four decimal places we want to find log 1.192. The characteristic is 0. For the mantissa, we have

$$\begin{aligned}
\text{Mantissa of log } 1.190 &= .0755 \\
\text{Mantissa of log } 1.200 &= \underline{.0792} \\
\text{Tabular difference} &= .0037
\end{aligned}$$

$$0.2 \times \text{tabular difference} = .00074 \text{ or } .0007 \text{ to four decimal places}$$

$$\text{Mantissa of log } 1.192 = .0755 + .0007 = .0762$$

Then log 1.1917 = 0.0762.

(See Probs. A3.6 and A3.7.)

A3.5 TO FIND THE NUMBER CORRESPONDING TO A GIVEN COMMON LOGARITHM

When the given mantissa is found in the table, read off the row number and the column heading and then point off using the characteristic rule. The resulting number is called the *antilogarithm* (antilog) of the given logarithm.

EXAMPLE A3.5   Antilog 1.8808 = 76.0.

The mantissa .8808 is found in the row opposite 76 and the column headed 0. Since the characteristic is 1, there are two digits to the left of the decimal point.

When the given mantissa is not found in the table, interpolation must be used.

EXAMPLE A3.6   Antilog 9.5657 − 10 = 0.3679.

$$
\begin{array}{ll}
\text{Mantissa of log 3670} = .5647 & \text{Given mantissa} = .5657 \\
\text{Mantissa of log 3680} = \underline{.5658} & \text{Next smaller mantissa} = \underline{.5647} \\
\text{Tabular difference} = .0011 & \text{Difference} = .0010
\end{array}
$$

$$
\text{Correction} = \frac{0.0010}{0.0011}(0.0010) = 0.00091 \quad \text{or} \quad 0.0009 \text{ to four decimal places.}
$$

Then antilog 9.5657 − 10 = 0.3670 + 0.0009 = 0.3679.

Note that the essential operation here is $\dfrac{10 \times 10}{12} = 9.1$ or 9.

(See also Prob. A3.8.)

3.6 COLOGARITHMS

The cologarithm of a positive number N (written colog N) is the logarithm of its reciprocal $1/N$. Thus, colog $N = \log(1/N) = \log 1 - \log N = -\log N$.

EXAMPLE A3.7   Colog 38.38 = 8.4159 − 10.

$$
\text{colog } 38.38 = \log \frac{1}{38.38} = \log 1 - \log 38.38
$$

$$
\log 1 = 10.0000 - 10
$$

$$
(-)\log 38.38 = \frac{1.5841}{8.4159 - 10}
$$

Note that colog N may be obtained by subtracting each digit (starting at the left) of log N from 9 except the last significant digit, which is subtracted from 10, and affixing − 10 when N is greater than 1. For example:

(a)   log 3163 = 3.5001; colog 3163 = 6.4999 − 10.
(b)   log 0.0399 = 8.6010 − 10; colog 0.0399 = 1.3990.

(See also Probs. A3.12 and A3.13.)

Solved Problems

A3.1   Prove the laws of logarithms.

Restricting the proofs to common logarithms, let $P = 10^p$ and $Q = 10^q$; then log $P = p$ and log $Q = q$.

I. Since $P \cdot Q = 10^p \cdot 10^q = 10^{p+q}$,   then $\log (P \cdot Q) = p + q = \log P + \log Q$.

II. Since $P/Q = 10^p/10^q = 10^{p-q}$,   then $\log (P/Q) = p - q = \log P - \log Q$.

III. Since $P^n = (10^p)^n = 10^{np}$,   then $\log P^n = np = n \log P$.

IV. Since $\sqrt[n]{P} = (10^p)^{1/n} = 10^{p/n}$ then $\log \sqrt[n]{P} = \dfrac{p}{n} = \dfrac{1}{n} \log P$.

A3.2 Express the logarithm of the given expression in terms of the logarithms of the individual letters or numbers involved.

(a) $\log \dfrac{P \cdot Q}{R \cdot S} = \log (P \cdot Q) - \log (R \cdot S) = (\log P + \log Q) - (\log R + \log S)$

$$= \log P + \log Q - \log R - \log S$$

(b) $\log \dfrac{\sqrt[3]{P}}{Q^4} = \log \sqrt[3]{P} - \log Q^4 = \dfrac{1}{3} \log P - 4 \log Q$

(c) $\log \dfrac{34(104)^2}{(49)^3} = \log 34 + 2 \log 104 - 3 \log 49$

(d) $\log \dfrac{(34.2)^2 \sqrt[3]{1.06}}{(9.8)^3 \sqrt{2.33}} = 2 \log 34.2 + \dfrac{1}{3} \log 1.06 - 3 \log 9.8 - \dfrac{1}{2} \log 2.33$

A3.3 Given $\log 2 = 0.3010$ and $\log 3 = 0.4771$, find the logarithm of:
(a) 30,   (b) 200,   (c) 25,   (d) 120,   (e) 2.5,   (f) $\sqrt{6}$,   (g) $\sqrt[3]{24}$

(a) $30 = 3 \times 10$; $\log 30 = \log 3 + \log 10 = 0.4771 + 1.0000 = 1.4771$

(b) $200 = 2 \times 10^2$; $\log 200 = \log 2 + 2 \log 10 = 0.3010 + 2.0000 = 2.3010$

(c) $25 = 10^2/2^2$; $\log 25 = 2 \log 10 - 2 \log 2 = 2.0000 - 0.6020 = 1.3980$

(d) $120 = 2^2 \cdot 3 \cdot 10$; $\log 120 = 2 \log 2 + \log 3 + \log 10 = 0.6020 + 0.4771 + 1.0000 = 2.0791$

(e) $2.5 = 10/2^2$; $\log 2.5 = \log 10 - 2 \log 2 = 1.0000 - 0.6020 = 0.3980$

(f) $\sqrt{6} = (2 \times 3)^{1/2}$; $\log \sqrt{6} = \frac{1}{2}(\log 2 + \log 3) = \frac{1}{2}(0.7781) = 0.3890$

(g) $\sqrt[3]{24} = \sqrt[3]{2^3 \times 3} = 2 \sqrt[3]{3}$; $\log \sqrt[3]{24} = \log 2 + \dfrac{1}{3} \log 3 = 0.3010 + \dfrac{1}{3} (0.4771) = 0.4600$

A3.4 Given $\log 2 = 0.3010$ and $\log 3 = 0.4771$, find the logarithm of:
(a) 0.2,   (b) 0.003,   (c) 0.5,   (d) $(0.02)^3$,   (e) $\sqrt[4]{0.006}$

(a) $0.2 = 2/10$; $\log 0.2 = \log 2 - \log 10 = 0.3010 - 1.0000 = -1 + 0.3010$. We shall write this $9.3010 - 10$.

(b) $0.003 = 3/10^3$; $\log 0.003 = \log 3 - 3 \log 10 = -3 + 0.4771 = 7.4771 - 10$

(c) $0.5 = 1/2$; $\log 0.5 = \log 1 - \log 2 = 0.0000 - 0.3010$
$$= (10.0000 - 10) - 0.3010 = 9.6990 - 10$$

(d) $(0.02)^3 = (2/10^2)^3$; $\log (0.02)^3 = 3 \log 2 - 6 \log 10$
$$= 0.9030 - 6.0000$$
$$= (10.9030 - 10) - 6.0000 = 4.9030 - 10$$

(e) $\sqrt[4]{0.006} = \sqrt[4]{2 \times 3/10^3}$; $\log\sqrt[4]{0.006} = \frac{1}{4}(\log 2 + \log 3 - 3\log 10)$
$$= \frac{1}{4}(0.3010 + 0.4771 - 3.0000)$$
$$= \frac{1}{4}(7.7781 - 10) = \frac{1}{4}(37.7781 - 40) = 9.4445 - 10$$

A3.5 Determine the characteristic of the common logarithm of each of the following numbers:

(a) 3864   (c) 8.746   (e) 0.3874   (g) 0.07295   (i) 2.3567   (k) 0.44636

(b) 286   (d) 982,600   (f) 0.00826   (h) 0.000023   (j) 88.725   (l) 0.00072358

The characteristics are:

(a) 3   (c) 0   (e) 9−10   (g) 8−10   (i) 0   (k) 9−10

(b) 2   (d) 5   (f) 7−10   (h) 5−10   (j) 1   (l) 6−10

A3.6 Verify each of the following logarithms. (Use Table 4, Appendix 2.)

(a) $\log 38.64 = 1.5870$ (e) $\log 2.356 = 0.3722$

(b) $\log 286 = 2.4564$ (f) $\log 88.72 = 1.9480$

(c) $\log 0.3874 = 9.5881\text{-}10$ (g) $\log 0.4463 = 9.6496\text{-}10$

(d) $\log 0.00826 = 7.9170\text{-}10$ (h) $\log 0.0007235 = 6.8594\text{-}10$

A3.7 Verify each of the following. (Use Table 4, Appendix 2.)

(a) $\log (0.07324 \times 0.0006235) = \log 0.07324 + \log 0.0006235$
$$= 8.8647 - 10 + 6.7948 - 10 = 15.6595 - 20 = 5.6595 - 10$$

(b) $\log (8.7633 \times 0.0074288) = \log 8.763 + \log 0.007429$ (round each number to four digits)
$$= 0.9426 + 7.8709 - 10 = 8.8135 - 10$$

(c) $\log 34.72/5.384 = \log 34.72 - \log 5.384$
$$= 1.5406 - 0.7311 = 0.8095$$

(d) $\log 7218/0.0235 = \log 7218 - \log 0.0235$
$$= 3.8584 - (8.3710 - 10) = 13.8584 - 10 - (8.3710 - 10) = 5.4873$$

(e) $\log (24.56)^3 = 3 \log 24.56 = 3(1.3902) = 4.1706$

(f) $\log (0.4893)^4 = 4 \log 0.4893 = 4(9.6896 - 10) = 38.7584 - 40 = 8.7584 - 10$

(g) $\log \sqrt{876.4} = \frac{1}{2} \log 876.4 = \frac{1}{2}(2.9427) = 1.4714$

(h) $\log \sqrt[3]{66.75} = \frac{1}{3} \log 66.75 = \frac{1}{3}(1.8244) = 0.6081$

(i) $\log \sqrt{0.9494} = \frac{1}{2} \log 0.9494 = \frac{1}{2}(9.9775 - 10) = \frac{1}{2}(19.9775 - 20) = 9.9888 - 10$

A3.8 Verify each of the following. (Use Table 4, Appendix 2.)

(a)   Antilog 2.5615 = 3.64.3.

(b)   Antilog 5.6900 = 489800.

(c)   Antilog 8.8135 − 10 = 0.06509. From Prob. A3.7(b), 8.7633 × 0.0074288 = 0.06509.

(d)   Antilog 1.4365 = 27.32.

(e)   Antilog 8.6915 − 10 = 0.04915.

(*f*)   Antilog 4.1706 = 14,810. From Prob. A3.7(*e*), (24.56)³ = 14,810.

(*g*)   Antilog 1.4714 = 29.61. From Prob. A3.7(*g*), $\sqrt{876.4}$ = 29.61.

Calculate each of the following using logarithms. (Use Table 4, Appendix 2).

A3.9 $N = 36.23 \times 2.674 \times 0.007175$

$$
\begin{aligned}
\log 36.23 &= 1.5591 \\
(+) \log 2.674 &= 0.4271 \\
(+) \log 0.007175 &= \underline{7.8558 - 10} \\
\log N &= 9.8420 - 10 \\
N &= 0.6950
\end{aligned}
$$

A3.10 $N = \dfrac{47.75 \times 8.643}{6467}$

$$
\begin{aligned}
\log 47.75 &= 1.6790 \\
(+) \log 8.643 &= \underline{0.9366} \\
& \quad\; 12.6156 - 10 \quad (2.6156 = 12.6156 - 10) \\
(-) \log 6467 &= \underline{3.8107} \\
\log N &= 8.8049 - 10 \\
N &= 0.06381
\end{aligned}
$$

A3.11 $N = \sqrt[3]{0.4847}$.

$$
\log N = \tfrac{1}{3} \log 0.4847
$$

$$
\begin{aligned}
\log 0.4847 &= 9.6854 - 10 \\
&= 29.6854 - 30 \\
\log N &= 9.8951 - 10 \\
N &= 0.7854
\end{aligned}
$$

A3.12 Solve Prob. A3.10 using cologarithms.

$$
N = 47.75 \times 8.643 \times 1/6467.
$$

$$
\begin{aligned}
\log 47.75 &= 1.6790 \\
(+) \log 8.643 &= 0.9366 \\
(+) \operatorname{colog} 6467 &= \underline{6.1893 - 10} \qquad (\log 6467 = 3.8107) \\
\log N &= 8.8049 - 10 \\
N &= 0.06381
\end{aligned}
$$

A3.13 $\dfrac{74.72}{\sqrt{8.394}\sqrt[3]{0.002877}} = N.$

$$\log N = \log 74.72 + \tfrac{1}{2} \operatorname{colog} 8.394 + \tfrac{1}{3} \operatorname{colog} 0.002877$$

$$
\begin{aligned}
\log 74.72 &= 1.8734 \\
(+)\ \tfrac{1}{2}\operatorname{colog} 8.394 &= 9.5380 - 10 \\
(+)\ \tfrac{1}{3}\operatorname{colog} 0.002877 &= 0.8470 \\
\log N &= 12.2584 - 10 \\
&= 2.2584
\end{aligned}
$$

$$
\begin{aligned}
\log 8.394 &= 0.9240 \\
\operatorname{colog} 8.394 &= 9.0760 - 10 \\
\log 0.002877 &= 7.4590 - 10 \\
\operatorname{colog} 0.002877 &= 2.5410
\end{aligned}
$$

$$N = 181.3$$

Supplementary Problems

Use Table 4, Appendix 2.

A3.14 Find:

(a) $\log 211 = 2.3243$

(b) $\log 9.17 = 0.9624$

(c) $\log 0.00466 = 7.6684 - 10$

(d) $\log 0.6754 = 9.8295 - 10$

(e) $\log 32.86 = 1.5167$

(f) $\log 264.7 = 2.4227$

(g) $\log 7.177 = 0.8559$

(h) $\log 0.9663 = 9.9851 - 10$

(i) $\log 4287. = 3.6322$

(j) $\log 0.005555 = 7.7447 - 10$

(k) $\log 0.09714 = 8.9874 - 10$

(l) $\log 2.122 = 0.3267$

(m) $\log 66.98 = 1.8260$

(n) $\log 781.5 = 2.8930$

(o) $\log 2348 = 3.3707$

(p) $\log 0.09123 = 8.9602 - 10$

A3.15 Find:

(a) antilog $1.9864 = 96.91$

(b) antilog $0.7500 = 5.624$

(c) antilog $8.6208 - 10 = 0.04176$

(d) antilog $1.0970 = 12.50$

(e) antilog $2.6561 = 453$

(f) antilog $0.9182 = 8.283$

(g) antilog $8.1184 - 10 = 0.01313$

(h) antilog $3.6662 = 4637$

(i) antilog $1.1207 = 13.20$

(j) antilog $2.6282 = 424.8$

(k) antilog $0.9584 = 9.086$

(l) antilog $9.6129 - 10 = 0.4101$

(m) antilog $2.2395 = 173.6$

(n) antilog $1.2225 = 16.69$

(o) antilog $4.8403 = 69230$

(p) antilog $2.6718 = 469.7$

A3.16 Evaluate:

(a) $\dfrac{819(748)}{3670} = 166.9,$

(b) $\dfrac{827.6}{518.3} = 1.596,$

(c) $\dfrac{48.62}{77.65} = 0.6260,$

(d) $787.9(0.003323) = 2.618$

(e) $\dfrac{(227.3)^2 \sqrt[3]{0.007764}}{(86.35)^3 \sqrt{0.3848}} = 0.02563$

(f) $\sqrt[3]{\dfrac{781.5(3.434)}{852.7(586.7)}} = 0.1751$

Index